VxWorks 嵌入式实时操作系统设备驱动与 BSP 开发设计

王　洋　主编

宋彦国　但　果　沈福生　副主编

北京航空航天大学出版社

内容简介

本书深入而详细地讲解了 VxWorks 操作系统下 BSP 开发内容。全书共分 10 章，从 Tornado 开发环境的安装和设置，到工程的建立、BSP 移植、驱动程序和应用程序的开发都做了详细的讲解。本书以 S3C2410 处理器为例，完整介绍了 VxWorks 下 TTY 设备驱动开发，串口驱动开发，块设备驱动开发，Nand FLASH 和 Nor FLASH 设备驱动开发，网络设备驱动开发，LCD 液晶驱动开发，WindML 开发，I^2C 设备驱动开发，普通字符设备驱动开发，并设计了多例字符型设备驱动程序设计实验和多例 VxWorks 应用程序设计实验。

本书所有驱动程序和例程都在 MagicARM2410 实验箱上进行过实测，适合高校相关专业学生学习参考，也可供嵌入式开发人员和系统设计人员参考使用。

图书在版编目(CIP)数据

VxWorks 嵌入式实时操作系统设备驱动与 BSP 开发设计 / 王洋主编. --北京 ：北京航空航天大学出版社，2012.7

ISBN 978-7-5124-0861-6

Ⅰ. ①V… Ⅱ. ①王… Ⅲ. ①实时操作系统—软件开发 Ⅳ. ①TP316.2

中国版本图书馆 CIP 数据核字(2012)第 153078 号

VxWorks 嵌入式实时操作系统设备驱动与 BSP 开发设计

王 洋 主编

宋彦国 但 果 沈福生 副主编

责任编辑 苗长江 王 彤

*

北京航空航天大学出版社出版发行

北京市海淀区学院路 37 号(邮编 100191) http://www.buaapress.com.cn

发行部电话：(010)82317024 传真：(010)82328026

读者信箱：emsbook@gmail.com 邮购电话：(010)82316936

涿州市新华印刷有限公司印装 各地书店经销

*

开本：710×1 000 1/16 印张：23.25 字数：496 千字

2012 年 7 月第 1 版 2012 年 7 月第 1 次印刷 印数：4 000 册

ISBN 978-7-5124-0861-6 定价：45.00 元

前　言

伴随着嵌入式行业的迅猛发展，嵌入式操作系统已经被越来越多的人和企业关注。在所有的嵌入式操作系统中，VxWorks始终因为优秀的实时性能和稳定性，牢牢占据航空、通信、国防、工业控制、网络设备、医疗设备、消费电子等应用领域。尤其像华为、中兴这样一流的国际通信设备厂商，都以VxWorks作为通讯产品中的嵌入式操作系统。

本书的编写，没有过多地介绍VxWorks系统任务等基本概念，而是直接以开发基于S3C2410的设备驱动和BSP为目标，完整介绍了VxWorks下TTY设备驱动开发，串口驱动开发，块设备驱动开发，Nand FLASH和Nor FLASH设备驱动开发，网络设备驱动开发，LCD液晶驱动开发，WindML开发，I^2C设备驱动开发，普通字符设备驱动开发，并设计了多例字符型设备驱动程序设计实验和多例VxWorks应用程序设计实验。本书所采用的开发平台是广州致远公司设计的MagicARM2410实验箱，书中给出了该平台的所有设备与S3C2410之间的电路原理图，因此读者在阅读和学习书中各章节代码时不会因为身边硬件开发平台的差异而无法进行实验。书中所有的例程都经过作者在实验平台上的测试，保证代码的可用性，例如串口驱动、网口驱动、文件系统等，让学习者尽量减少摸索的时间。

本书引用的部分资料和图片都是书中讲述部分所需要的，无侵权意图，特此声明。另外，书中所有例程的源代码可在北京航空航天大学出版社网站的“下载专区”下载。

本书主要面向从事嵌入式开发领域的广大开发设计人员、工程技术人员，也可作为高等院校相关专业本科生、研究生的参考学习资料。

最后感谢我们身边的各位同事和研究生们，他们为本书的编写和校对提出了许多宝贵建议。

我们力争整本书的编写都做到实用性和实战性，但水平有限，难免有错漏不足之处，欢迎广大读者批评指正。

作者

2012年5月

目 录

第1章

VxWorks 实时操作系统介绍

1.1 嵌入式实时操作系统

1.1.1 实时操作系统特点

实时系统可以分为实时控制系统和实时信息处理系统。所谓实时系统是对未来事件在限定时间内能做出反应的系统。限定时间所指的范围很广,可以从微秒级(如信号处理)到分钟级(如联机查询系统)。在实时控制系统中,计算机通过特定的外围设备(以下简称为外设)与被控对象发生联系,被控对象的信息经过加工后,通过显示屏幕向控制人员显示或通过外设向被控对象发出指示,实现对被控对象的控制;在实时信息处理系统中,用户通过终端设备向系统提出服务请求,系统完成服务后通过终端回答给用户。

在实时系统中,主要有 3 个指标来衡量系统的实时性:响应时间(Response Time)、生存时间(Survival Time)、吞吐量(Throughput)。

① 响应时间:是计算机从识别一个外部事件到做出响应的时间。在控制应用中,它是最重要的指标,如果时间不能及时地处理,系统可能就会崩溃。响应时间不是一定要最快,只要能满足过程的时间要求即可。比如,对于一些慢变化的过程,具有几分钟甚至更长的响应时间都可以认为是实时的;对于快速过程,其响应时间可能要求达到毫秒、微秒、纳秒级甚至更短。

② 生存时间:是指数据有效等待时间,在这段时间里,数据是有效的。

③ 吞吐量:是在给定时间段内,系统可以处理的事件总数。例如通信控制器用每秒钟处理的字符数来表示吞吐量。吞吐量可能是平均响应时间的倒数,但它通常

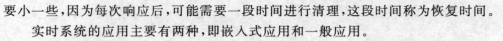

要小一些，因为每次响应后，可能需要一段时间进行清理，这段时间称为恢复时间。

实时系统的应用主要有两种，即嵌入式应用和一般应用。

1. 嵌入式应用

嵌入式系统(Embedded System)是由硬件和软件相结合组成的具有特定功能、用于特定场合的独立系统。嵌入式系统硬件主要由嵌入式微处理器、外围硬件设备组成；其软件主要由底层系统软件和用户应用软件组成。嵌入式应用系统广泛地应用于办公自动化、消费、通信、汽车、工业和军事等领域。其中，办公自动化、消费和通信领域占据的份额最大，具体来说有：

① 过程控制，即对生产过程中各种动作流程的控制。这种控制是在对被控制对象和环境进行不断观测的基础上做出及时的、恰当的反应。在控制的过程中，计算机扮演着中心角色。它通过传感器从外部接收有关过程的信息，对这些信息进行加工处理，然后对执行设备发出控制指令。

② 通信设备，如程控交换机、路由器、移动通信基站、桥接器、集线器、ADSL 等。

③ 智能仪器，如频谱仪、示波器、医疗仪器等。

④ 消费产品，如机顶盒、空调、洗衣机、微波炉、电视机、游戏机等。

⑤ 计算机外设设备，如蓝牙设备、打印机、PCM 卡设备等。

⑥ 军事设备，如无人侦察机、导弹、穿戴式电脑、定位仪器、战机、战舰等。

2. 一般应用

实时系统的一般应用，通常指的是类似于普通计算机操作系统，如 Windows XP、Linux 等带有人机界面的使用环境。该系统支持符合标准的外设，如键盘、显示器、磁盘等，支持应用程序。用户可以根据需要在操作系统支持下开发各种应用，其典型应用如下：

① 测控计算机。在大型控制系统中，作为上位机，与其他计算机和设备相连，进行系统的监测、控制、协调及数据处理等工作。

② 交互式系统。通常指实时信息查询系统，如快递单查询系统、高考成绩查询系统、股票交易系统等。这些系统的响应时间要求不高，只要在人可接受的范围内即可。

1.1.2 嵌入实时操作系统特点

根据 IEEE 的实时 Unix 分委会定义，实时操作系统应具备以下特点：

① 异步的事件响应。实时系统为能在系统要求的时间内响应异步的外部事件，要求有异步 I/O 和中断处理能力。I/O 响应时间常受内存访问、硬盘访问和处理机总线速度所限制。

② 切换时间和中断延迟时间确定。

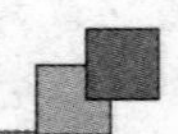

③ 优先级中断和调度。必须允许用户定义中断优先级和被调度的任务优先级，并指定如何服务中断。

④ 抢占式调度。为保证响应时间，实时操作系统必须允许高优先级任务准备就绪后，可以马上抢占 CPU，打断低优先级任务的执行。

⑤ 内存锁定。必须具有将程序或部分程序锁定在内存的能力。锁定在内存的程序减少了为获取该程序而访问盘的时间，从而保证了快速的响应时间。

⑥ 连续文件。应提供存取盘上数据的优化方法，使得存取数据时查找时间最短。通常要求把数据存储在连续文件上。

⑦ 同步。具有同步和协调共享数据使用和时间执行的手段。

嵌入式实时操作系统是指在嵌入式系统中采用的实时操作系统，它既是嵌入式操作系统，又是实时操作系统。作为一种嵌入式操作系统，它具有嵌入式软件共有的可裁剪、低资源占用、低功耗等特点；而作为一种实时操作系统，它必须满足实时系统的特点。总体来说，嵌入式实时操作系统的特点如下。

1. 确定性和可预测性

实时软件对于外部事件的响应时间必须是实时的、确定的和可以重复实现的，不管当时系统内部状态如何，都是可预测的。测量操作系统确定性能力的其中一个指标就是从一个高优先级设备中断到开始被服务之间的最大延迟。在非实时操作系统中，这个延迟可能是几十到数百毫秒，并且不可预测；而在嵌入式实时操作系统中有一个明确的上限，从几微秒到几毫秒不等。

2. 响应性

响应性是与确定性相关但又不同的特征。确定性考虑在应答一个中断前，操作系统延迟时间；而响应性是在应答中断后，操作系统服务中断时间。响应性包含以下几个方面：

- 初始化中断处理和开始执行中断服务程序(ISR)需要的时间。如果是要求一个过程切换的 ISR 执行，那么比在当前进程上下文中的 ISR 执行延迟更长时间。
- 执行 ISR 需要的时间。这通常依赖于硬件平台的处理能力。
- 中断嵌套的作用。如果一个 ISR 可能因另一个中断的到达而被中断，那么它的服务将被延迟。

确定性和响应性共同构成了对外部事件的响应时间。响应时间对于嵌入式实时操作系统是至关重要的，这是由于系统必须满足系统外部个人、设备或者数据流的实时要求。

3. 用户控制

用户控制是嵌入式实时操作系统一个重要特点。在一个典型的非实时操作系统

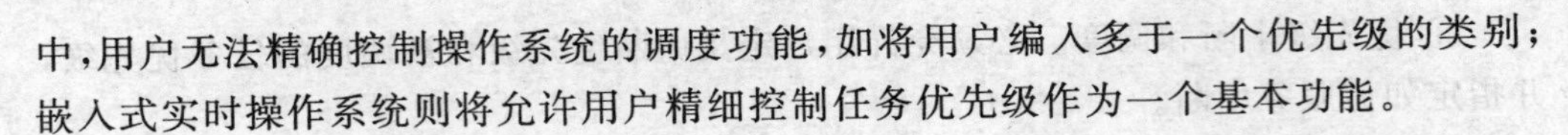

中，用户无法精确控制操作系统的调度功能，如将用户编入多于一个优先级的类别；嵌入式实时操作系统则将允许用户精细控制任务优先级作为一个基本功能。

4. 可靠性

可靠性是衡量嵌入式实时操作系统最为重要的指标之一。在非实时系统中，一个瞬时错误只需要简单地通过重新引导系统就可以解决。但是，实时系统需要实时响应和控制事件，性能的丧失或降低可能造成灾难性的后果，比如导弹不能及时修正目标等。

5. 软失败操作

当嵌入式实时操作系统中某个进程或者任务失败后，操作系统将试图纠正这个问题或者将它的影响最小化，同时继续运行系统。典型情况下，系统通知用户或者用户进程，它将试图进行纠正动作，然后继续操作，或降低服务级别。非实时系统遇到此种情况，则只是尽可能地保留数据，然后失败。例如，一个典型的传统 Unix 系统，当它检测到内核数据的误用，在系统控制台上发出失败消息，将内存内容存储到磁盘用于以后的失败分析，同时终止系统的执行。

1.1.3 VxWorks 操作系统特点

VxWorks 操作系统是美国 WindRiver 公司于 1983 年设计开发的一种嵌入式实时操作系统(RTOS)，具有良好的可持续发展能力、高性能的内核以及友好的用户开发环境，在嵌入式实时操作系统领域占据非常重要的地位。VxWorks 操作系统以其良好的可靠性和卓越的实时性被广泛地应用在通信、军事、航空、航天等高精尖技术及实时性要求极高的领域中，如卫星通信、军事演习、弹道制导、飞机导航等。

VxWorks 操作系统是实时多任务操作系统，能在确定的时间内执行其功能，并对外部的异步事件做出响应的计算机系统。VxWorks 系统提供多处理器间和多任务间高效的信号量、消息队列、管道、网络透明的套接字，提高事件的响应速度。

VxWorks 实时操作系统具体特征如下：

1. 具有高性能的操作系统内核 Wind

VxWorks 的内核 Wind 是一个具有较高性能的、标准的嵌入式实时操作系统内核，主要特点包括快速多任务切换、抢占式任务调度、任务间通信手段多样化等。该内核具有任务切换时间短、中断延迟小、网络流量大的特点，与其他嵌入式实时操作系统相比有一定的优势。

首先，VxWorks 的任务调度策略为以可抢占式调度(Preemptive Scheduling)为基础，辅以时间片轮转调度算法。这一调度算法使得 VxWorks 能够及时地响应高优先级的任务。而同级任务间则可选择时间片轮转法使多个同优先级的任务并发

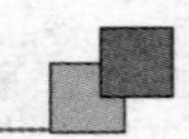

执行。

其次，VxWorks 采用中断处理与普通任务分别在不同的栈中处理的中断处理机制。这使得中断的产生只会引发一些关键寄存器的存储而不会导致任务的上下文切换，从而减小了中断延迟。同时，VxWorks 的中断处理程序在最小时间内处理中断中必须要处理的任务，而将其他的非实时处理尽量放入被引发的中断服务程序中来完成，从而进一步缩小了中断延迟。VxWorks 在内核中普遍采用互斥信号量(Semaphores)而非通过关闭中断来实现互斥访问的方法对缩小中断延迟也有一定的贡献。

2. 开发调试环境友好

VxWorks 具有友好的开发调试环境，便于操作、配置和开发调试应用程序。嵌入式系统的本身特点通常使其开发和调试过程较一般系统更为复杂。一个友好的开发环境能够大大地降低嵌入式系统开发的难度，提高开发效率。

VxWorks 的开发环境 Tornado 就是一个友好的开发环境，能够运行在多种主机上，包括 Sun、HP、IBM - rs6000、Dec、MIPS 等；主机操作系统则支持 Unix、Windows NT、Windows XP 等。系统的各项配置由于使用了较为流行的图形界面(如对话框、列表、选项、按钮等)，并且具有可视化图形界面的调试工具，符合大多数用户的计算机使用习惯，方便用户调试。

VxWorks 支持应用程序的动态链接和动态下载，使开发者省去了每次调试都将应用程序与操作系统内核进行链接和下载的步骤，缩短了编辑调试的周期。VxWorks 提供的目标机仿真器 VxSim，使开发者可独立于硬件环境，先行开发应用程序，从而节省了新产品的研发时间和硬件方面的开销。

3. 具有良好的兼容性

VxWorks 是最早兼容 POSIX1003.1b 标准的嵌入式实时操作系统之一，同时是 POSIX 组织的主要会员。

VxWorks 的 TCP/IP 协议栈部分与 BSD4.4 版本的 TCP/IP 除了在实时性方面有较大差别外，其他方面基本兼容。这使得基于 BSD4.4 Unix Socket 的应用程序可以很方便地移植到 VxWorks 中。

不仅如此，VxWorks 还是第一个通过 Windows NT 测试的，是可以在 Windows NT 平台进行开发和仿真的嵌入式实时操作系统。VxWorks 同时支持 ANSI C 标准，并通过了 ISO9001 的认证。

VxWorks 良好的兼容性，使其在不同运行环境间可以方便地移植，从而使用户在开发和培训方面所做的工作得到保护，减少了开发周期和经费。

4. 支持多种开发和运行环境

VxWorks 的应用范围和领域比较广泛。VxWorks 开发环境支持的主机包括

Sun、HP、IBM - rs6000、Dec、MIPS 等。系统运行环境支持 PowerPC、68K、CPU32、SPARC、i960、x86、MIPS 等众多 CPU 及支持 RISC、DSP 等技术。VxWorks 同世界诸多硬件厂商有着良好的合作关系，对于多种硬件的有效支持是 VxWorks 得以流行的一个重要原因。

5. 快速支持新技术

VxWorks 对于出现的各种新技术采取跟踪和及时改进的策略，始终能够满足客户的最新需求。它是最早实现捆绑集成交叉开发环境的嵌入式实时操作系统之一，也是最早在其内核中加入 TCP/IP 网络协议的嵌入式实时操作系统。此后，VxWorks 又率先宣布支持网络文件系统 NFS，在系统集成最新网络协议方面一直保持良好的势头。VxWorks 还是最先支持 RISC 处理器的嵌入式实时操作系统。从这可以看出 VxWorks 具有较强的开发革新能力及较短的更新周期。

1.2 VxWorks 操作系统基本结构

VxWorks 操作系统的基本构成模块包括以下部分。

(1) 高效的实时内核。VxWorks 实时内核（Wind）主要包括基于优先级的任务调度、任务同步和通信、中断处理、定时器和内存管理。

(2) 兼容实时系统标准 POSIX。VxWorks 提供接口来支持实时系统标准 P.1003.1b。

(3) I/O 系统。VxWorks 提供快速灵活的与 ANSI C 相兼容的 I/O 系统，包括 Unix 的缓冲 I/O 和实时系统标准 POSIX 的异步 I/O。VxWorks 包括以下驱动：

- 网络——网络设备（以太网、共享内存）；
- 管道——任务间通信；
- RAM——驻留内存文件；
- SCSI——SCSI 硬盘、磁碟、磁带；
- 键盘——PC x86 键盘（BSP 仅支持 x86）；
- 显示器——PC x86 显示器（BSP 仅支持 x86）；
- 磁碟——IDE 和软盘（BSP 仅支持 x86）；
- 并口——PC 格式的目标硬件。

(4) 本机文件系统。VxWorks 针对不同的设备提供了多种文件系统。VxWorks的文件系统与 MS - DOS、RT - 11、RAM、SCSI 等相兼容。针对块设备，VxWorks 提供了兼容 MS - DOS 的文件系统及 RT - 11 文件系统。所支持的 Raw File System 的文件系统能够将整个硬盘视为单个文件来操作。VxWorks 还提供支持磁带设备和 CD - ROM 设备的文件系统。

(5) I/O 系统。VxWorks 的 I/O 系统提供了操作系统与各硬件设备的接口。

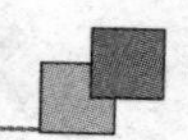

VxWorks的I/O系统主要包括字符设备、块设备、虚拟设备(管道、Socket)、监控设备和网络设备等。在外部接口上,VxWorks的I/O系统提供标准的C函数库支持Basic I/O和Buffered I/O。其中,Basic I/O与Unix标准兼容,Buffered I/O与ANSI C兼容。在内部实现上,VxWorks的I/O系统采用独立的机制来实现I/O系统的快速性和灵活性。

运行在VxWorks上的应用程序通过VxWorks的文件系统访问各设备接口。每个设备都有一个同普通文件一样的文件名,并对它像一个文件一样存取。VxWorks的I/O系统还支持异步I/O(Asynchronous Input/Output)。异步I/O使得I/O操作与调用该操作的程序可以异步执行,从而避免了I/O操作速度对应用程序执行的影响。异步I/O通过调用符合POSIX标准的aioPxLib中的接口来实现。

(6) 网络特性。VxWorks网络能与许多运行其他协议的网络进行通信,如TCP/IP、4.3BSD、NFS、UDP、SNMP、FTP等。VxWorks可通过网络允许任务存取文件到其他系统中,并对任务进行远程调用。关于网络部分会在网卡驱动设计部分进行具体分析。

(7) 虚拟内存(可选单元VxVMI)。VxVMI主要用于对指定内存区的保护,如内存块只读等,加强了系统的健壮性。

(8) 共享内存(可选单元VxMP)。VxMP主要用于多处理器上运行的任务之间的共享信号量、消息队列、内存块的管理。

(9) 驻留目标工具。Tornado集成环境中,开发工具工作于主机侧。驻留目标外壳、模块加载和卸载、符号表都可进行配置。

(10) Wind基类。VxWorks系统提供对C++的支持,并构造了系统基类函数。

(11) 工具库。VxWorks系统向用户提供丰富的系统调用,包括中断处理、定时器、消息注册、内存分配、字符串转换、线性和环形缓冲区管理,以及标准ANSI C程序库。

(12) 性能优化。VxWorks系统通过运行定时器来记录任务对CPU的利用率,从而进行有效地调整,合理安排任务的运行,给定适宜的任务属性。

(13) 目标代理。目标代理可使用户远程调试应用程序。

(14) 板级支持包(Board Support Package,BSP)。VxWorks采用模块化设计方法,把依赖于硬件环境的函数和信息分离出来,放入所谓BSP的组件中。而BSP Libraries向上层软件提供一致的接口。VxWorks在Target上运行需要相应的BSP的支持。BSP中包含硬件环境中CPU的初始化及系统各项硬件资源的安装和配置,包括RAM、Clock、网络接口、中断控制器等。要使VxWorks在某个硬件环境下运转起来,必须首先进行BSP的相关设计开发。

(15) VxWorks仿真器(VxSim)。可选产品VxWorks仿真器,能模拟VxWorks目标机的运行,用于应用系统的分析。

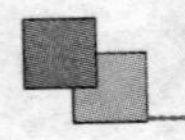

1.3 VxWorks 操作系统开发流程

通常情况下，VxWorks 操作系统开发指系统的应用程序开发和 BSP 开发。其中，VxWorks 操作系统的开发最难最主要的就是 BSP 的开发。BSP 的英文全称是 Board Support Package，即板级支持包，其作用是针对特殊的硬件平台，为 VxWorks 内核提供操作的接口。它与内核、驱动程序以及应用程序的关系如图 1.1 所示。

(1) 在通电后，完成硬件初始化；

(2) 支持 VxWorks 和硬件驱动通信；

(3) 使与硬件相关(Hardware - dependent)和独立于硬件(Hardware - independent)的两大部分能够在 VxWorks 系统中良好地结合，共同为系统服务。

BSP 是用“make”来编译连接生成的，而不是采用 Tornado 工具完成编译。BSP 的设置包括以下驱动：中断控制器(Interrupt controller)、定时器(Timer)、串行口(Serial)、以太网卡、LCD、键盘、触摸屏等。其中，前面 3 个是 BSP 的基础部分，以太网卡驱动是最为重要的一个 BSP 驱动。这是因为 BSP 默认的下载 VxWorks RAM Image 的方式是使用以太网连接方式，而串行口电缆需要用来和开发板(COM1)通信，使用的是 WDB 通信协议。

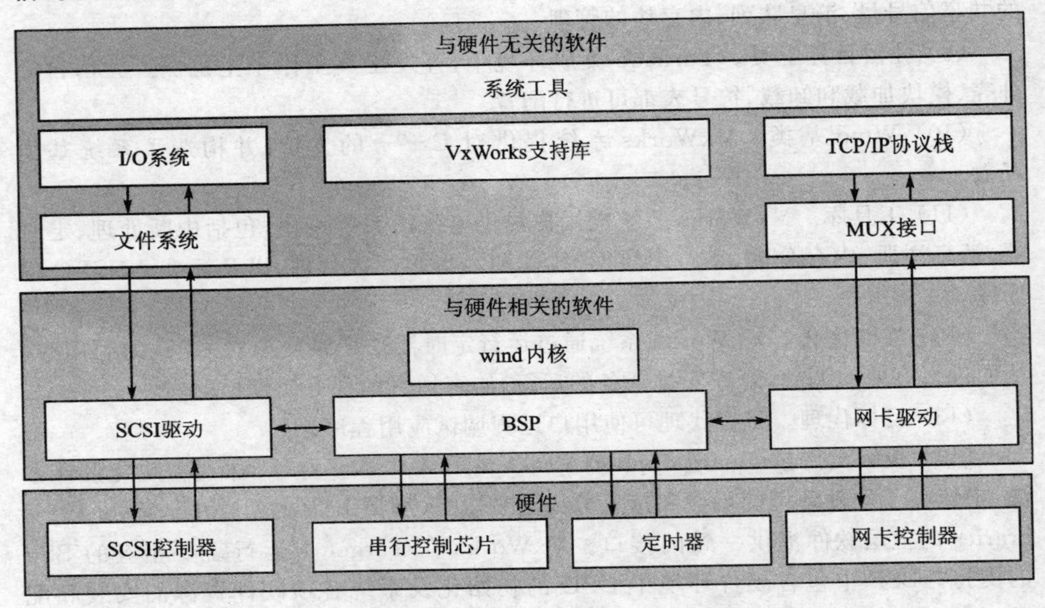

图 1.1 VxWorks BSP 与内核、驱动程序以及应用程序关系图

从图 1.1 中可以看出，BSP 的作用实际上就是在内核和硬件之间建立一个桥梁，以便进行数据交换。BSP 的开发总体上分为 5 个阶段：

（1）配置开发环境和选择调试工具；

（2）编写 BSP 初始化代码；

（3）明确硬件的具体情况，配置一个最小化内核；

（4）启动并测试上一步创建的最小内核；

（5）编写其他驱动（如网络驱动、文件系统驱动）。

第 2 章

Tornado 交叉开发环境

2.1 Tornado 基本介绍

Tornado 2.2 是嵌入式实时领域里最新一代的开发调试环境。Tornado 给嵌入式系统开发人员提供了一个不受目标机资源限制的开发和调试环境。Tornado 包含 3 个高度集成的部分：

- 运行在宿主机和目标机上的交叉开发工具和实用程序；
- 运行在目标机上的高性能、可裁剪的实时操作系统 VxWorks；
- 集成开发环境(Integrated Development Environment, IDE)，支持连接宿主机和目标机的多种通信方式，如以太网、串口、ICE 或 ROM 仿真器等。

集成开发环境是一种更直观的自动化环境，用于项目的生成和管理，建立和维护主机与目标机间的通信，运行、调试和监视 VxWorks 的应用程序，具有如下特性：

- 1 个源代码编辑器；
- 1 个 C 和 C++ 的编译器；
- CrossWind，1 个源代码级的图形增强的调试器；
- Wind Sh，1 个 C 语言的 Shell，用于控制目标机；
- VxSim，1 个仿真器。

Tornado Facilities 主要在主机上运行，它的各部件之间的关系如图 2.1 所示。在主机上运行的开发工具和 VxWorks 之间的通信由目标服务器(Server)和目标代理(Agent)来负责。

开发过程中，有两个主要的通信协议：

① WTX 协议(Wind River Tool eXchange)，用于开发机内部 Tornado 工具与 Target Server 间通信。

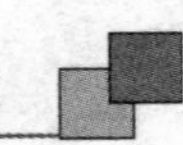

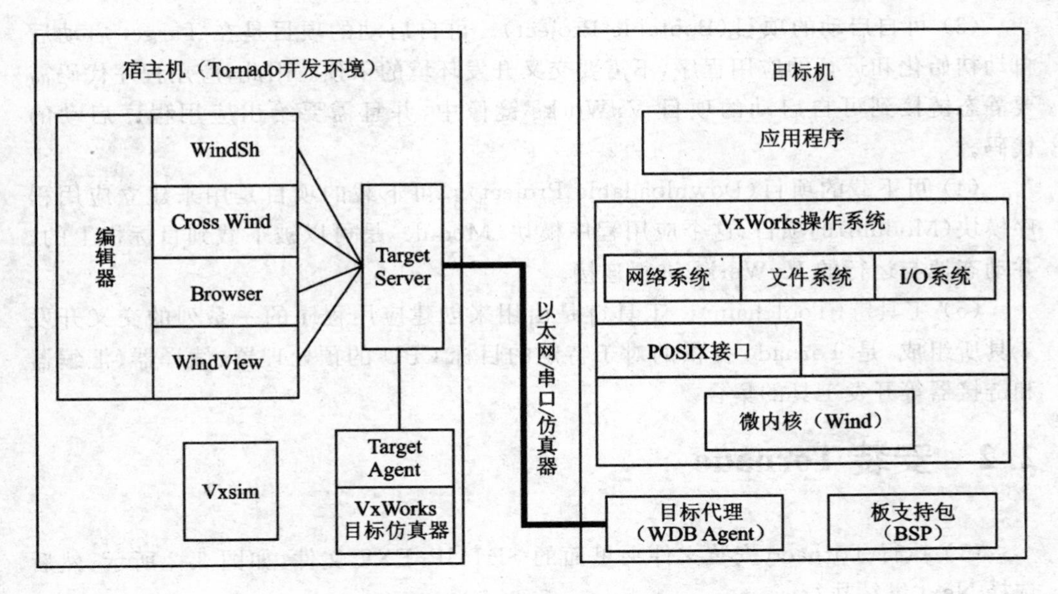

图 2.1　(宿)主机与目标机

② WDB 协议(Wind DeBug),用于主机 Target Server 与目标机之间的通信。

由图 2.1 所示,一个完整的 VxWorks 操作系统只是 Tornado(或目标系统)中的一个组成部分。VxWorks 操作系统包含一个多任务的内核;运行库包含 POSIX、内部任务通信、外部网络连接、文件系统支持等。

Tornado 2.0 具有如下主要特点:

- 集成工具 VxWorks 目标仿真器(VxSim);
- 集成工具 WindView 逻辑分析器;
- 新的项目设备环境(Project Facility);
- 新的调试引擎和图形用户界面。

其中,目标仿真器(VxSim)允许用户在没有硬件平台的条件下,使用 Tornado 开发硬件模块程序;WindView 逻辑分析器提供了在仿真器上的应用程序执行情况的动态分析和可视化;项目环境采用了与其他流行开发环境类似的人机界面,用户在管理项目文件,配置 VxWorks,建立应用时感觉方便易懂;同样,调试程序界面的设计跟常用的软件开发环境类似,无须用户重新学习。

在学习使用 Tornado 之前,首先介绍一下相关概念:

(1) 项目(Project)。项目是由建立应用程序所需要的一组源文件、二进制文件、以及编译设置的集合所组成的。

(2) 工作空间(Workspace)。工作空间是一个或几个相关项目的集合。创建了项目和工作空间后,Tornado 环境会显示一个工作空间窗口,该空间窗口能够显示它所包含的所有项目的信息。

(3) 可自启动的项目(Bootable Project)。可自启动的项目是在 Target 启动后自动初始化和运行的应用程序,不需要交叉开发环境的干预。因此,应用程序代码需要静态链接到可自启动的项目 VxWorks 镜像中,并且需要给出应用程序启动的代码。

(4) 可下载的项目(Downloadable Project)。可下载的项目是用来建立应用程序模块(Module)的项目,这个应用程序模块(Module)是可以被下载到目标机上的,并动态地与运行的 VxWorks 进行连接。

(5) 工具链(Toolchain)。工具链是由用来创建应用程序的一系列的交叉开发工具所组成,是 Tornado 提供的对于特定的目标 CPU 的预处理器、编译器、汇编器和链接器等开发工具的集合。

2.2 安装 Tornado

(1) 执行 Tornado 安装文件夹里面的 SETUP. EXE 文件,如图 2.2 所示,然后选择 Next 继续执行。

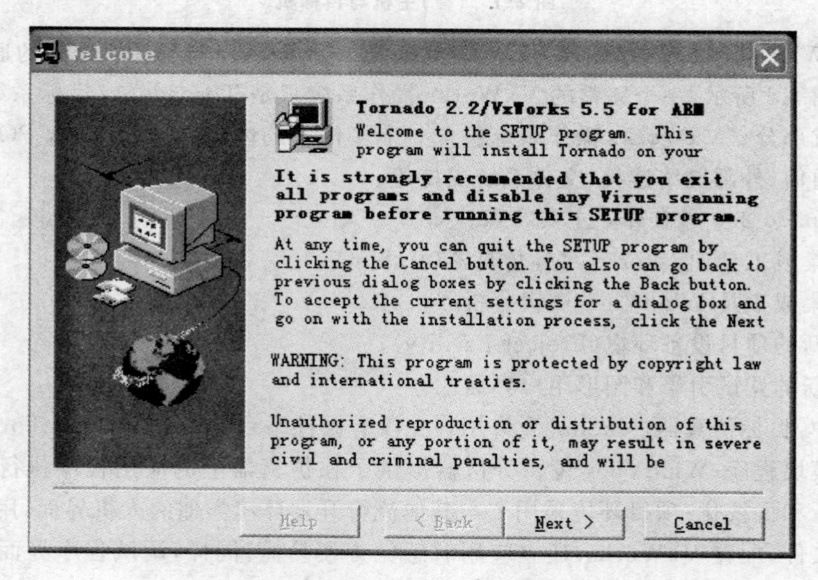

图 2.2 执行 SETUP. exe 后提示框

(2) 如图 2.3 所示,继续执行 Next。

(3) 选择 Next 后,如图 2.4 所示,选择 Accept 后,再执行 Next。

(4) 如图 2.5 所示,输入 Name 和 Company 名字,在 Install 里面输入注册码,执行 Next。

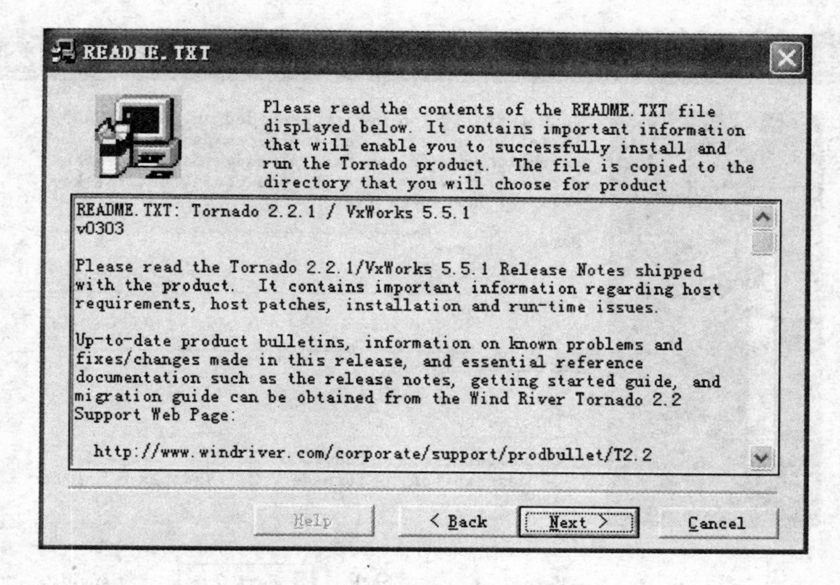

图 2.3 README.TXT 提示框

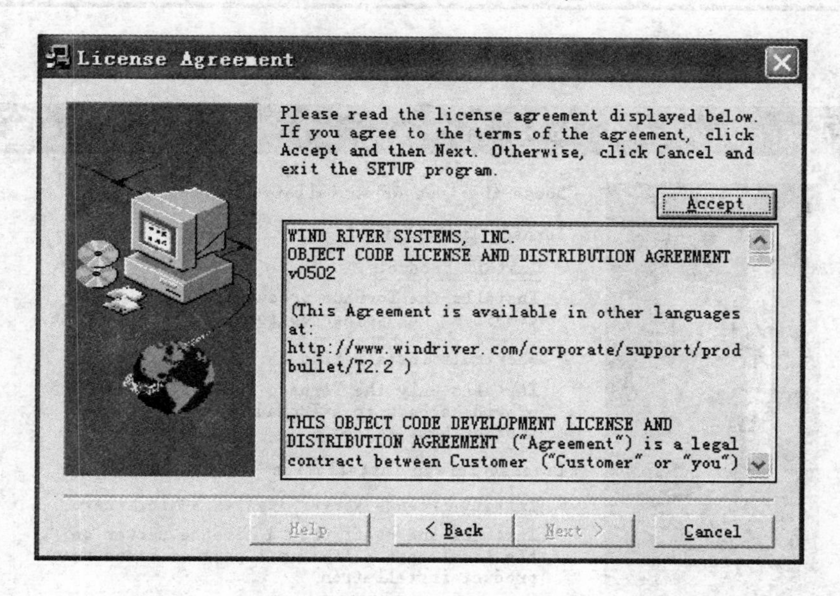

图 2.4 License Agreement 提示框

(5) 如图 2.6 所示,选择 Install Product,然后执行 Next。

(6) 如图 2.7 所示,选择安装的目录,然后执行 Next。

(7) 如图 2.8 所示,选择编译器,这里默认即可,执行 Next。

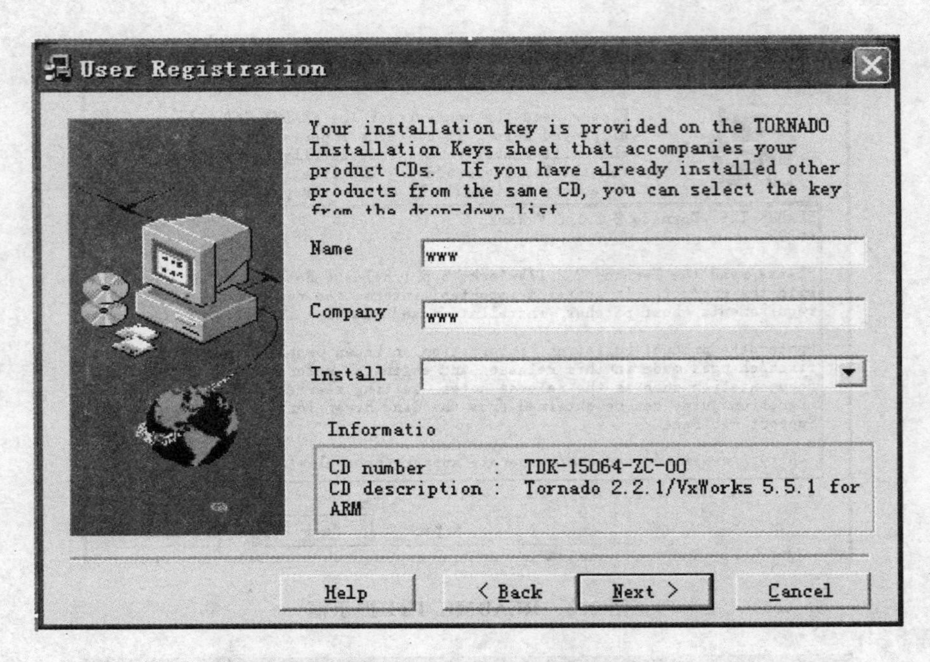

图 2.5　User Registration 对话框

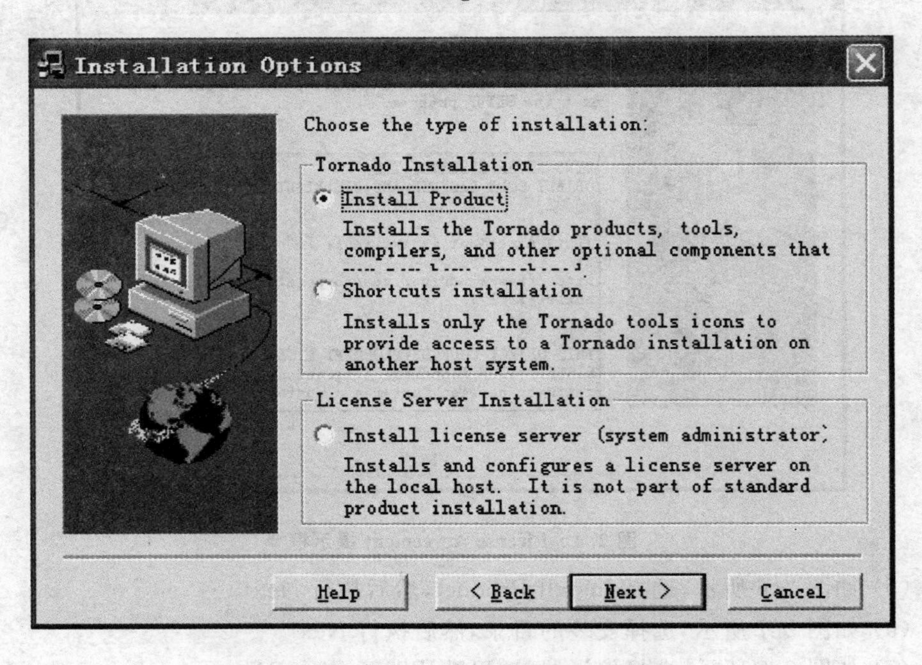

图 2.6　Installation Options 对话框

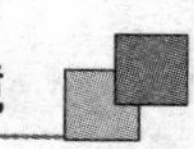

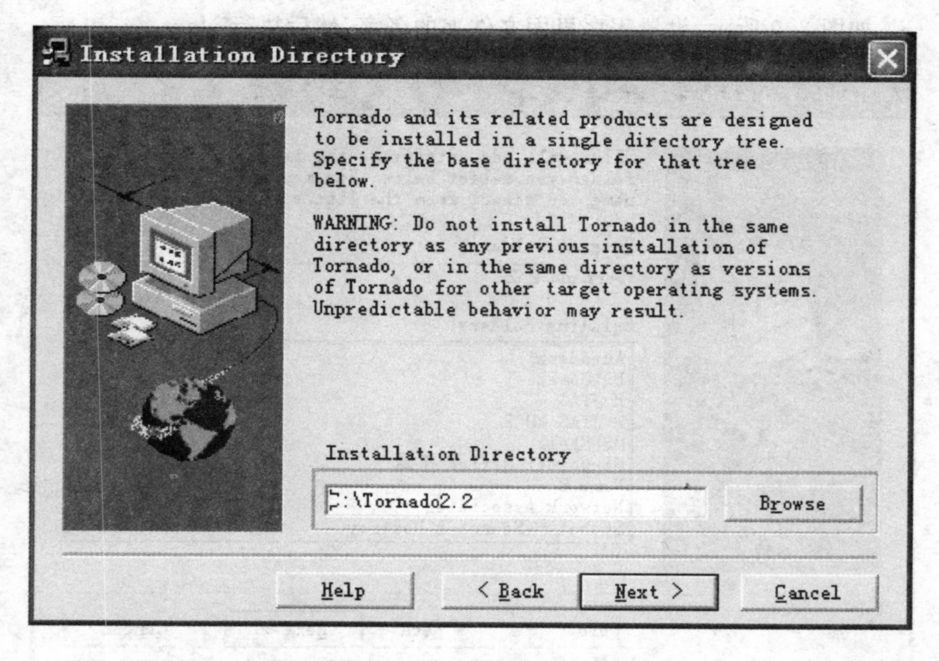

图 2.7 Installation Directory 对话框

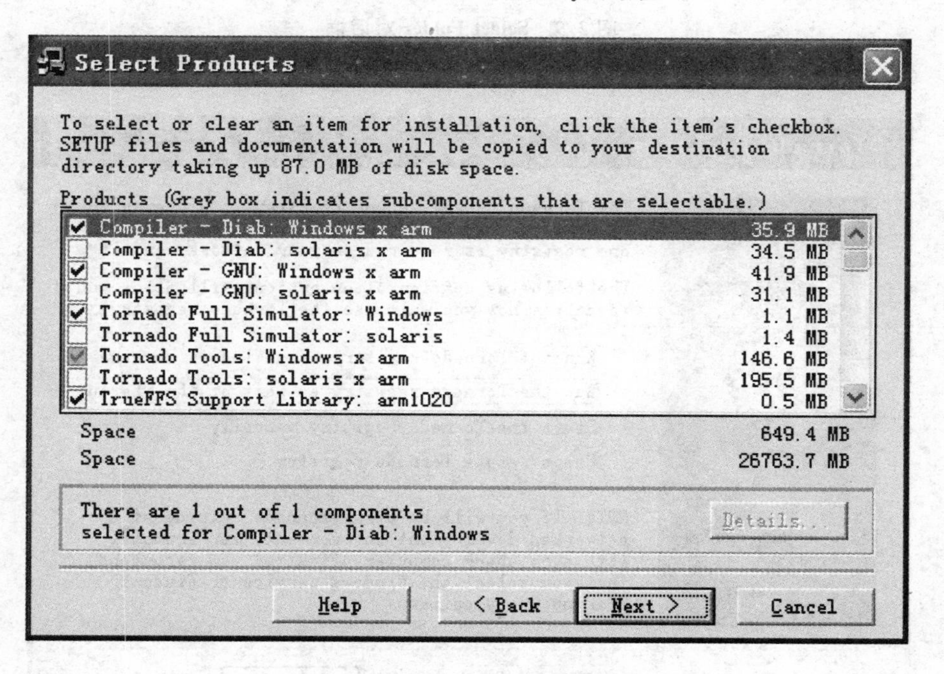

图 2.8 Select Products 对话框

(8) 如图 2.9 所示,设置程序里面文件夹的名字,然后执行 Next。

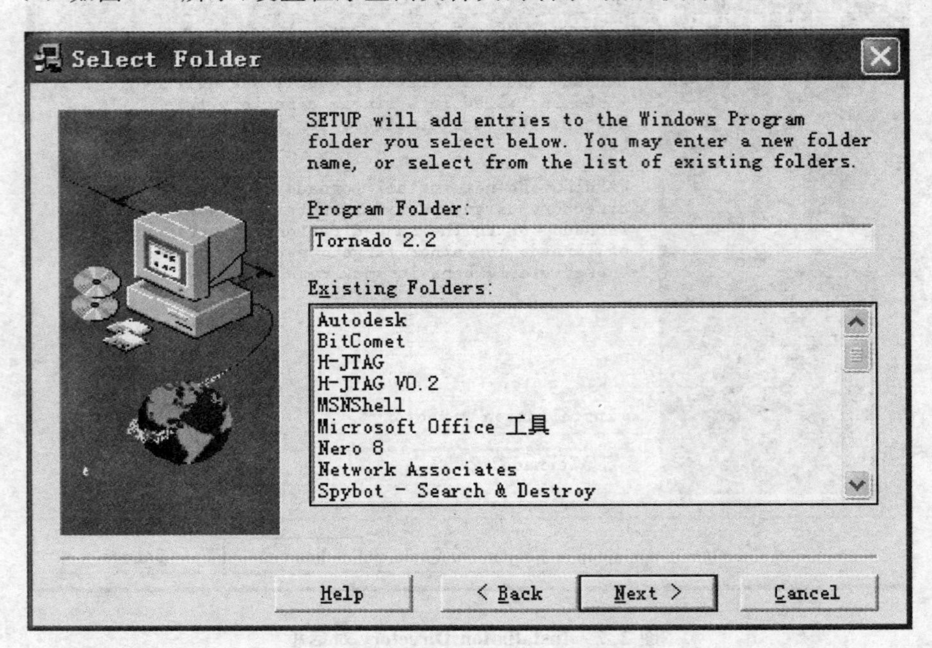

图 2.9 Select Folder 对话框

(9) 如图 2.10 所示,选择第 2 项,执行 Next。

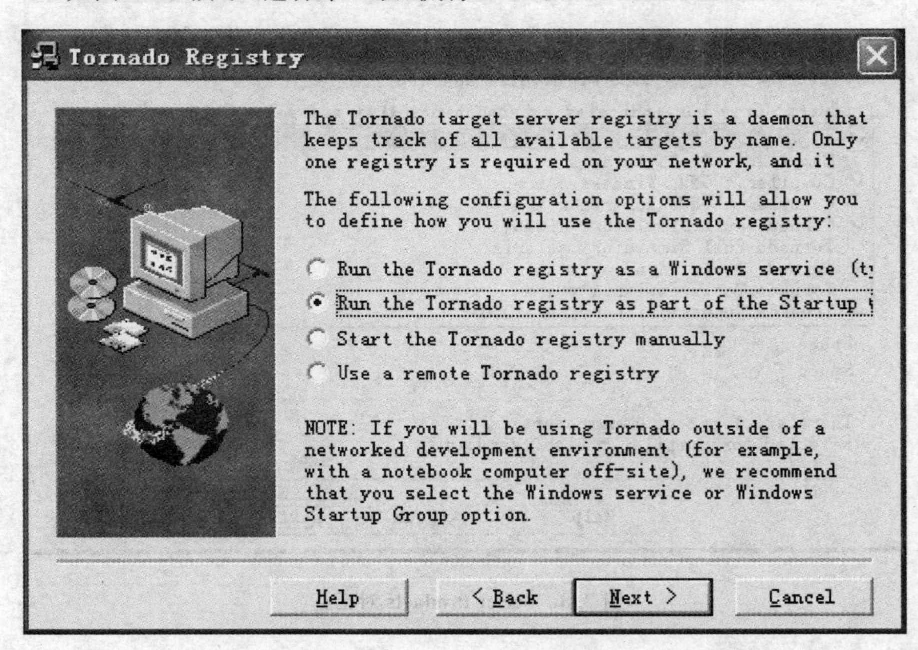

图 2.10 Tornado Registry 对话框

(10) 如图 2.11 所示,选择安装 T1.0.1 工具的适配器,执行 Next。

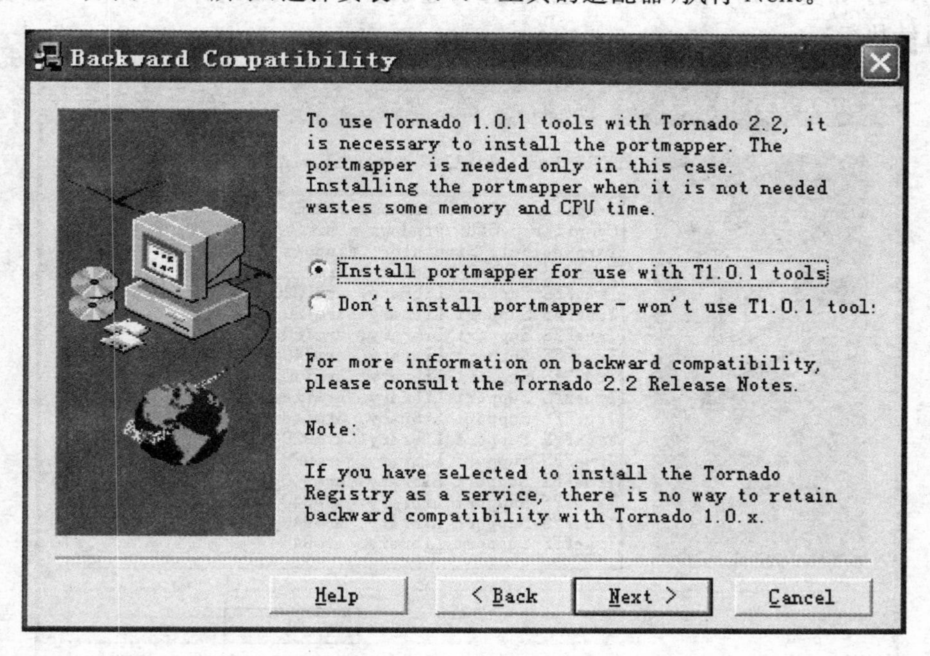

图 2.11 Backward Compatibility 对话框

(11) 如图 2.12 所示,默认选择即可,执行 Next。

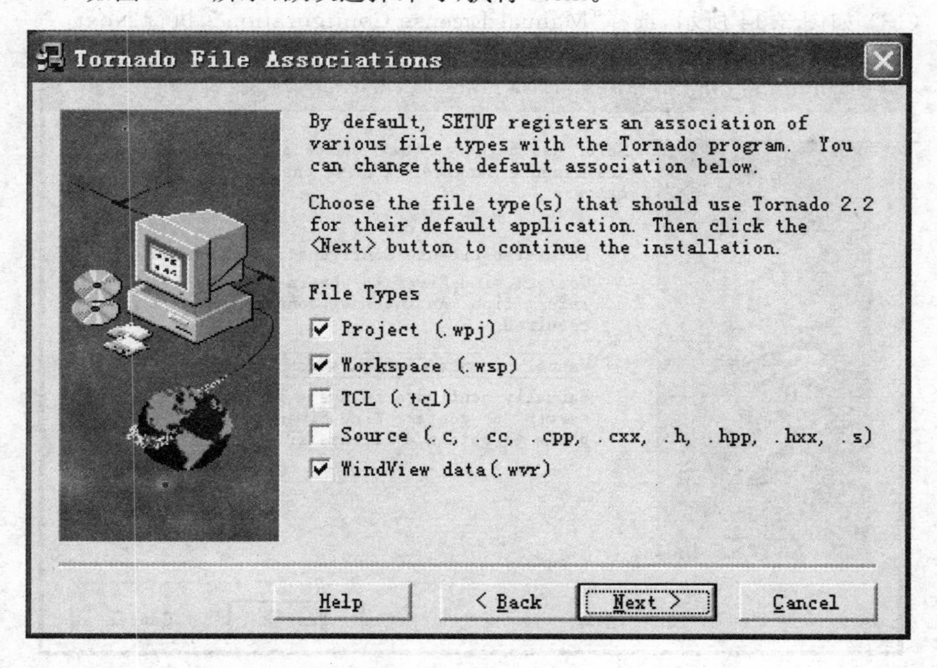

图 2.12 Tornado Files Associations 对话框

(12) 如图 2.13 所示，安装设置完成后，显示安装的产品组件，单击 OK，安装程序继续执行。

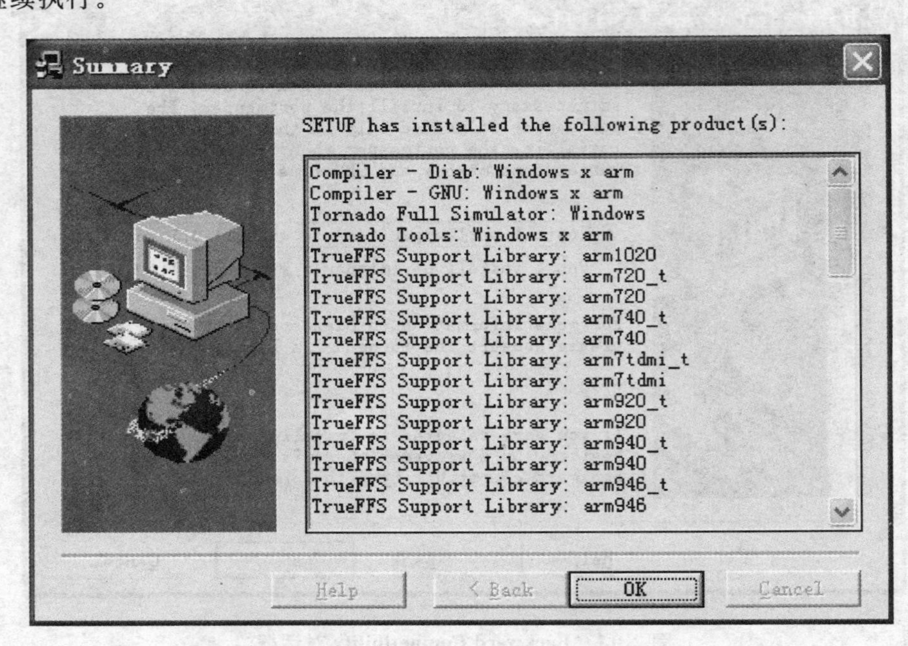

图 2.13 Summary 对话框

(13) 如图 2.14 所示，选择“Manual License Configuration”，执行 Next。

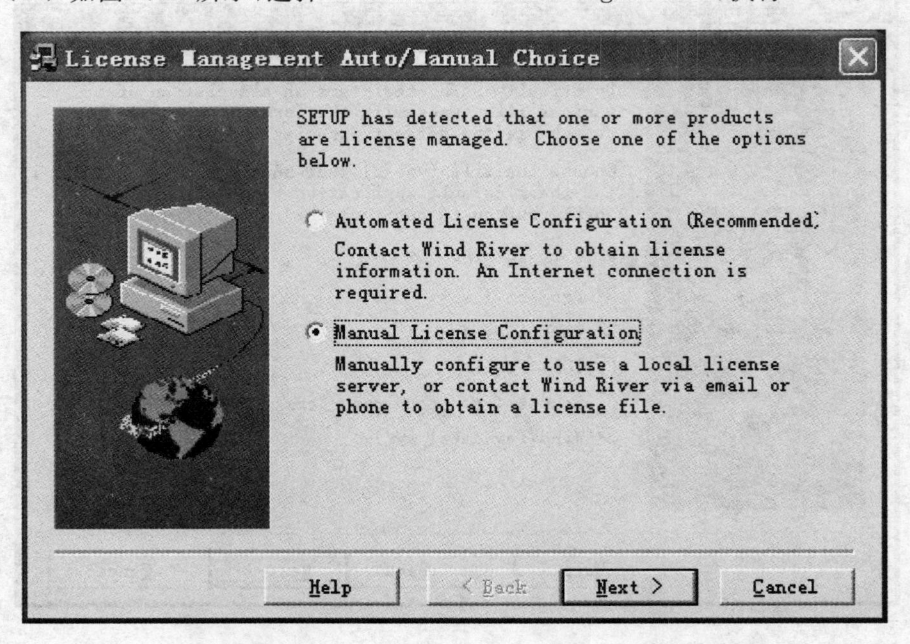

图 2.14 License Management Auto/Manual Choice 对话框

(14) 如图 2.15 所示,选择 Request a node—locked license file 项,执行 Next。

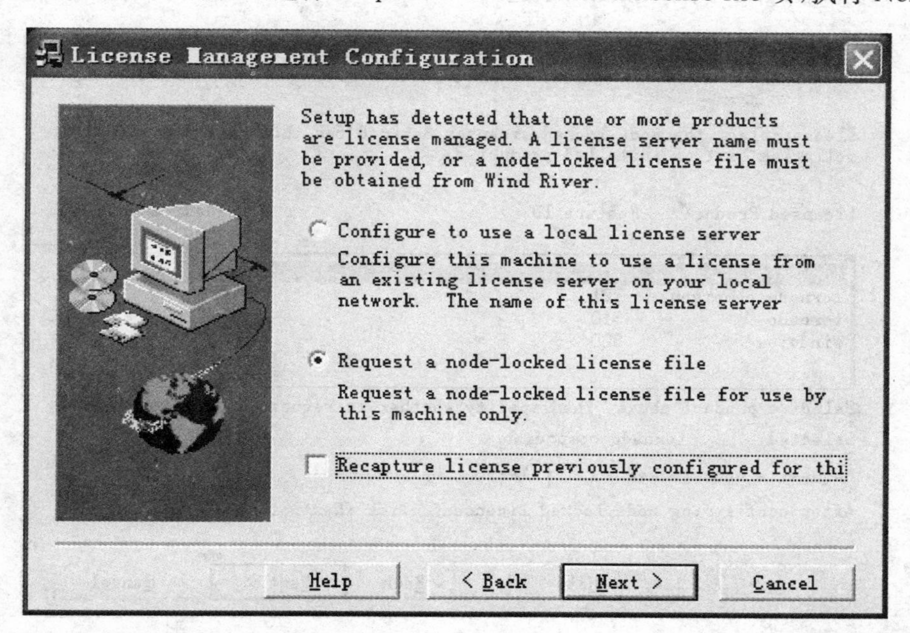

图 2.15 License Management Configuration 对话框

(15) 如图 2.16 所示,选择第二项(Phone or Fax),执行 Next。

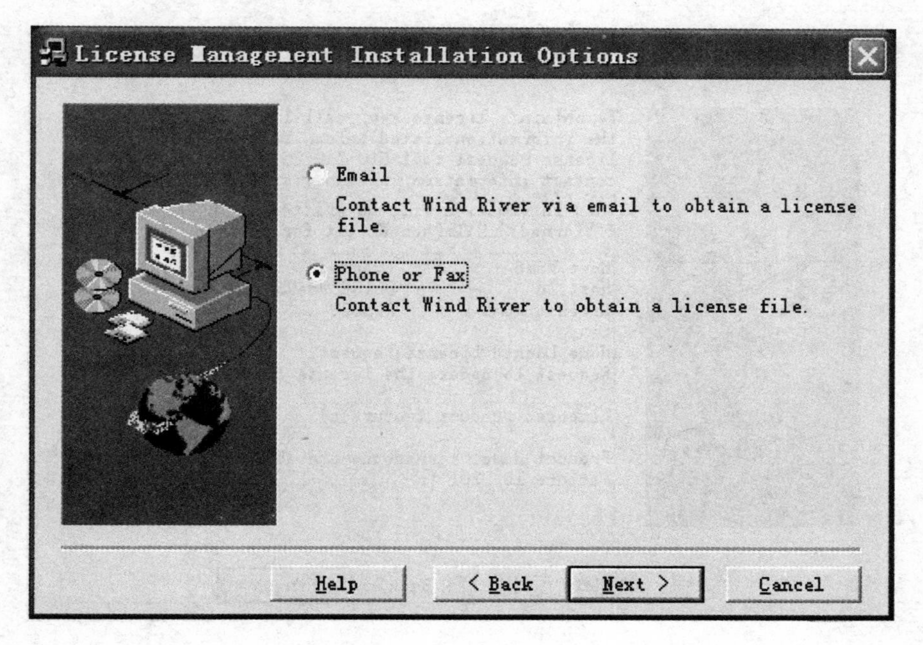

图 2.16 License Management Installation Options 对话框

(16) 如图 2.17 所示,按图所示选择 tornado - comp - diab 后,执行 Next。

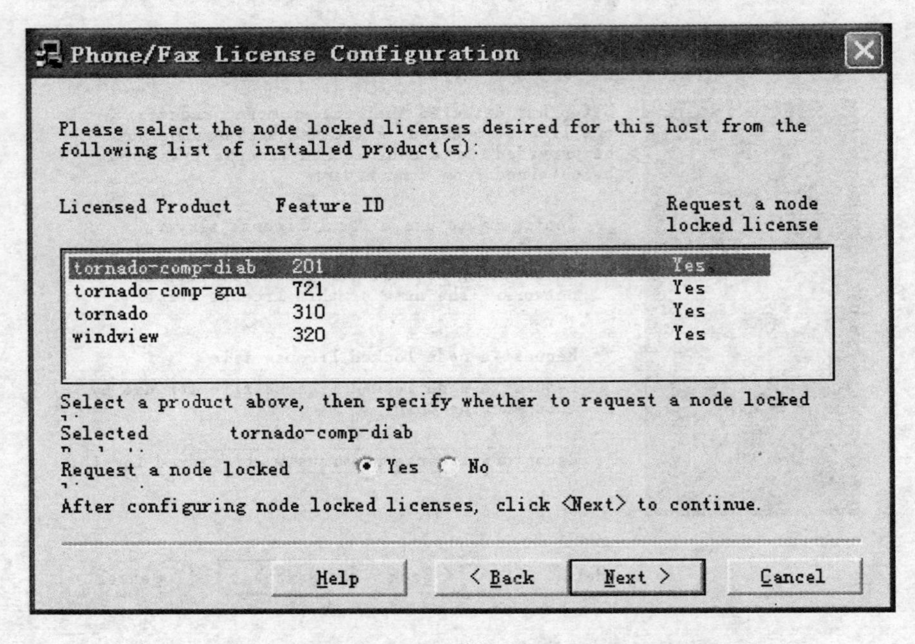

图 2.17 Phone/Fax License Configuration 对话框

(17) 如图 2.18 所示,执行 Finish,完成全部安装。

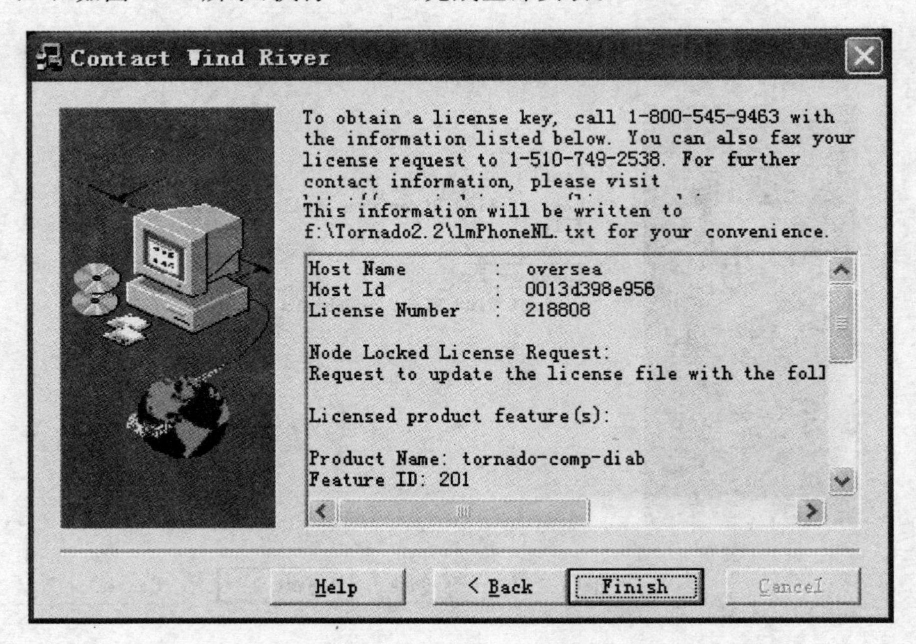

图 2.18 Contact Wind River 对话框

2.3 安装 WindML3.0

(1) 执行 WindML 安装文件夹里面的 SETUP.EXE，如图 2.19 所示，执行 Next。

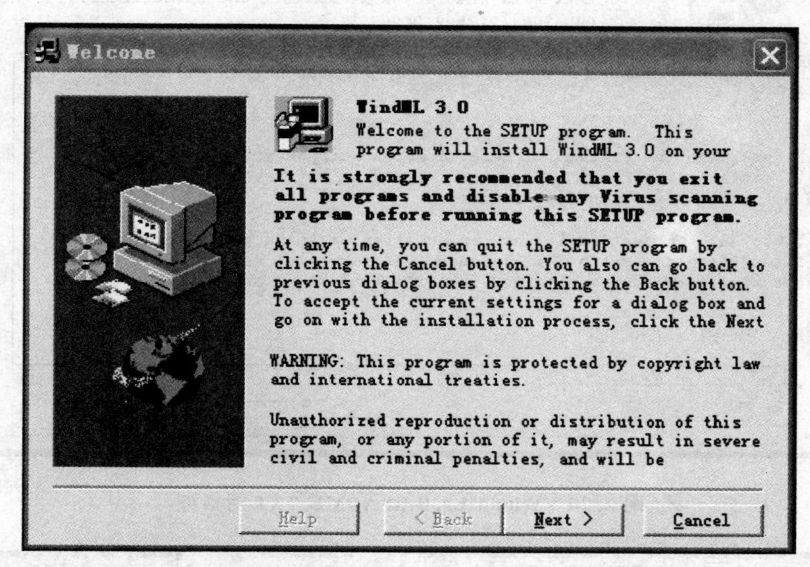

图 2.19 WindML3.0 安装对话框

(2) 如图 2.20 所示，继续执行 Next。

(3) 如图 2.21 所示，选择 Accept，执行 Next。

(4) 如图 2.22 所示，输入 Name 和 Company 名字，在 Install 里面输入注册码，执行 Next。

(5) 如图 2.23 所示，选择 Product installation(typical)，执行 Next。

(6) 如图 2.24 所示，填写序列号，工程名等，执行 Next。

(7) 如图 2.25 所示，选择 Tornado 安装路径，执行 Next。

(8) 如图 2.26 所示，选择针对性产品，执行 Next。

(9) 如图 2.27 所示，选择存放文件夹，执行 Next。

(10) 如图 2.28 所示，选择文件关联类型，执行 Next。

(11) 如图 2.29 所示，选择 OK，完成全部安装。

安装完成后，打开 Tornado 开发环境，如图 2.30 所示，在开发环境中工具栏中出现了 WindML 组件图标。

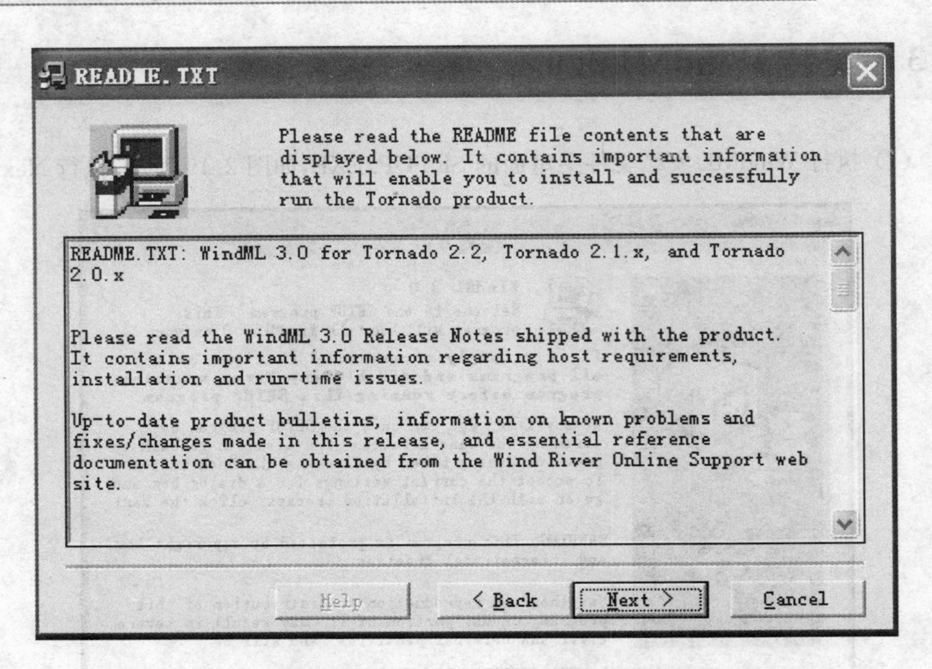

图 2.20 WindML3.0 README.TXT 对话框

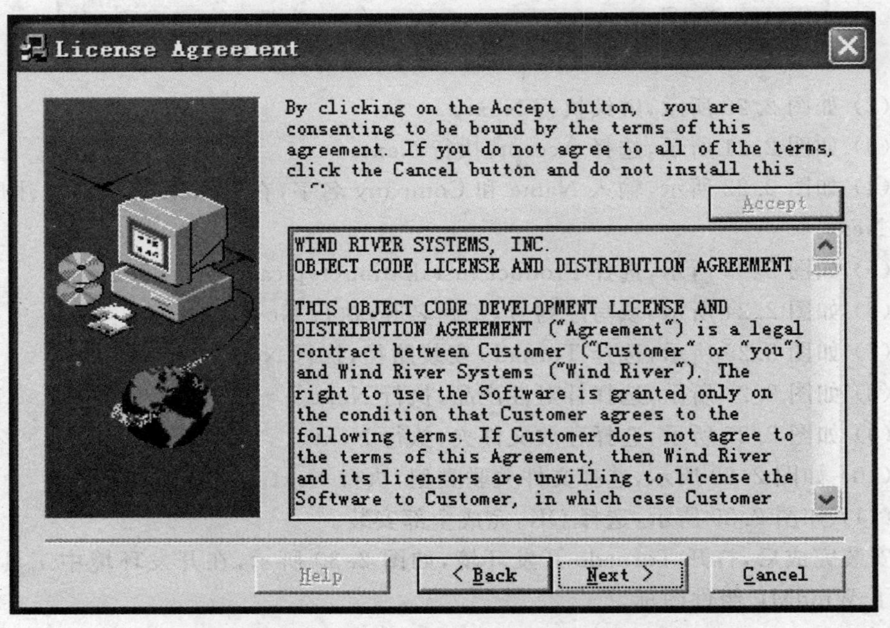

图 2.21 WindML3.0 License Agreement 对话框

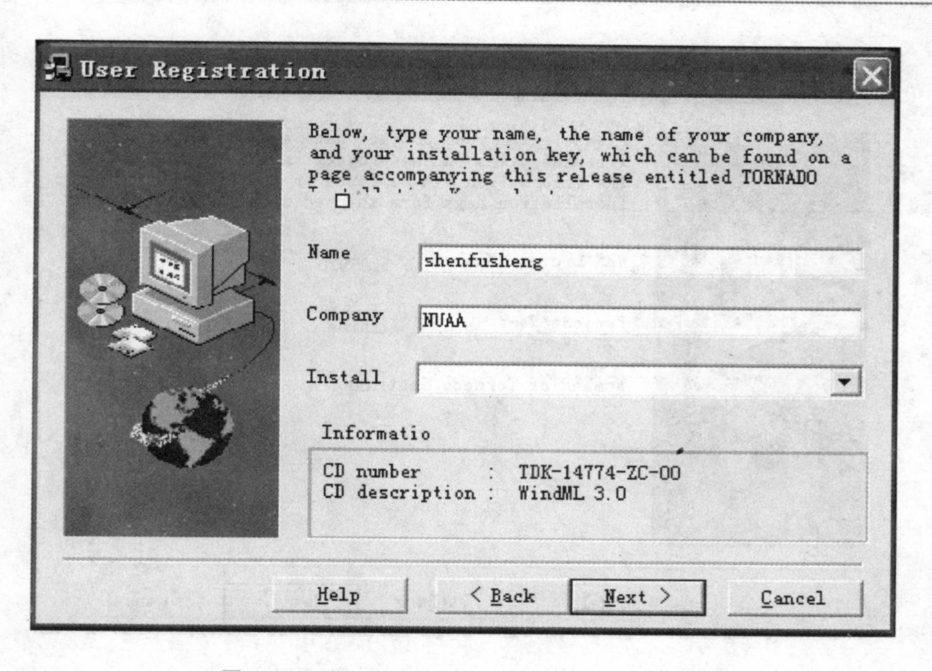

图 2.22　WindML3.0 User Registration 对话框

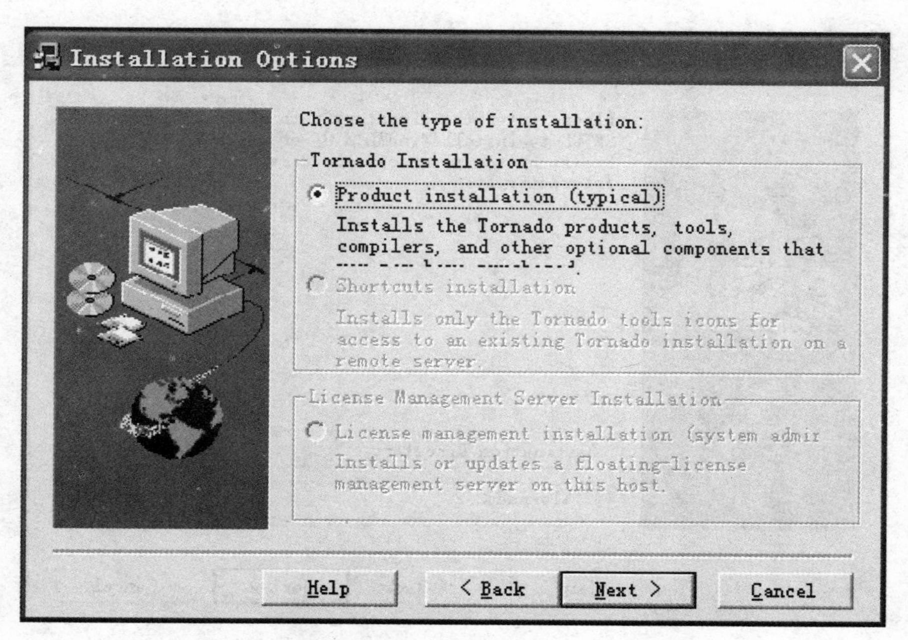

图 2.23　WindML3.0 Installation Options 对话框

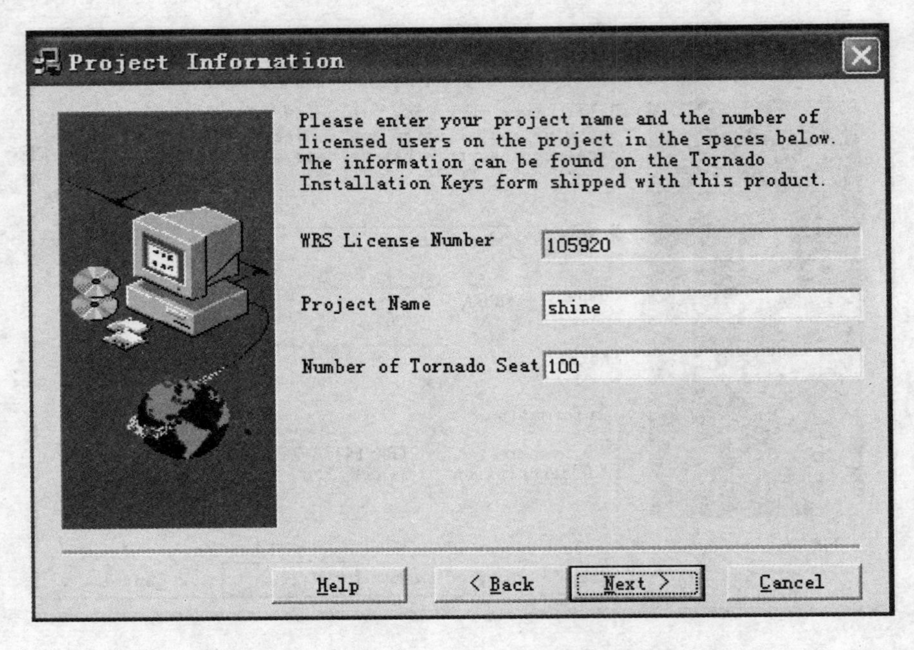

图 2.24　WindML3.0 Project Information 对话框

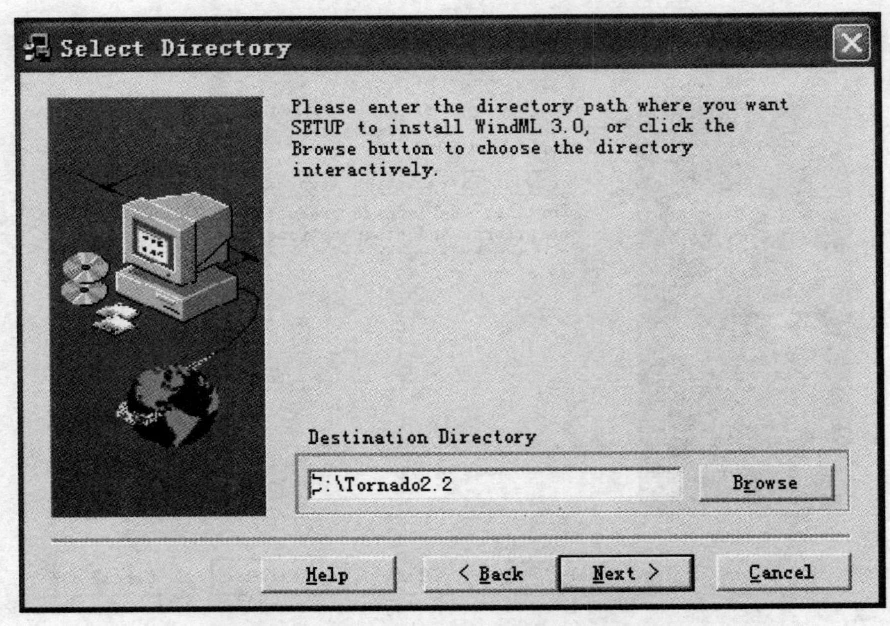

图 2.25　WindML3.0 Select Directory 对话框

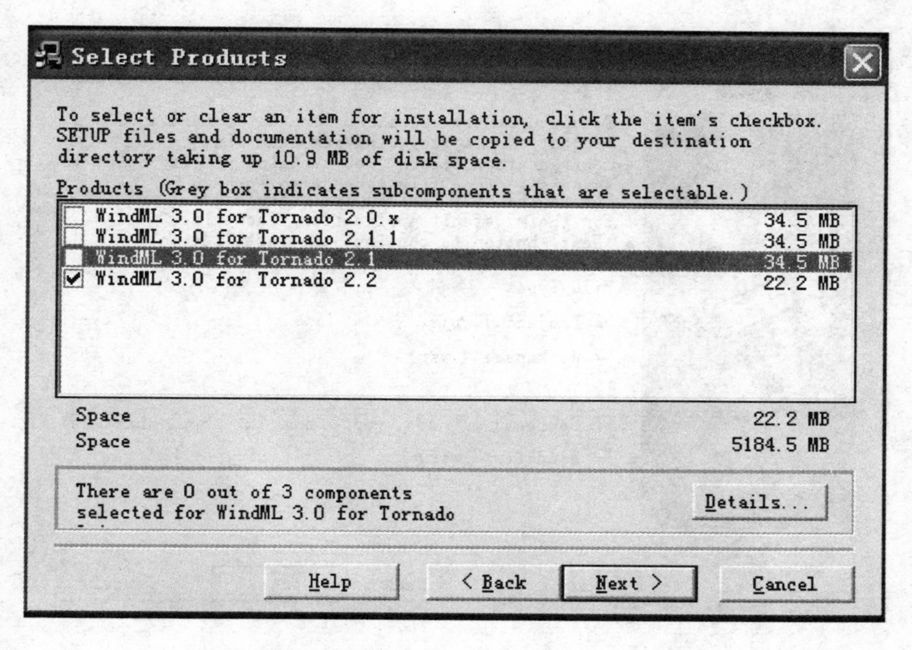

图 2.26 WindML3.0 Select Products 对话框

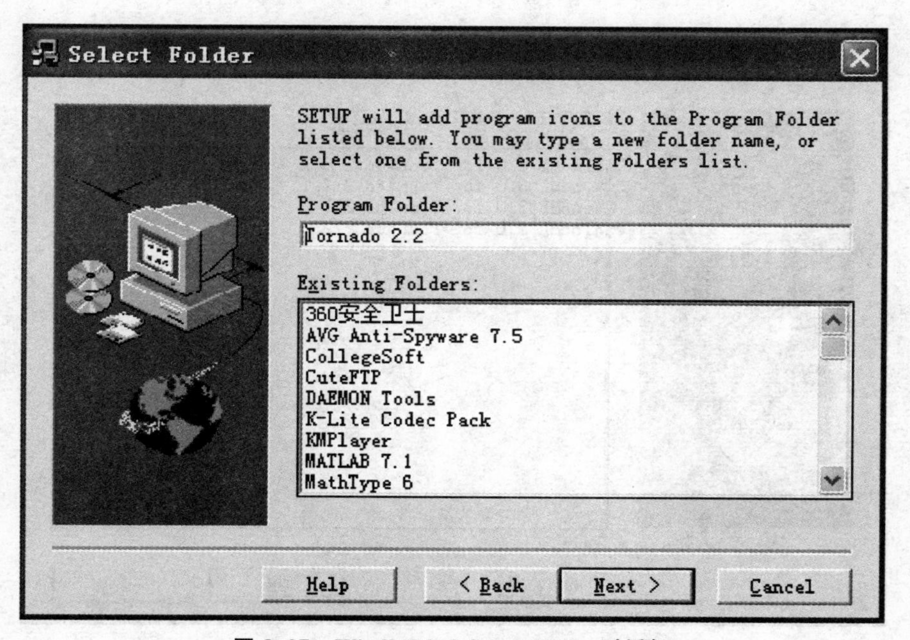

图 2.27 WindML3.0 Select Folder 对话框

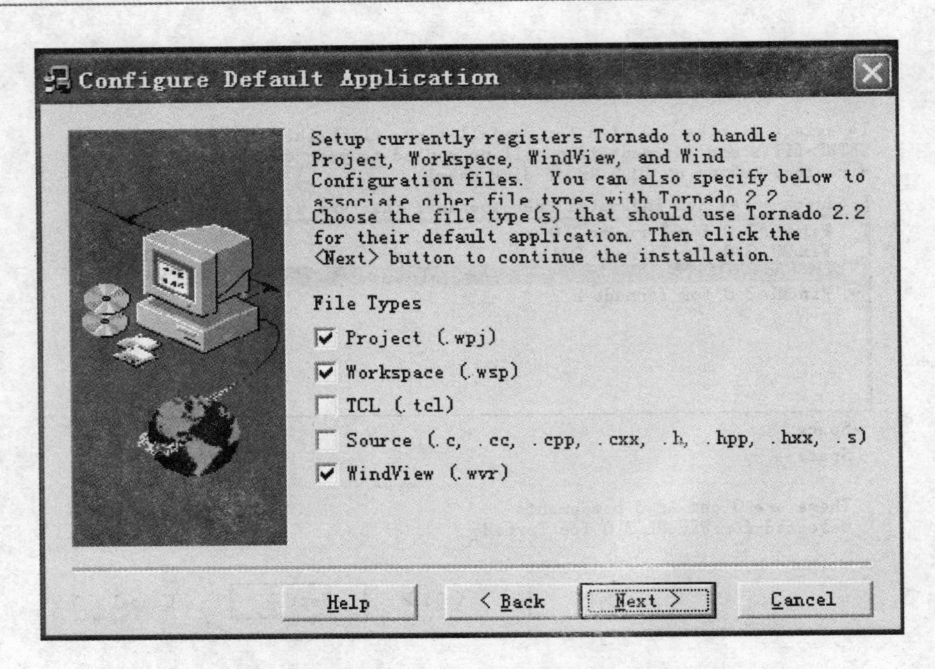

图 2.28　WindML3.0 Configure Default Application 对话框

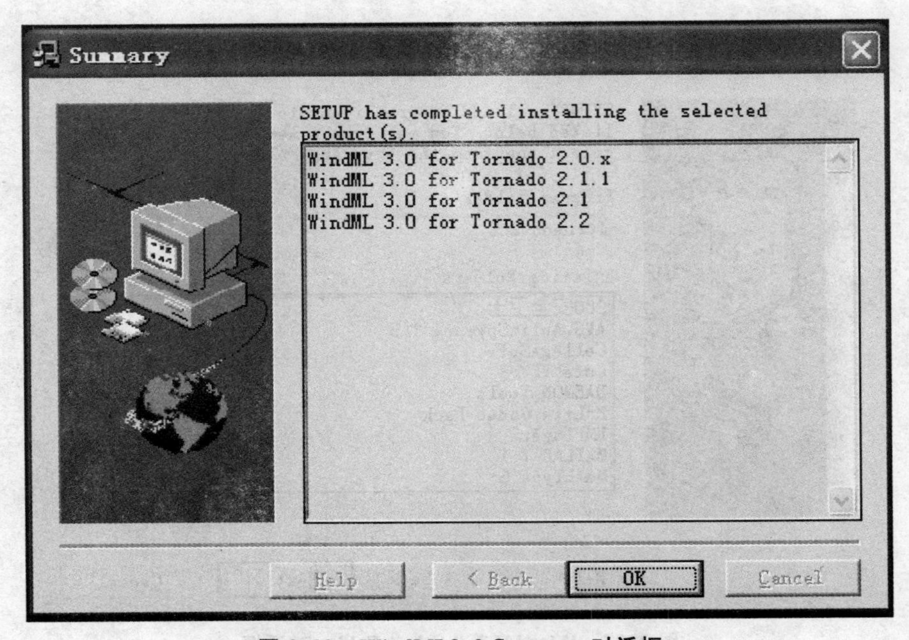

图 2.29　WindML3.0 Summary 对话框

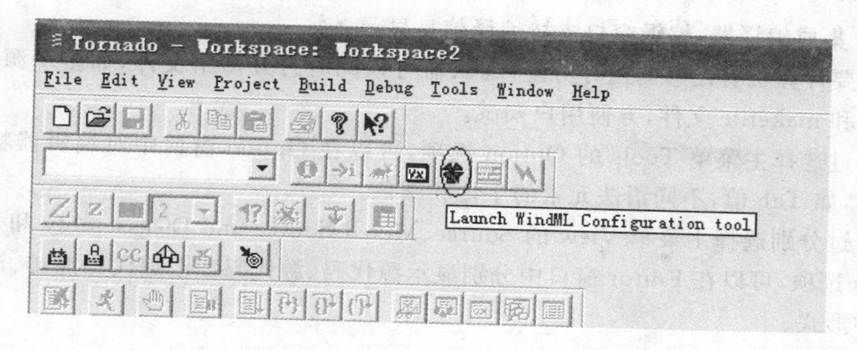

图 2.30　WindML3.0 在 Tornado 开发环境的图标

2.4　Tornado 工具包介绍

2.4.1　集成编辑器

如图 2.31 所示，Tornado 的源码编辑器包含标准的文本处理功能，采用 Windows 标准的编辑用法，同时具有以下特征：

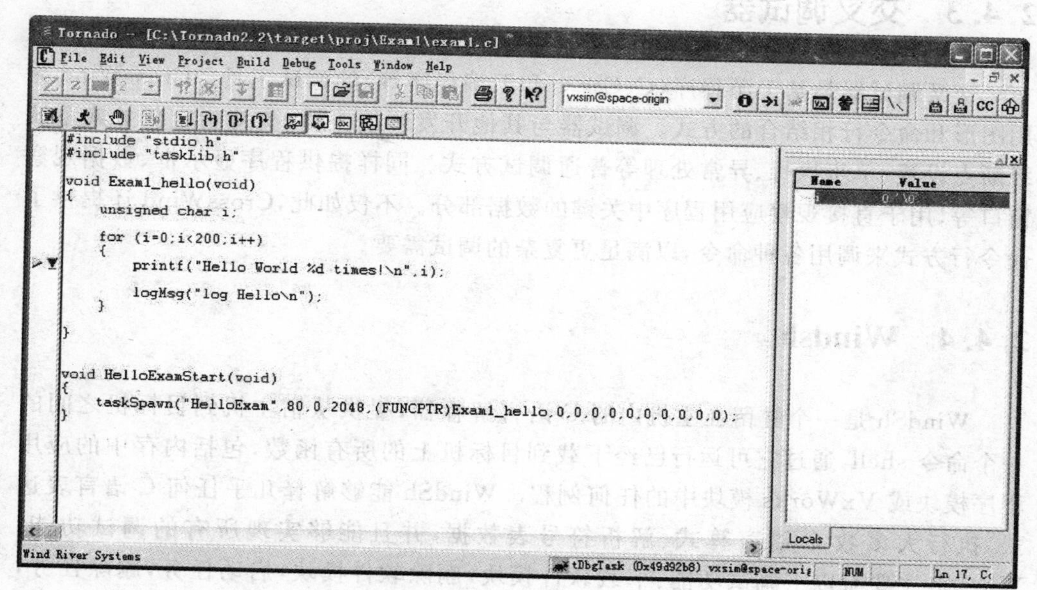

图 2.31　Tornado 开发环境集成编辑器

- 支持 C\C++语法元素不同颜色字体显示；
- 集成调试器，编辑窗口支持代码执行跟踪；

● 集成编译器,编辑窗口支持编译信息显示。

只要计算机有足够的内存分配,编辑器可以同时打开多个文件,如多个源文件,头文件和 makefile 文件,方便用户调试。

通过选择主菜单 Tools 的 Option 选项,可以在 Option 窗口中对编辑器特征进行设置,如 Tab 值,不同语法元素的字体和颜色。

通过分别选择主菜单 View 的 Source、Mixed Source and Disassembly 和 Disassembly 选项,可以在 Editor 窗口中分别显示源代码、源代码和汇编代码混排、汇编代码 3 种形式。

2.4.2 集成仿真器

集成仿真器,不依赖于硬件设备,能够独立在宿主机上实现对应用程序进行仿真分析。仿真器能够在宿主机上模拟目标机的 VxWorks 操作环境,实现目标机上大多数 VxWorks 系统功能。仿真器与目标机的主要区别在于:在 VxSim 里,内存 Image 是作为宿主机(Windows or Unix)的一个进程执行,不与目标硬件进行交互,无法仿真设备驱动运行结果。

2.4.3 交叉调试器

交叉调试器是基于源程序(C、C++和汇编程序等)的调试工具。CrossWind 采用图形和命令行相结合的方式。调试器与其他开发环境一样,支持指定任务或系统级断点设置、单步执行、异常处理等普通调试方式。同样提供程序显示框、数据观察窗口等,用于直接观察应用程序中关键的数据部分。不仅如此,CrossWind 还提供了命令行方式来调用各种命令,以满足更复杂的调试需要。

2.4.4 Windsh

WindSh 是一个驻留在主机内的 C 语言解释器,提供从宿主机到目标机之间的一个命令 shell,通过它可运行已经下载到目标机上的所有函数,包括内存中的应用程序模块或 VxWorks 模块中的任何例程。WindSh 能够解释几乎任何 C 语言表达式,执行大多数 C 语言算式、解析符号表数据,并且能够实现所有的调试功能。WindSh 主要有以下调试功能:下载软件模块,删除软件模块,启动任务,删除任务,设置断点,删除断点,运行、单步运行、继续执行程序,查看内存、寄存器、变量,修改内存、寄存器、变量,查看任务列表、内存使用情况、CPU 利用率,查看特定的对象(任务、信号量、消息队列、内存分区等),复位目标机等。

2.4.5　目标机代理(Target Agent)

目标代理遵循 WBD(Wind Debug)协议,允许目标机(Target)与主机(Host)上的 Tornado 开发工具相连。在目标代理的缺省设置中,目标代理是以 VxWorks 的一个任务(tWdbTask) 的形式运行的,如图 2.32 所示。

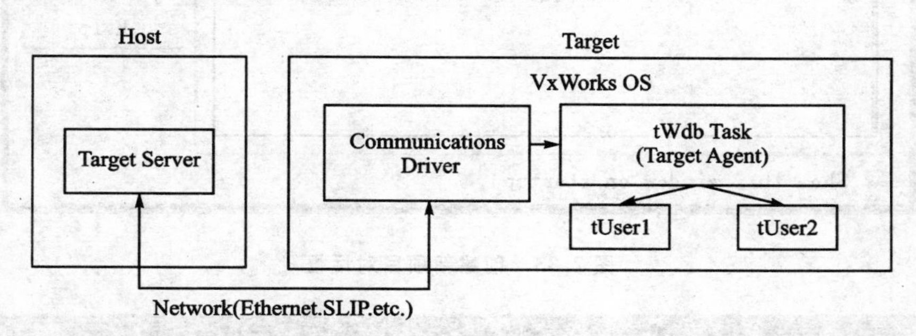

图 2.32　Tornado 目标机与主机通信关系图

Tornado 目标服务器(Target Server)向目标代理发送调试请求。调试请求通常决定目标代理对系统中其他任务的控制和处理。在缺省状态下,目标服务器与目标代理通过网络进行通信,但是用户也可以改变通信方式,比如串口方式等。

2.5　创建和管理工程

如图 2.33 所示,选择"Creat a bootable Vxworks image",单击 OK,创建可启动工程。

如图 2.34 所示,按照向导的提示做如下修改:

- 工 程 名　　MagicARM2410
- 工程目录　　C:\Tornado2.2\target\proj\MagicARM2410
- 工程描述　　<任意>
- 工程空间　　C:\Tornado2.2\target\proj\MagicARM2410.wsp

单击如图 2.34 中的"Next"进入下一步向导,如图 2.35 所示;选择"A BSP",从下拉列表中选择"MagicARM2410",单击"Next"进入确认选择界面,如图 2.36 所示;单击"Finish"出现工作空间窗口,如图 2.37 所示。新项目生成后,文件管理器窗口,VxWorks 操作系统组件管理器窗口、开发窗口,项目生成管理窗口,分别如图 2.38、图 2.39、图 2.40、图 2.41 所示。

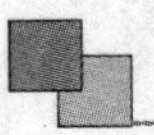

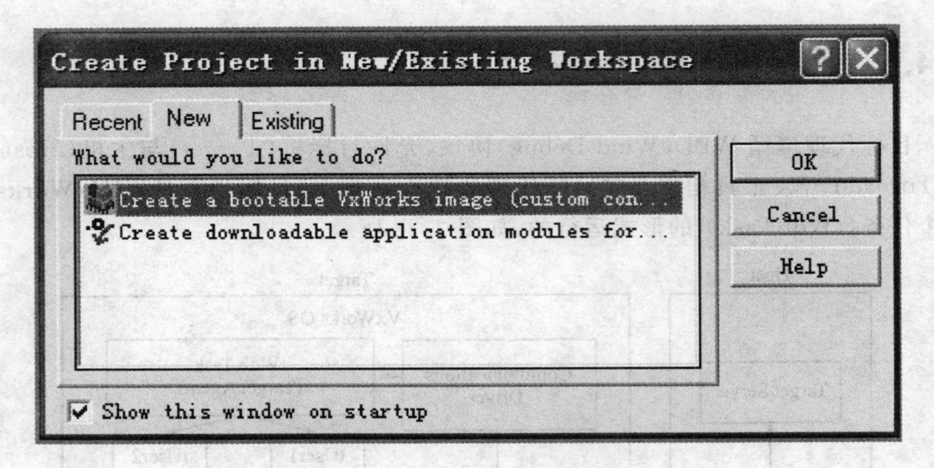

图 2.33 创建新项目对话框

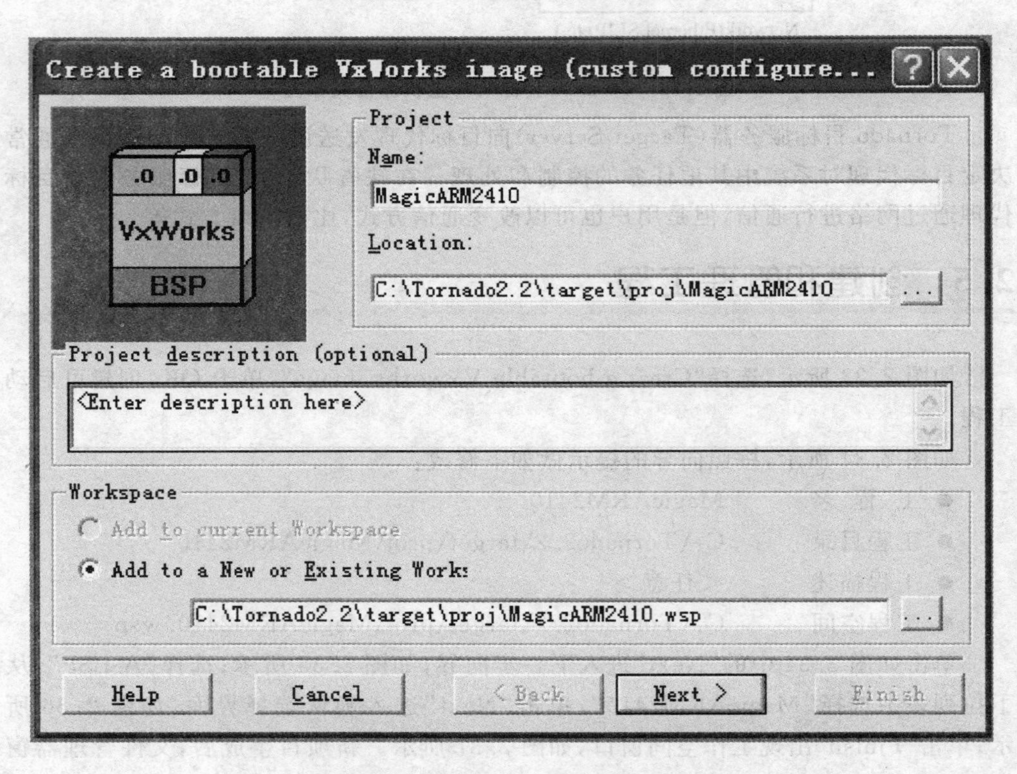

图 2.34 新项目配置信息对话框步骤 1

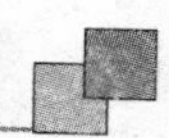

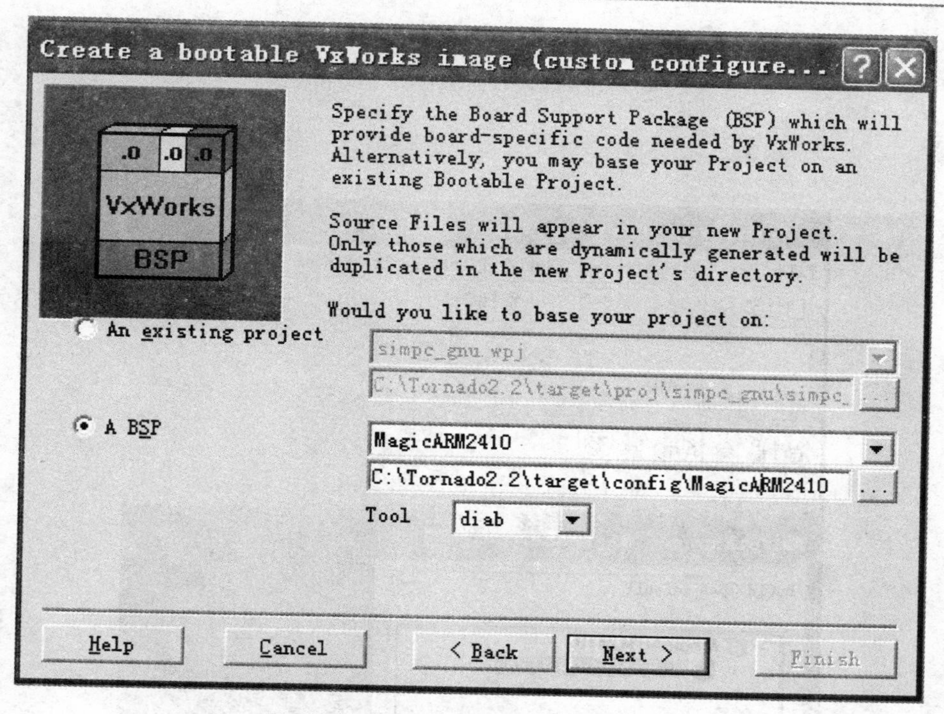

图 2.35　新项目配置信息对话框步骤 2

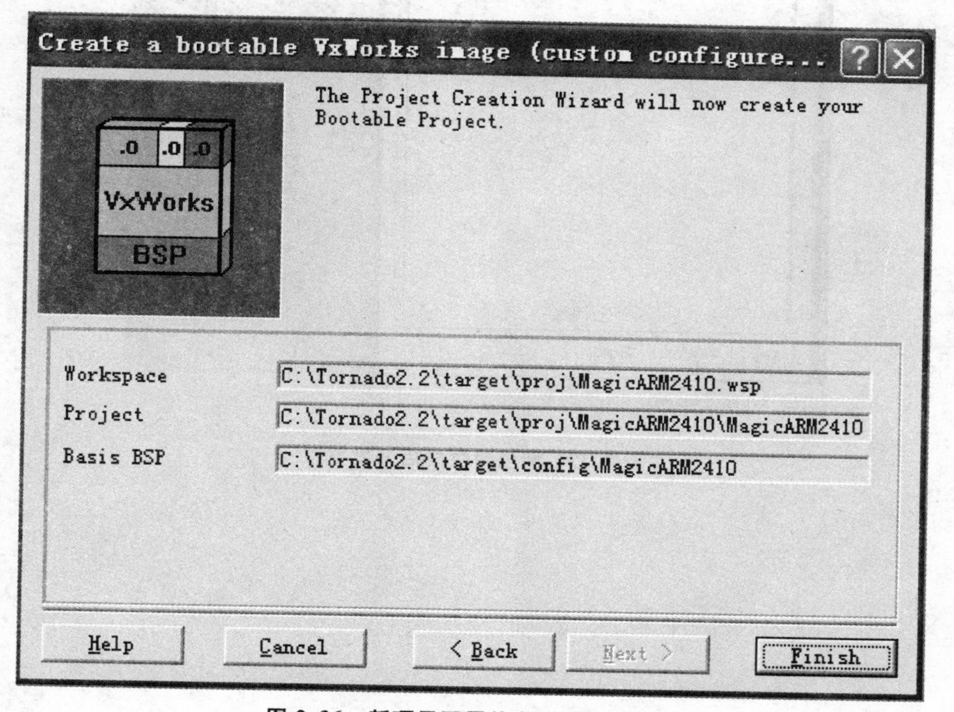

图 2.36　新项目配置信息对话框步骤 3

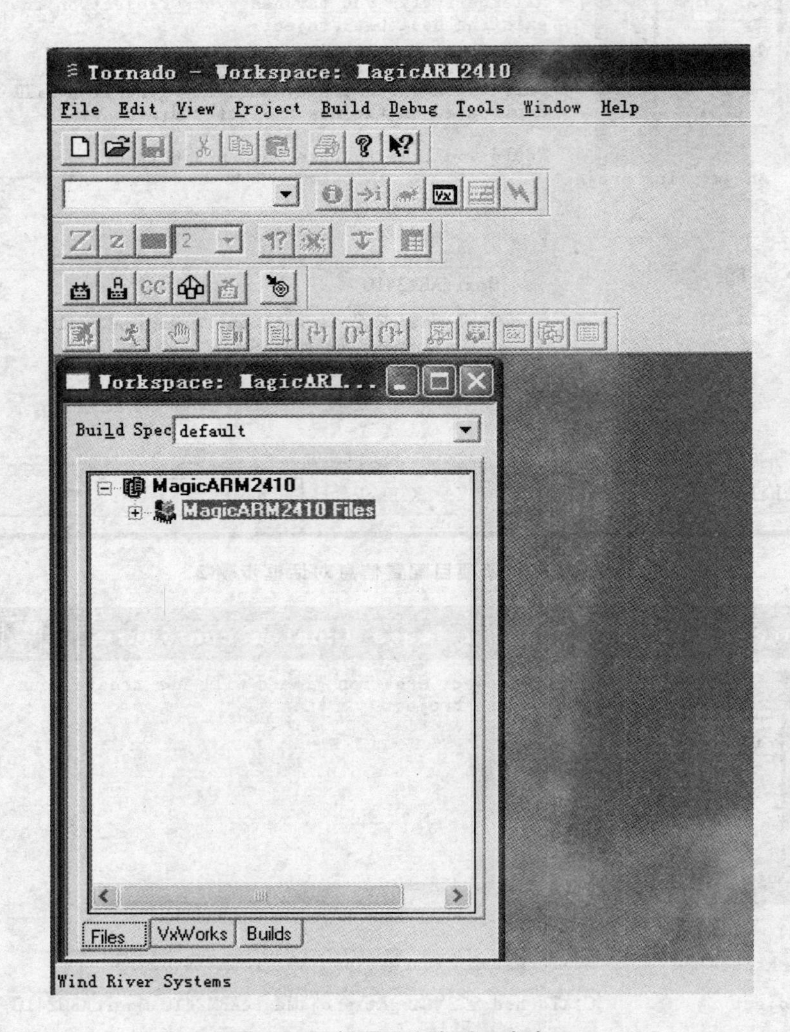

图 2.37　新项目生成后开发窗口

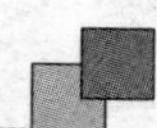

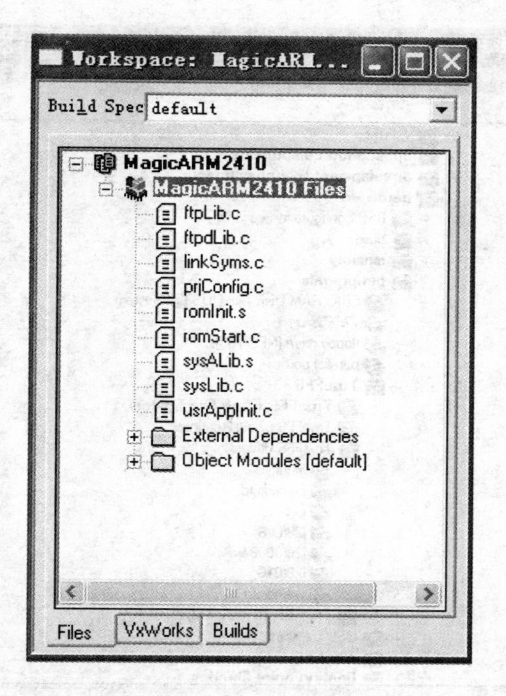

图 2.38 新项目生成后文件管理器

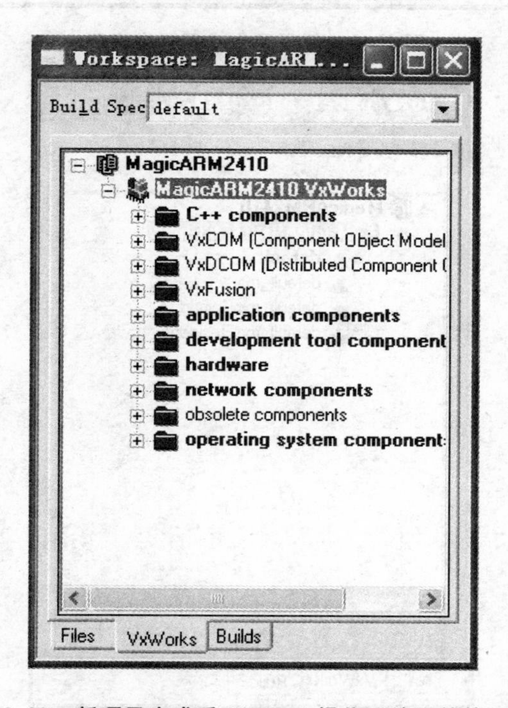

图 2.39 新项目生成后 Vxworks 操作系统组件管理器

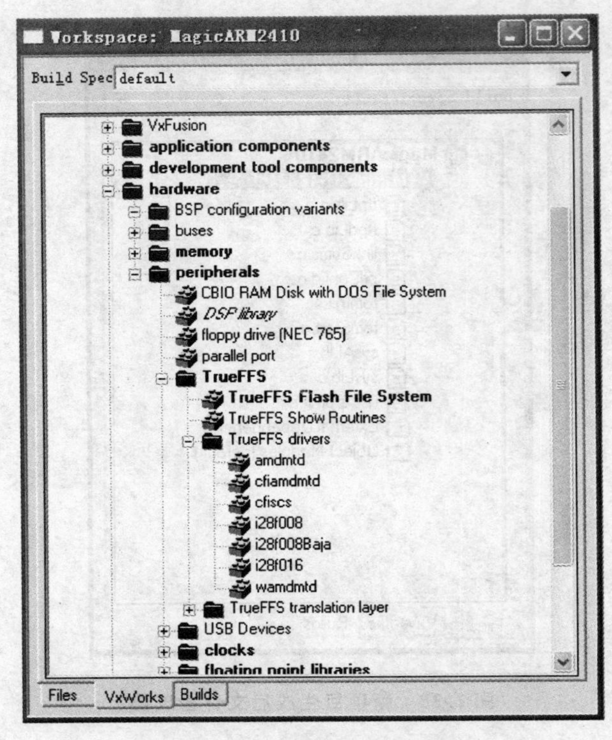

图 2.40　新项目生成后开发窗口

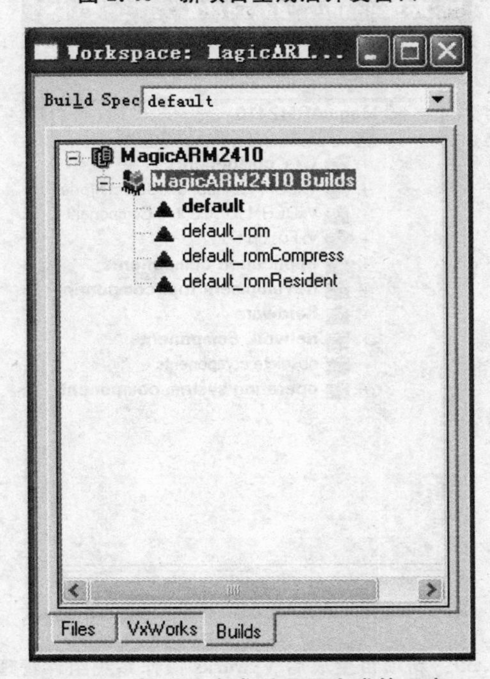

图 2.41　新项目生成后项目生成管理窗口

2.6 内核配置和裁剪

VxWorks 开发可以通过配置来增加或者删减组件，实现自由裁剪功能。如图 2.42所示，在项目的“Workspace”窗口的“VxWorks”标签页中，可以看到 VxWorks 支持的各种组件。此时，如果要添加“TrueFFS”组件下的“TrueFFS Show Routines”，则在选择该选项后单击鼠标右键，在弹出的菜单中，能够看到有关的选择。其中，如果选择“Include ‘TrueFFS Show Routines’”选项，则可以将该组件包含进内核当中；选择“Exclude ‘TrueFFS Show Routines’”，则是删除该组件。

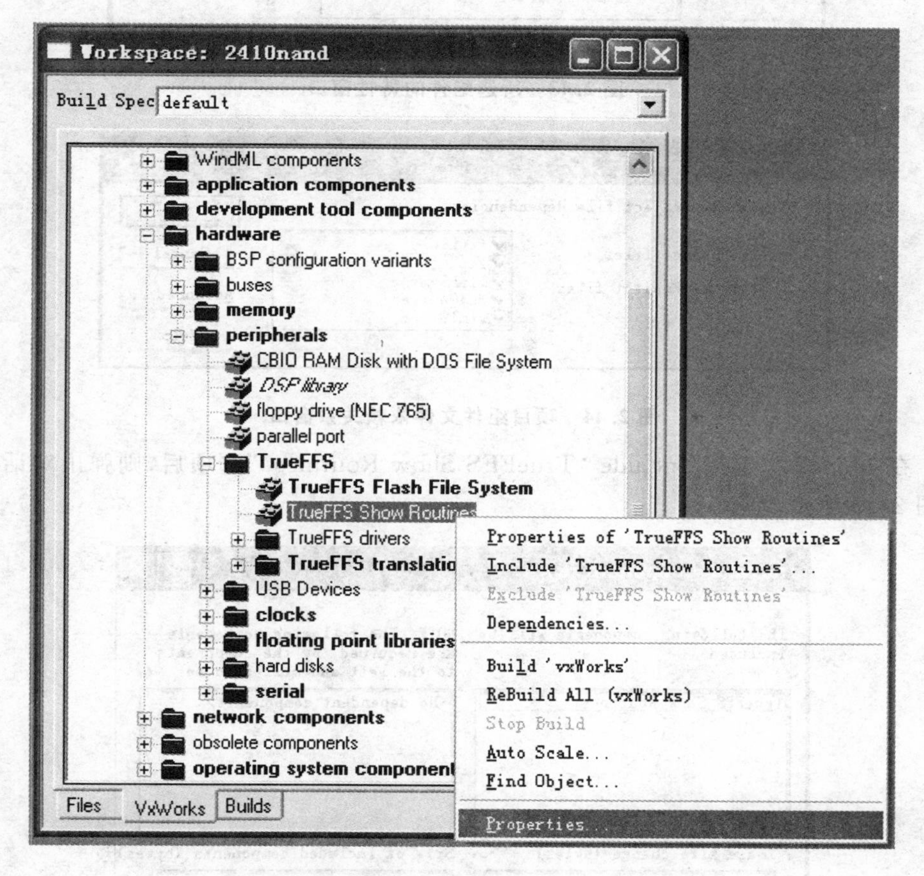

图 2.42 项目生成后组件管理窗口组件特性

在图 2.42 中选择“Properties”，则弹出“TrueFFS Show Routines”组件特性对话框，如图 2.43 所示。

在图 2.42 中选择“Dependencies”，则弹出项目的文件依赖关系对话框，如图2.44 所示。

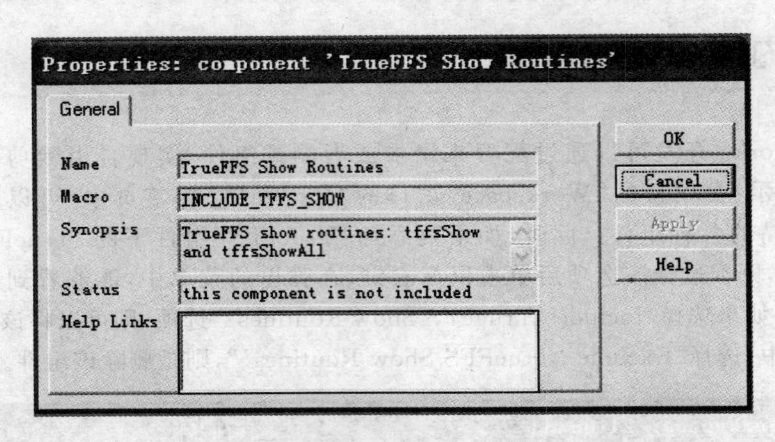

图 2.43　所选组件的特性窗口

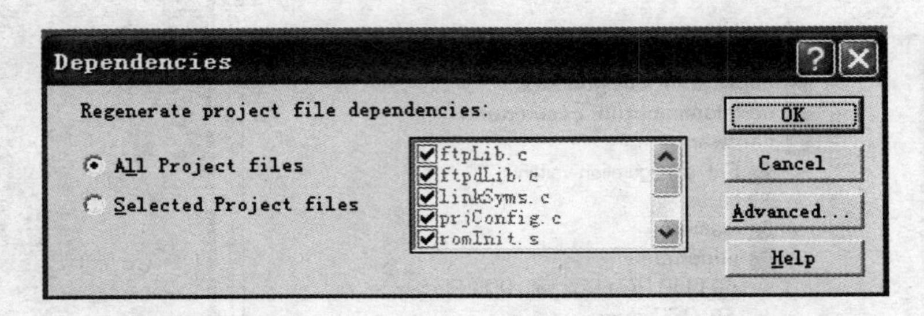

图 2.44　项目组件文件依赖关系窗口

在图 2.42 中选择“Include ‘TrueFFS Show Routines’”选项后，则弹出对话框，如图 2.45 所示。

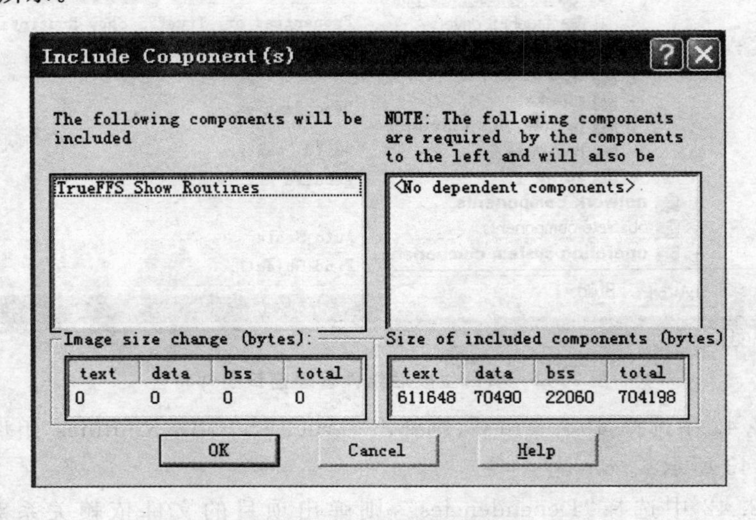

图 2.45　增加组件对话框窗口

如果需要在已有的项目中增加文件，则需要在“Workspace”窗口的“Files”标签页中选择项目名称，然后单击鼠标右键弹出对话框，如图 2.46 所示，此时选择“Add Files”项，就可以增加需要增加的文件。选择该项后，则弹出对话窗口，如图 2.47 所示，在该窗口完成文件的选择工作。

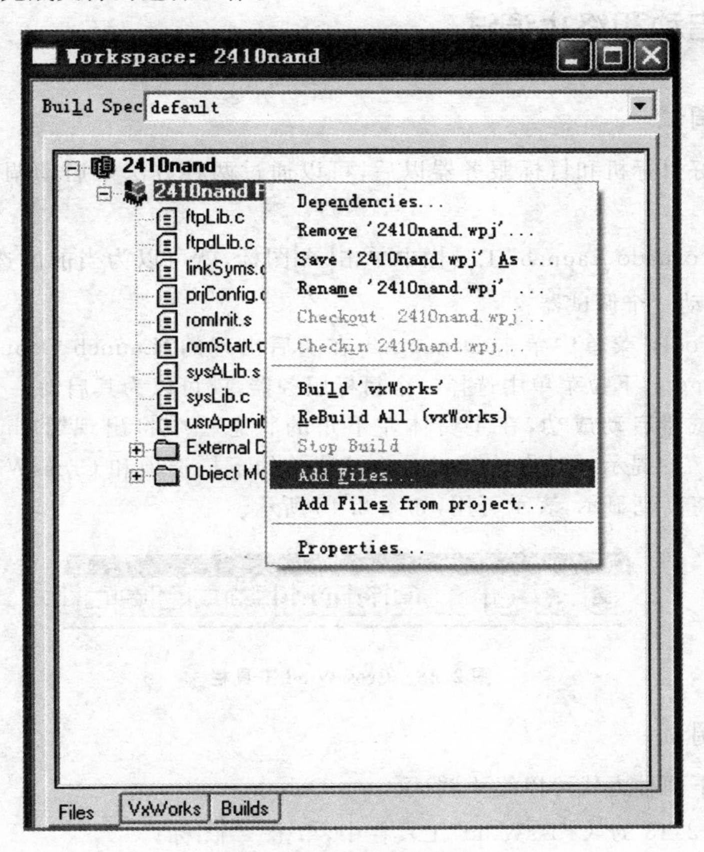

图 2.46　增加文件菜单

图 2.47　选择增加文件窗口

2.7 WDB 调试程序方法

2.7.1 启动和终止调试

1. 启动调试

当配置好目标机和目标服务器以后，可以通过两种方法来启动调试（Start Debugging）：

- 在“Tornado Launch”工具栏中单击图标，就可以为当前所选的目标服务器启动一个调试器；
- 在“Tools”菜单中单击 Debugger，在随后出现的“Launch Debugger”窗口的“Targets”下拉菜单中选择一个目标服务器，就可以为其启动一个调试器。

如果调试器启动成功，在主窗体左下角的状态栏中将出现“Debugger started successfully.”的提示。对应的“Debugger”菜单的下拉选项和 CrossWind 工具栏中的快捷图标将高亮显示，表示可用，如图 2.48 所示。

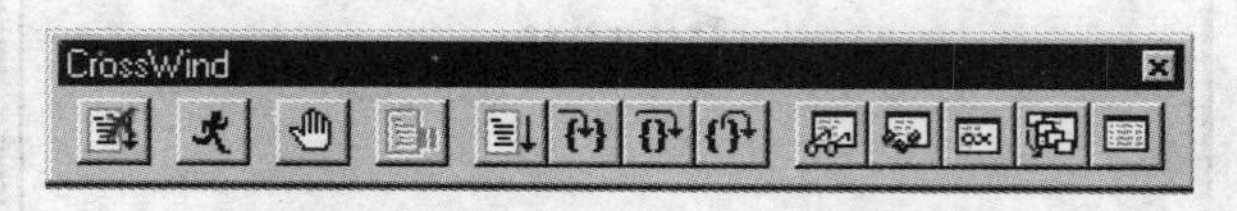

图 2.48 CrossWind 工具栏

2. 终止调试

通过以下两种方式可以终止调试（Stop Debugging）：

- 在图 2.48 的“CrossWind”工具栏中，单击图标；
- 在“Debug”下拉菜单中单击“Stop Debugging”选项关闭调试器，相应的调试工具选项将变成灰色，如需进行调试，需重新启动调试器。

3. 中断调试

单击“CrossWind”工具栏中的图标；或者选择“Debug”下拉菜单中的“Interrupt Debugger”选项可以中断程序的执行。若当前调试的任务正处于全速运行的状态，可以中断其执行。

2.7.2 运行程序

单击“CrossWind”工具栏中的图标或选择“Debug”下拉菜单中的“Run”选

项，就会出现“Run Task”窗口，如图 2.49 所示。

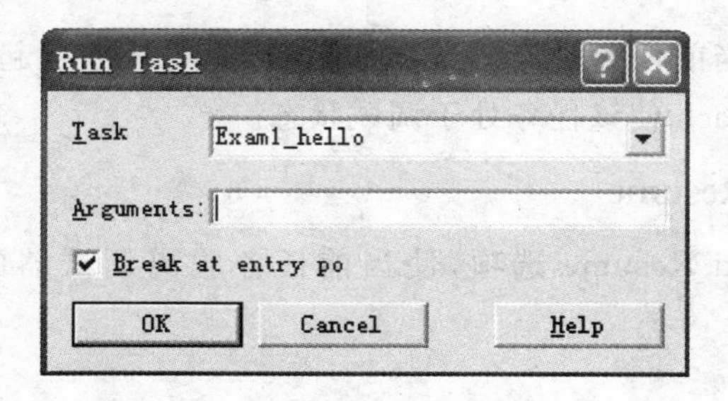

图 2.49 Run Task 窗口

利用“Run Task”窗口指定需要运行的函数和函数参数。函数参数之间以空格隔开。参数列表必须是整数或地址，不能是浮点或双精度值、函数调用。选中“Break at entry point”框可以在函数的第一条语句处设置一个临时断点，这样程序一运行就会停在第一条语句处，用户可以执行单步、跳过子函数调用或恢复执行。

2.7.3 Attach 和 Detach 一个任务

1. Attach 任务

选择 Debug 下拉菜单中的 Attach 选项，能够实现将一个已经运行的任务设置成调试状态。如果在此之前正在调试另一个任务，之前的任务就会被释放掉，脱离调试器的控制，并保持其当前状态（运行或中止）。如图 2.50 所示，Attach 窗口中列出运行于目标板上的所有任务的滚动列表。用户可以在任务列表中选择一个任务，也可以通过在“Attach to”框中键入任务名称（或任务 ID）选择一个任务。Attach 一个任务以后，调试器立即挂起该任务。

如图 2.50，Attach 窗口的第一项为 System，选择此项可以进入系统调试模式。如果 BSP 配置不支持系统模式，将会显示出错信息。

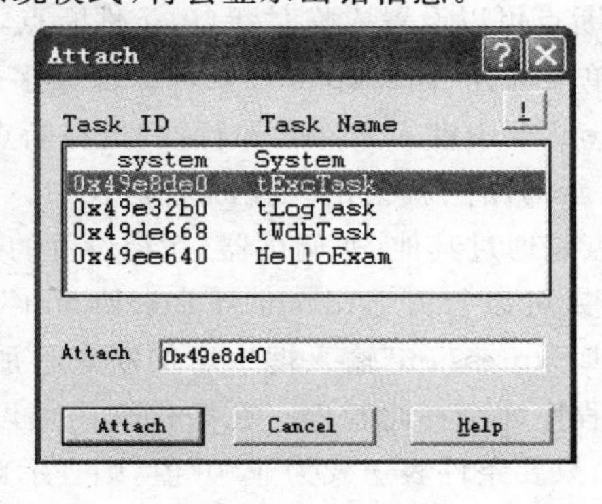

图 2.50 Attach 窗口

2. Detach

Detach 选项使当前任务脱离调试器的控制,并将任务挂起,在以后需要的时候仍可以通过选择 Attach 使该任务处于调试状态。

3. Detach and Resume

单击 Detach and Resume 选项,使当前任务脱离调试器的控制,任务将继续执行。

2.7.4 断 点

1. 断点类型

① 任务级断点。任务级断点,仅对当前调试任务有效,设置时单击菜单命令“Debug|Toggle BreakPoint”,或者将光标放在源文件处,点击图标即可。

② 全局断点。全局断点对所有任务都有效,设置全局断点,采用菜单命令“Debug|Toggle Global Breakpoint”。在任务模式下,断点只对当前被调试的任务有效,但如果想了解当前被调试任务的动作是否会对另一个任务有影响,而同时又只能调试一个任务,就必须使用全局断点。全局断点对任何任务都有效,一旦程序执行到此处,当前调试任务都会直接进入到“Suspend”状态,然后系统切换到所设全局断点所在任务进行调试。

③ 临时断点。设置临时断点,选择菜单命令“Debug|Toggle temp BreakPoint”,临时断点仅中止程序一次。一旦程序在此中止,“Debugger”会自动删除它。临时断点的标识是一个中空的倒三角,与其他类型断点有区别。

④ 条件断点。只有当条件满足时,条件断点才起作用。

2. 设置断点

任务级断点和全局断点可以设置成临时断点,条件断点,或临时条件断点。

在“Debug”下拉菜单中选择“Breakpoint”就可以设置多个不同类型的断点。如图 2.51 所示,在“Location”框中键入文件名和行数,选择断点类型(任务级断点或全局断点),单击“Add”键,新的断点就会出现在断点列表中。如果选中了“Externally managed”框,表示该断点是通过其他(非调试器)途径设置的,如“Tornado Shell”。

单击“Advanced”按钮可以打开“Advanced Breakpoint”窗口,如图 2.52 所示。允许用户在“Conditional Expression”输入框中增加断点附加条件,只有在此条件满足时,断点才会起作用,程序才会在此暂停。在条件框内可以输入一个整型表达式,或是一个变化的内存值,只要条件表达式为非“0”值,则表示此条件成立。

“Number of times to skip”框指定在导致程序暂停之前允许经历的断点次数。

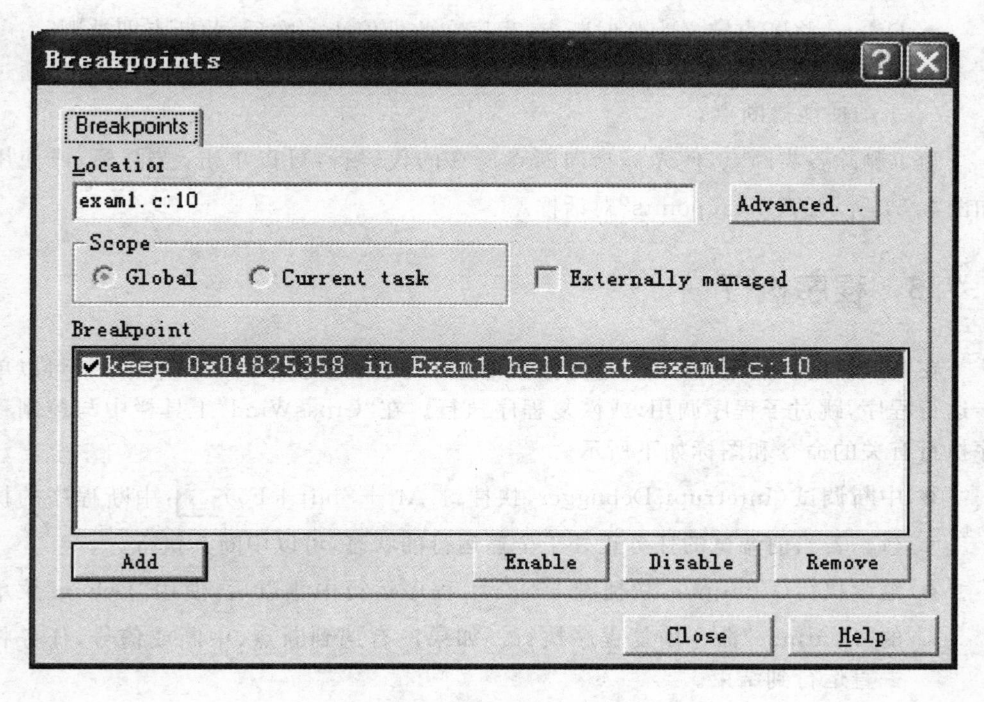

图 2.51　断点设置工具栏

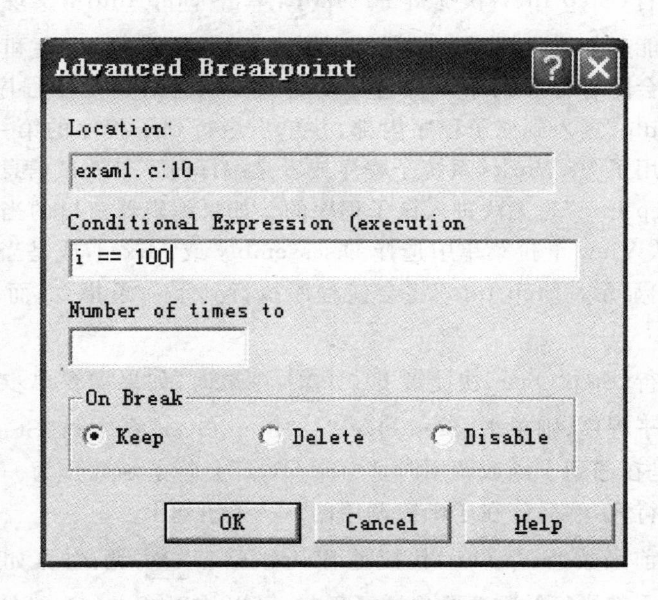

图 2.52　断点设置高级选项工具栏

"On Break"选项指定了如何处理一个断点：

- Keep，将断点定义为永久断点；

- Delete，将断点定义为临时断点，程序在此处停止一次后，此断点即被删除；
- Disable，将断点定义为临时断点，程序在此处停止一次后即关闭它，之后通过手动使能该断点。

为了删除各类断点，使光标指向断点所在的代码行，可以单击图标，或使用如图 2.51 所示的"Breakpoints"对话框。

2.7.5 程序执行

在调试器的控制下，一旦一个任务被中止了(通常是遇到了一个断点)，就可以单步运行程序，跳过子程序调用，或恢复程序执行。在"CrossWind"工具栏中与控制程序执行有关的命令和图标如下所示。

- 中断调试(Interrupt Debugger，快捷键 Alt+Shift+F5)：中断程序的执行。若当前调试的任务正处于全速运行的状态，可以中断其执行。
- 继续执行(Continue，快捷键 F5)：程序运行中止以后，使用"Debug"菜单的"Continue"命令恢复程序执行。如果没有遇到断点、中断或信号，任务将一直运行到结束。
- 单步执行(Step Into，快捷键 F11)：单击"Step Into"，实现单步执行程序调试功能。如果打开了调试器的观察窗口(检查数据、内存和堆栈)，窗口中的数值会随着程序的单步执行自动更新。执行语句是子程序调用的话，则"Step Into"进入到该子程序内部，并单步运行到子程序的第一行。但是，如果是调用了 VxWorks 系统子程序或者是编译时不带调试信息的应用子程序时，"Step Into"是无法进入该子程序的。如果编辑器窗口的当前视图以汇编代码(从 View 下拉菜单中选择 Disassembly 或 Mixed，或是当前代码没有调试符号)显示，"Step Into"将会使程序执行到下一条指令，而不是下一条源代码。
- 单步执行(Step Over，快捷键 F10)：调试时，如果需要单步执行程序而不进入其子程序，则单击"Step Over"。"Step Over"命令与"Step Into"命令类似，只是在遇到子函数调用时，"Step Over"会将子函数作为一条普通的指令一次执行完，然后停在子函数调用的下一条语句上。
- 跳出当前函数(Step Out，快捷键 Shift+F11)：调试时，如果是在一个子程序的内部执行并需要跳出该子程序，可以使用"Step Out"命令继续执行程序直到当前子函数结束。跳出后，程序停在子函数调用的下一条语句，调试器重新获得控制权。
- 执行到光标处(Run to Cursor)：调试时，需要程序执行到一个指定的位置，可

将光标放在所需的代码行，单击鼠标右键，在弹出的菜单中选择“Run to Cursor”，而不需要设置断点。

2.7.6 观察运行信息

在调试状态下当程序处于暂停状态时，可以通过调试器的辅助窗口来观察局部变量和全局变量、函数参数、寄存器、目标板内存和执行函数栈。辅助窗口有两种显示方式，“docked”方式和“free－floating”方式。在“Tools|Options|Debugger”窗口中取消“Docking views”选择框，将辅助窗口设为浮动的，就可以随意调整窗口大小和位置。

在CrossWind工具栏中，与观察各类信息行有关的图标和定义如下所示：

:打开/关闭“Watch”窗口（快捷键Alt＋3），观察全局变量、表达式或其他符号的值。“Watch”窗口显示了程序运行过程中变量的当前值。“Watch”窗口有4页，将逻辑相关的变量组成一页，方便观察。在编辑窗口选择所需的变量单击鼠标右键，在弹出的菜单中选择“Add to Watch”选项就可以将该变量添加到“Watch”窗口了，同时允许手工修改变量的值。

:打开/关闭Variables窗口（快捷键Alt＋4），单击“Debug”下拉菜单的“Variable”命令或工具栏中的“Variable”图标，则可以打开显示局部变量的“Variable”窗口，能够查看局部变量的值。“Variable”窗口总是显示当前执行函数的局部变量的值。如果程序运行跳到另一个函数，新函数的局部变量就会取代旧函数的局部变量显示在“Variable”窗口中。“Variable”窗口中的局部变量是由Tornado自动加入或者删除的，无需用户干预，支持用户手工修改局部变量的值。

:打开/关闭Registers窗口（快捷键Alt＋5），单击“Debug”下拉菜单的“Register”命令或工具栏中的“Register”图标，则可以打开显示寄存器值的“Register”窗口，用来查看寄存器的值。窗口中的内容取决于目标机CPU的架构。在“Tools|Options|Debugger”窗口中取消“Docking views”选择框，将窗口设为浮动（free－floating）的，窗口的标题就会显示出与芯片结构有关的信息，如对于PowerPc系列显示为“ppc”。但是，Tornado的“Register”窗口只能显示出目标机CPU有限的几个通用寄存器和特殊功能寄存器，用户可以手动修改寄存器的值。如果想要访问更多的寄存器，则需要安装其他的专用组件实现。

:打开/关闭Back Trace窗口（快捷键Alt＋7），用来查看当前函数的调用栈情况；利用“Backtrace”窗口可以检查程序运行到当前函数之前的函数调用顺序，即函数调用栈。在“Backtrace”窗口中双击任一函数，则能将编辑窗口的光标移至程序调用该函数处。这种操作不改变程序的运行流程，只是方便用户查阅程序之前的函数调用情况。此时光标颜色不同，以区别程序正常运行时的光标颜色。

▣:打开/关闭 Memory 窗口(快捷键 Alt+6),用来查看指定地址的内存内容。"Memory"窗口显示出从指定起始地址开始的一部分目标内存值。调试器将起始域中键入的每一个起始地址保存下来,可以从下拉列表中选择一个以前显示过的地址。单击 ! 按钮可以更新内存显示。"Memory"窗口中显示的内存值不能手动修改。如果想修改某一地址的内存值可通过 Shell 命令"m"来完成。在"Tools|Options|Debugger"窗口中修改"Memory Window"选项可以改变"Memory"窗口中的内存值的显示方式。

需要注意的是,以上窗口可以同时打开。每次程序在调试器的控制下暂停时,窗口中的内容将会更新。更新时只有变化了的值会高亮。

2.7.7 调试方法

VxWorks 支持两种调试模式:任务模式调试(Task Mode Debug)和系统模式调试(System Mode Debug)。

1. 任务模式调试

任务模式调试是指对系统中单个任务进行调试,而系统其他任务和中断正常运行不受影响。系统调试缺省设置模式为任务模式,选择菜单"Debug|Attach",然后选中所要调试的任务就可以对此任务进行调试。一个调试器同时只能调试一个任务,如果要对其他任务进行调试,则需通过"Attach"窗口选择该任务,调试器会停下原任务调试而进行新选任务调试。在任务调试模式下,"Target Server"和"Target Agent"通过中断方式进行通信,因此无法调试中断服务程序。

在调试过程中,用户能够像在其他系统开发环境下一样进行断点设置,断点类型如下:

- 任务级断点。在任务级调试模式下,任务级断点仅对当前调试任务有效,程序运行中若遇到非调试任务中的任务级断点,该断点将不生效。
- 全局断点。全局断点对所有任务都有效。一旦程序执行到全局断点处,不管系统在调试哪个任务,包含此全局断点的任务都进入"Suspend"状态;如果所调试任务非当前任务,则可通过"Attach"命令将调试器切换至此任务调试。

2. 系统模式调试

系统模式下调试把整个系统当作一个任务进行调试,因此可以调试中断。断点对所有的任务都起作用。在"Attach"窗口选择"System"可以进入系统级调试模式,进入系统级调试模式将中止整个目标系统:所有的任务、内核和 ISR。此时"Target Server"和"Target Agent"通过轮询方式进行通信。因此,系统模式调试需要支持 END 模式的网络或者串口通信模式。

需要注意的是，调试器能够工作在何种模式下是需要与目标代理的配置相一致的。目标代理具有 3 种配置模式：

- 任务模式。该模式下，目标代理作为一个 VxWorks 的任务运行。调试是在一个任务的基础上进行的，该任务被独立出来而不影响目标系统的其余任务运行；
- 系统模式。该模式下，目标代理运行于 VxWorks 之外，所调试的应用程序与 VxWorks 系统可被看做为一个单线程的任务。在此模式下，当目标运行遇到一个断点，VxWorks 和应用程序都会停止并锁住中断。该模式下，支持单步调试中断服务程序；
- 双模式。该模式下，同时配置两种代理，任务模式代理和系统模式代理。系统一次只能激活一个代理模式，通过调试器（选择 Attach | tTask、Attach | system）或者 Shell 下输入命令（sysResume/sysSuspend）实现两种模式之间切换。其中，END(Enhanced NetWork Driver)模式网络支持双模式调试。

要支持系统模式代理，目标通信方式必须工作在轮询方式（因为系统挂起时，外部代理仍需要和主机进行通信）。

在配置 VxWorks 映象时可以同时配置目标代理。如图 2.53 所示，配置支持双模式的目标代理，需要将工作台的 VxWorks 窗口中“development tool components | select WDB connection”选项设为“WDB END driver connection”，并选中“development tool components | select WDB mode”中的“WDB system debugging”和“WDB task debugging”。

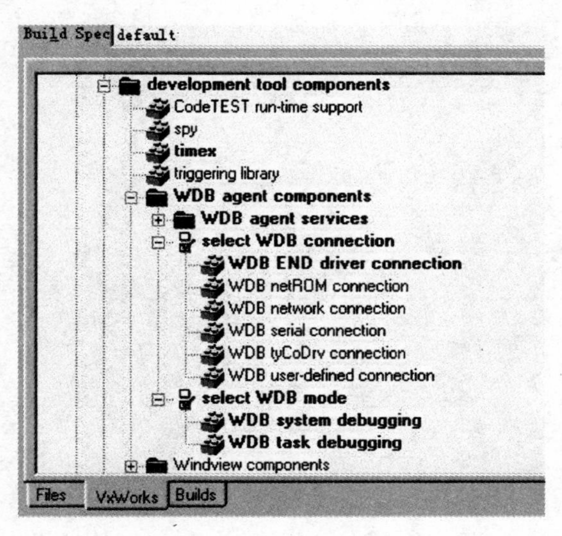

图 2.53　配置双模式目标代理

只有支持 END 驱动器（具有一个轮询接口）的目标板才支持双模式代理。在

END 连接下，代理直接使用 END 驱动器而不是 UDP/IP 协议栈。

调试器支持双模式的前提条件是：

- 下载配置双模式代理的 VxWorks 映象；
- 目标板使用 END 驱动器。

采用仿真器进行应用程序调试时，缺省仿真器的 VxWorks 映象的“WDB agent connection”为“WDB simulator pipe connection”，支持双模式调试。

在系统级模式下调试，程序断点对所有任务都有效。值得注意的是，在系统级调试模式下启动一个程序不能使用“run”指令，而要采用 Shell 里面的命令“sp”启动。使用“run”命令启动程序，系统将会自动运行在任务级调试模式。

在 Shell 下，系统级调试常用命令有：

① sysSuspend，中止目标系统进入系统模式；

② sysResume，恢复目标系统的执行，返回任务模式；

③ agentModeShow，显示当前处于何种调试模式(任务级或系统级)；

④ sysStatusShow，显示系统上下文状态(挂起或运行)；

⑤ b，设置系统级断点，无论在哪个任务、内核或 ISR 中遇到了断点，整个系统都停止运行；

⑥ c，恢复整个系统运行，但仍处于系统调试模式；

⑦ i，显示系统上下文状态和代理模式；

⑧ sp，启动一个任务至执行队列中，在系统模式下只能通过 sp 来启动新任务，而不能运行 Debugger 的 run 命令。

第3章

VxWorks BSP在MagicARM2410上的移植

3.1 BSP文件结构

3.1.1 BSP文件组成

BSP主要由以下几部分的文件组成。

① BSP实现文件：BSP的实现文件位于目录InstallPath\target\config\bspname\下，是与目标机相关的部分，包括源程序、include文件、makefile、生成的中间目标代码、文档文件等。

② BSP相关文件：BSP的相关文件位于目录InstallPath\target\config\all和Installpath\target\h\make下。其中，InstallPath\target\config\all目录是所有BSP的公共部分。

③ 设备驱动程序：设备驱动程序位于目录InstallPath\target\src\drv和InstallPath\target\h\drv下。

④ 其他：BSP还用到InstallPath\target\config\comps\vxworks和InstallPath\target\config\comps\src目录。其中，InstallPath\target\config\comps\vxworks包括基本模块描述文件(*.cdf)，InstallPath\target\config\comps\src包括模块配置文件(被usrConfig.c使用)。

3.1.2 BSP 主要文件目录及文件作用

1. InstallPath\target\config\all 目录

这个目录下的文件是所有 BSP 文件共享的，尽量不要修改里面的任何文件。该目录下主要有如下几个文件。

① bootConfig.c 文件：此文件完成 BootROM Image 的初始化和控制，是所有 Boot ROM 的主要初始化与控制文件。bootConfig.c 是 usrConfig.c 的一个子集，所以 Boot Image 只能提供完整 VxWorks Image 中的部分功能。Boot Image 不使用 MMU 库(SPARC 除外)。在集成环境下修改工程相关的配置信息不会影响 Boot Image，只有直接修改 config.h、configAll.h、bootConfig.c 和 bootInit.c 文件才会影响到 Boot Image。

② bootInit.c 文件：此文件在 romInit.s 运行之后执行，它定义了 romstart()函数，用于完成 Boot ROM 的第二步初始化。程序从 romInit.s 中的 romInit()函数跳到这个文件中的 romstart()函数，来执行必要的解压缩和 ROM Image 的放置(即把 text 和 data 段从 ROM 复制到 RAM 中)。

③ usrConfig.c：包含 VxWorks Image 的主要初始化代码。使用命令行编译时才会用到此文件。

④ configAll.h 文件：此文件默认定义了所有 VxWorks 的设置。如果不用默认的设置，可在 BSP 目录下的 config.h 文件中用“#define”或“#undef”方式来更改设置。

2. InstallPath\target\config\comps\src 目录

此目录涉及系统核心的组件，主要由 target\config\All 中 usrconfig.c 中的函数调用。

3. InstallPath\target\config\bspname 目录

此目录包含系统或硬件相关的 BSP 文件，主要有如下几个文件：

① makefile 文件：用于命令行下编译。此文件包含一些命令行控制 Images 的生成。其中定义了一些宏，如 ROM_TEXT_ADRS、ROM_WARM_ADRS、ROM_SIZE、RAM_LOW_ADRS 和 RAM_HIGH_ADRS。

② config.h 文件：含所有的地址头文件和与 CPU 相关的宏定义，如对 Cache 和 MMU 的配置，RAM 和 ROM 的定位以及大小配置，外部总线映射及 PCI 地址定义。

③ romInit.s：该文件包括 Boot Rom 和 Rom-based Vxworks Image 的初始化入口汇编代码，即实现了 romInit()。目标板一上电就开始执行 romInit()，主要完成 3 部分工作：

- 屏蔽中断，初始化 CPU。
- 配置内存系统，通常包括关闭 Cache，初始化内存控制器。
- 初始化堆栈指针和其他寄存器以执行 C 语言，然后调用 romstart()。

注意，romInit()函数必须设计成与内存地址无关的代码(PIC)，romInit. s 只执行一些必要的初始化，其余的初始化工作可放到函数 sysHwInit()中实现。

① sysALib. s：此文件是汇编语言文件，程序员可以把自己的汇编函数放在该文件里，在上层调用。VxWorks Image 的入口点_sysInit 函数在此文件，是程序在 RAM 中执行的第一个函数。sysALib. s 类似于 romInit. s 完成的工作，但它不必设计成 PIC，且可调用其他库中的函数。

② sysLib. c：包含目标板或与系统相关的 C 语言函数，提供板组接口，sysLib. c 文件包含 src\drv 目录下的驱动程序子文件，驱动程序的安装和初始化在子文件中完成。

③ sysLib. c 中应该实现以下几类函数。

- 系统时钟中断相关的函数：sysClkConnect()、sysClkDisable()、sysClkEnable()、sysClkInt()、sysClkRateGet()、sysClkRateSet()。
- 系统硬件初始化函数：sysHwInit()、sysHwInit2()。
- 内存相关的函数：sysMemTop()、sysNvRamGet()、sysNvRamSet()。
- 串行相关的函数：sysSerialHwInit()、sysSerialHwInit2()、sysSerialChanGet()。
- 杂项函数：sysBspRev()、sysModel()、sysToMonitor()。
- 其他可选函数：如辅助时钟中断相关函数及总线相关函数等。

④ sysSerial. c 文件：此文件是可选文件，用于所有的串行口设置和初始化。

⑤ sysNet. c：网络接口设备的安装和初始化。

⑥ bspname. h：包含与板子相关的宏定义。若要把 src\drv 下的某驱动程序移植到目标板上，必须在 bspname. h 文件开始包含目录 target\h\drv 下该驱动程序相对应的头文件。bspname. h 应该定义中断向量或中断号、I\O 设备地址、设备寄存器位的定义、系统时钟和辅助时钟最大和最小时钟速率。

⑦ README 文件：此文件是记录 BSP 的发布纪录、版本等的文档。

⑧ configNet. h 文件：此文件是网络驱动的主要设置文件，主要对 END 驱动设置。

⑨ sysScsi. c 文件：此文件是可选文件，用于 scsi 设备设置和初始化。

⑩ BootROM. hex 文件：此文件是 ASIC 文件，包含 VxWorks BootROM 代码。

⑪ VxWorks 文件：此文件运行在目标机上，是完整的、连接后的 VxWorks 二进制文件。

⑫ Vxworks. sym 文件：此文件是完全的、连接后带有符号表的 VxWorks 二进制文件。

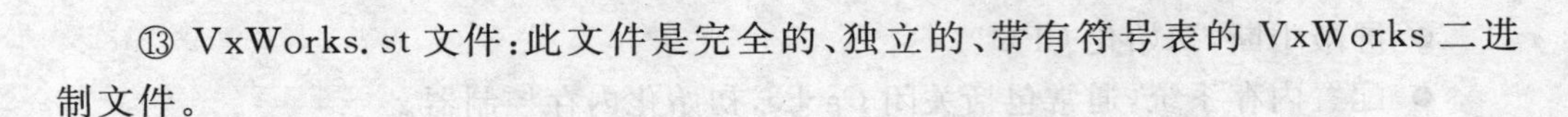

⑬ VxWorks.st 文件：此文件是完全的、独立的、带有符号表的 VxWorks 二进制文件。

3.2 BSP 配置文件

VxWorks BSP 主要配置文件有 config.h 和 makefile，下面分别解释。

3.2.1 config.h 文件

config.h 文件代码，分段解释说明如下。

```
/*
* This module contains the configuration parameters for the MagicARM2410 BSP.
*/
#ifndef    INCconfigh
#define    INCconfigh
/* BSP version/revision identification,before configAll.h */
#define BSP_VER_1_1       1          /* 1.2 is backwards compatible with 1.1 */
#define BSP_VER_1_2       1
#define BSP_VERSION       "1.0"
#define BSP_REV           "/1"       /* 0 for first revision */
#include "configAll.h"               /* 这个文件定义了 VxWorks 所有的默认设置语句 */

/*
* Default boot line
*/
#define DEFAULT_BOOT_LINE        "dmf(0,0)host:vxWorks " \
                                 "h = 192.168.18.5 " \
                                 "e = 192.168.18.18:ffffff00 " \
                                 "g = 192.168.18.1 " \
                                 "u = magic " \
                                 "pw = magic "      \
                                 "tn = vxWorks " \
                                 "o = dmf"/
```

DEFAULT_BOOT_LINE 参数定义了启动设备类型(dmf)、主机的 IP 地址(h)、目标机的 IP 地址(e)、网关 IP(g)、用户名(u)、登录口令(pw)、目标板名(tn)、从网络启动时需要指定的网络接口(o)等信息，对于配置网络、连通 Target server 及下载调试程序非常重要。

DEFAULT_BOOT_LINE 的功能是在没有 NVRAM 的 Target 设计的，用户在

每次系统启动时不需要再手工输入这些配置参数。

系统启动网络时，xxxEndLoad()函数会解释这一行并按这一行的定义进行加载。这一行中，dmf(0,0)表示启动设备，支持软盘、硬盘、PCMCIA 卡等其他的设备。例如：fd 为软盘，(0,0)表示第一个软驱，3.5 寸盘；dc 则表示从 DEC 21x4x 芯片启动，系统有 NVRAM 存在，目前此种方式基本淘汰；elpci 表示启动设备是 3COM EtherLink XL PCI 网卡；fei 表示启动设备是 Intel 82559 EtherExpress 网卡；ene 表示 NE2000 网卡；ELT 表示 3COM 以太网卡；dmf 表示 DM9000 以太网卡；ata 表示 ATA/IDE 硬盘；host 表示主机名；vxWorks 表示从主机加载的 vxWorks 文件；h=192.168.18.5 表示主机的 IP 地址；e=192.168.18.18 表示目标机的 IP 地址，若网络启动 Target Server 时，这个 IP 地址必须和主机上 Target Server 配置的 Target IP 地址一致，且设置 Back End 选项为 wdbrpc。

u=magic 表示用户名，pw=magic 表示登录口令，若通过网络加载调试时，主机的 ftp 服务器和目标机的用户名和密码必须相同；否则，无法加载。tn=vxWorks 表示目标机名称。

```
/*
#undef      LOCAL_MEM_AUTOSIZE
#define USER_RESERVED_MEM         0
#define LOCAL_MEM_LOCAL_ADRS      0x30000000            /* fixed at 0x30000000 */
#define LOCAL_MEM_BUS_ADRS        LOCAL_MEM_LOCAL_ADRS  /* fixed at 0x30000000 */
#define LOCAL_MEM_SIZE            0x04000000            /* 64 Mbytes */
#define LOCAL_MEM_END_ADRS        (LOCAL_MEM_LOCAL_ADRS + LOCAL_MEM_SIZE)
*/
```

以上代码作用是进行内存设置。如果定义了 LOCAL_MEM_AUTOSIZE 则 SDRAM 的大小会在 boot 时指定。

下面主要设置系统的内存分配定义。若分配不当，则系统不能正常加载和运行。

```
#define ROM_BASE_ADRS     0x00000000          /* base of Flash/EPROM */
#define ROM_TEXT_ADRS     ROM_BASE_ADRS       /* code start addr in ROM */
#define ROM_WARM_ADRS     0x00000004          /* code start addr in ROM */
#define ROM_SIZE          0x00100000          /* size of ROM holding VxWorks */

#define ROM_COPY_SIZE     ROM_SIZE
#define ROM_SIZE_TOTAL    0x00200000          /* total size of ROM */
```

ROM_BASE_ADRS、ROM_TEXT_ADRS、ROM_SIZE、RAM_HIGH_ADRS 和 RAM_LOW_ADRS 要在 config.h 和 makefile 文件中进行定义，并且必须保持一致。具体地址的定义数值一定要参照 VxWorks 加载执行过程、硬件手册、MMU 和 VxWorks 的大小进行。基本原则是保证 VxWorks Image 在 ROM 和 RAM 中拥有一定的运行空间且高效运行，可参见 VxWorks BSP 和启动过程。

接下来定义内存的地址变量。

```
#define RAM_LOW_ADRS      0x30010000      /* VxWorks image entry point */
#define RAM_HIGH_ADRS      0x33000000      /* RAM address for ROM boot */

/*
 * Include MMU BASIC and CACHE support for command line and project builds
 */
#define      INCLUDE_MMU
#define      INCLUDE_MMU_BASIC
#define INCLUDE_CACHE_SUPPORT

/*
 * I-cache mode is a bit of an inappropriate concept, but use this.
 */
#undef   USER_I_CACHE_MODE
#define USER_I_CACHE_MODE        CACHE_WRITETHROUGH

/* has to be this. */
#undef   USER_D_CACHE_MODE
#define USER_D_CACHE_MODE        CACHE_COPYBACK

/* Enhanced Network Driver (END) Support */
#ifndef INCLUDE_END
#define INCLUDE_END
#define INCLUDE_FTP_SERVER
#endif

/* WDB configuration. */
#define INCLUDE_NETWORK
#define INCLUDE_WDB
#undef   WDB_COMM_TYPE
#define WDB_COMM_TYPE    WDB_COMM_END

/* Optional timestamp support */
#undef     INCLUDE_TIMESTAMP        /* define to include timestamp driver */


#undef   INCLUDE_TFFS
#define INCLUDE_TFFS                    /* to include TrueFFS driver */
#ifdef  INCLUDE_TFFS
#define INCLUDE_SHOW_ROUTINES
```

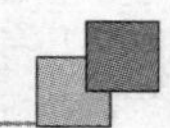

```
#define INCLUDE_TFFS_SHOW
#define INCLUDE_DOSFS
#define INCLUDE_DISK_UTIL
#define INCLUDE_TL_FTL
#endif
/* INCLUDE_TFFS */

#include "s3c2410x.h"

#undef BSP_VTS
#ifdef BSP_VTS

/*****************************************************************
 * Add these defines for the Validation Test Suite *
*****************************************************************/
#define INCLUDE_SHELL
#undef INCLUDE_RLOGIN
#undef INCLUDE_SHOW_ROUTINES
#undef INCLUDE_NET_SYM_TBL
#define INCLUDE_LOADER

#undef INCLUDE_PING
#undef INCLUDE_NET_SHOW
#endif /* BSP_VTS */

#define INCLUDE_SHELL
#ifdef  INCLUDE_SHELL
#define INCLUDE_SYM_TABLE
#define INCLUDE_STANDALONE_SYM_TABLE
#define INCLUDE_LOADER
#define INCLUDE_ELF
#define INCLUDE_STARTUP_SCRIPT
#define INCLUDE_UNLOADER
#endif /* INCLUDE_SHELL */

#undef INCLUDE_WINDML

#ifdef __cplusplus
}
#endif
#endif  /* INCconfigh */
```

```
#if defined(PRJ_BUILD)
#include "prjParams.h"
#endif
```

3.2.2 makefile 文件

makefile 文件代码如下。

```
# Makefile - make rules for MagicARM2410
#
# DESCRIPTION
# This file contains rules for building VxWorks for the MagicARM2410
# with an s3c2410x microprocessor module.
#
CPU            = ARMARCH4
TOOL           = diab
EXTRA_DEFINE       = -Wcomment -DCPU_920T \
            -DARMMMU=ARMMMU_920T -DARMCACHE=ARMCACHE_920T

TGT_DIR = $(WIND_BASE)/target

include $(TGT_DIR)/h/make/defs.bsp
#include $(TGT_DIR)/h/make/make.$(CPU)$(TOOL)
#include $(TGT_DIR)/h/make/defs.$(WIND_HOST_TYPE)

## Only redefine make definitions below this point, or your definitions will
## be overwritten by the makefile stubs above.

TARGET_DIR      = 2410diab
VENDOR          = NUAA
BOARD           = S3C2410X

RELEASE         += bootrom.bin
CONFIG_ALL      = $(TGT_DIR)\config\MagicARM2410\all

#
# The constants ROM_TEXT_ADRS, ROM_SIZE, and RAM_HIGH_ADRS are defined
# in config.h and Makefile.
# All definitions for these constants must be identical.
#
```

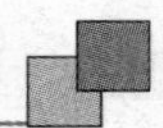

```
ROM_TEXT_ADRS      = 00000000 # ROM entry address
ROM_WARM_ADRS      = 00000004 # ROM warm address
ROM_SIZE     = 00100000 # number of bytes of ROM space

RAM_LOW_ADRS      = 30010000 # RAM text/data address
RAM_HIGH_ADRS      = 33000000 # RAM text/data address

MACH_EXTRA        = lv160mtd.o ftpdLib.o sysWindML.o

EXTRA_MODULES + = sysWindML.o
PRJ_LIBS          = C:\Tornado2.2\target\lib\arm\ARMARCH4\diab\libwndml.a

VMA_START      = 0x$(ROM_TEXT_ADRS)

 # Binary version of VxWorks ROM images, suitable for programming
 # into Flash using tools provided by ARM.  If other ROM images need to
 # be put into Flash, add similar rules here.

bootrom.bin: bootrom
    - @ $(RM) $@
    $(EXTRACT_BIN) -O binary bootrom $@

bootrom_res.bin: bootrom_res
    - @ $(RM) $@
    $(EXTRACT_BIN) -O binary bootrom_res $@

bootrom_uncmp.bin: bootrom_uncmp
    - @ $(RM) $@
    $(EXTRACT_BIN) -O binary bootrom_uncmp $@

vxWorks_rom.bin: vxWorks_rom
    - @ $(RM) $@
    $(EXTRACT_BIN) -O binary vxWorks_rom $@

vxWorks.st_rom.bin: vxWorks.st_rom
    - @ $(RM) $@
    $(EXTRACT_BIN) -O binary vxWorks.st_rom $@

vxWorks.res_rom.bin: vxWorks.res_rom
    - @ $(RM) $@
    $(EXTRACT_BIN) -O binary vxWorks.res_rom $@
```

```
vxWorks.res_rom_nosym.bin: vxWorks.res_rom_nosym
        - @ $(RM) $@
        $(EXTRACT_BIN) -O binary vxWorks.res_rom_nosym $@

## Only redefine make definitions above this point, or the expansion of
## makefile target dependencies may be incorrect.

include $(TGT_DIR)/h/make/rules.bsp
#include $(TGT_DIR)/h/make/rules.$(WIND_HOST_TYPE)
```

3.3 系统映像类型

系统的映像大致可以分为两类，一类是 VxWorks Image，另外一类是 BSP 引导映像。映像分类的根据是看映像是否包含用户 VxWorks 内核和组件代码。两种映像都包含了两种相关属性：①映像运行时是否驻留 ROM；②映像是否压缩。

通常驻留在 ROM 的映像可以减少对 RAM 的占用，运行时只将数据段装入 RAM 中，而代码段还是驻留在 ROM 中。如图 3.1 所示，给出了驻留在 ROM 的映像装入数据段到 RAM 时的位置。驻留在 ROM 的映像虽然节省了 RAM 占用率，但是却降低了运行速度，并且限制了访问 RAM 的数据宽度。

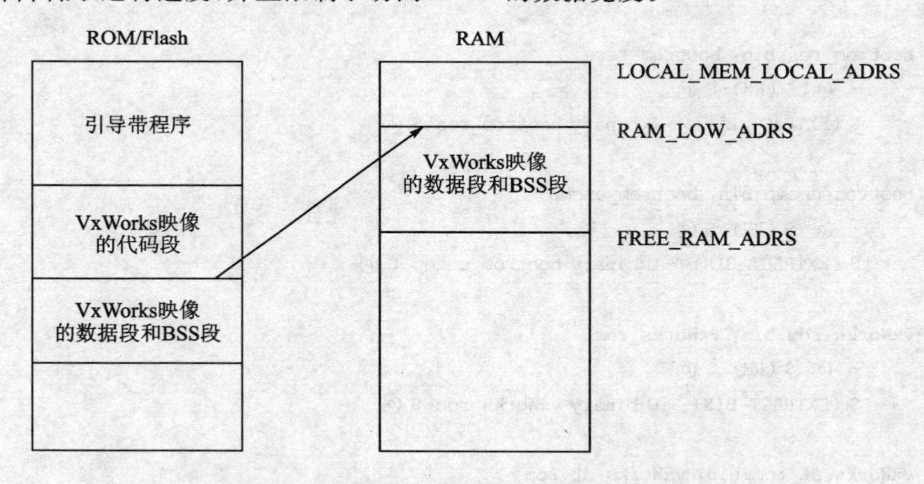

图 3.1 驻留在 ROM 的映像初始化逻辑示意

驻留 ROM 的 VxWorks Image 在 RAM 中的 RAM_LOW_ADRS 位置上开始装入数据段，BSP 引导映像则在 RAM_HIGH_ADRS 的位置上装入数据段。由于 BSP 引导映像的功能是装入 VxWorks Image，系统设定 RAM_LOW_ADRS 为 VxWorks

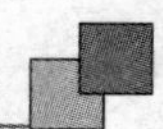

Image 位置，RAM 地址从 RAM_LOW_ADRS 到 RAM_HIGH_ADRS 作为 VxWorks 系统和应用程序地址空间。

3.3.1 VxWorks Image

VxWorks Image 包括 wind 多任务微内核，用户定制 VxWorks 组件，以及用户应用代码。VxWorks Image 同样支持在 ROM 中运行和在 RAM 中运行的两种模式。在启动过程中，映像是否是驻留在 ROM 上执行初始化根据映像的类型决定，包括 sysInit()函数的调用。

1. 在 ROM 中运行分析

VxWorks 能够在 ROM 中直接运行，是因为写入 ROM 中的 VxWorks 是非压缩的，不需要进行额外的解压工作。运行时，系统直接从 ROM 的首地址开始运行 VxWorks。需要注意的是，在 ROM 中运行的 VxWorks 并不支持所有主板，应以主板手册中的说明为准。

下面是在 ROM 中运行的 VxWorks 的启动顺序：

首先执行文件 romInit. s 中的 romInit()函数→执行文件 bootInit. c 中的 romStart()函数→执行文件 usrConfig. c 中的 usrInit()函数→执行 sysHwInit()函数→执行 usrKernelInit()函数→执行 KernelInit(usrRoot,…)函数。

romInit()函数是与目标系统 CPU 特性直接相关的汇编程序，在目标系统上电后被第一个执行，其功能是进行最基本的硬件初始化，以便正确地执行 romStart()函数。其函数主要功能包括：

- 屏蔽 CPU 中断，复位 CPU；
- 初始化 cache；
- 设置栈指针和其他寄存器，配置启动类型参数来调用 romStart()函数；
- 完成其他的 CPU 特定初始化。

通常来说，受到成本和体积限制，嵌入式主板的 RAM 一般较小，因此 VxWorks 在 ROM 中运行，可以节省主板上相对紧张的 RAM 空间，相应的用户应用程序运行空间就比较充足。运行时，只是将 VxWorks Image 的 DATA 段复制到 RAM 区域，而 TEXT 段驻留在 ROM 中并执行。相应地，在 ROM 中 VxWorks 运行的速度则比在 RAM 中慢。

2. 在 RAM 中运行分析

在 RAM 中运行 VxWorks，首先确定是哪种映像类型，如果是压缩的话，则需要先解压并复制所有的 TEXT 和 DATA 到 RAM 空间。下面 sysInit()函数主要是初始化 RAM 用的，系统直接跳到 RAM 的首地址，运行 VxWorks。

下面是在 RAM 中运行的 VxWorks 的启动顺序：

首先执行文件 romInit. s 中的 romInit. s()函数→执行文件 bootInit. c 中的 romstart()函数→执行 sysLib. s 中的 sysInit()函数→执行文件 usrConfig. c 中的 usrInit()函数→执行 sysHwInit()函数→执行 usrKernelInit()函数→执行 KernelInit(usrRoot,…)函数。

sysInit()函数是可装入映像的入口程序，其位置在 RAM 中的 RAM_LOW_ADRS 处。该函数执行映像的初始化，功能和上面的 romInit()函数基本相同，差别在于 sysInit()函数不进行内存初始化，其调用时假定已经完成了内存初始化。sysInit()函数是汇编程序，位于模块 sysLib. s。即在 FLASH 中首先执行 romInit. s，再跳到 bootInit. c 中的 romStart()函数，然后 romStart()函数中开始装载和解压缩 Image 到 RAM。sysaLib. s 是在 RAM 中执行的第一个文件。其中，usrRoot()函数是 VxWorks 启动的第一个任务，由它来初始化驱动程序、网络等。

3.3.2 BSP 引导映像

BSP 引导映像是由 BSP 的引导代码生成的映像，不含应用代码，只有基本的 wind 多任务微内核，能够引导目标系统，主要用于启动装载 VxWorks Image。引导映像被烧写在 ROM 或者装入 FLASH 中，执行时总从 ROM/FLASH 开始，差别在于是否驻留在 ROM/FLASH 上，以及是否进行压缩。

通常有以下 3 种映像：①bootrom 压缩的引导 ROM 映像；②bootrom_uncmp 未压缩的引导 ROM 映像；③bootrom_res 驻留 ROM 和引导 ROM 映像。与 VxWorks Image 的区别在于 BootROM 调用 bootConfig. c，而 VxWorks 调用 usrConfig. c。

下面是 BSP 引导映像的启动顺序：

执行文件 romInit. s 中的 romInit()函数→执行文件 bootInit. c 中的 romstart()函数→执行文件 bootConfig. c 中的 usrInit()函数→执行 sysHwInit()函数→执行 usrKernelInit()函数→执行 KernelInit(usrRoot,…)，usrRoot()→bootCmdLoop(void)命令行选择，或 autobooting→bootLoad(pLine,&entry)加载模块到内存(网络，TFFS，TSFS…)→netifAttach()→go(entry)→从入口开始执行，不返回。BootROM Image 初始化逻辑示意如图 3.2 所示。

其中/target/config/all/bootConfig. c 只为了装入系统映像，因此提供的功能主要包括取得引导参数，然后下载映像并开始执行。

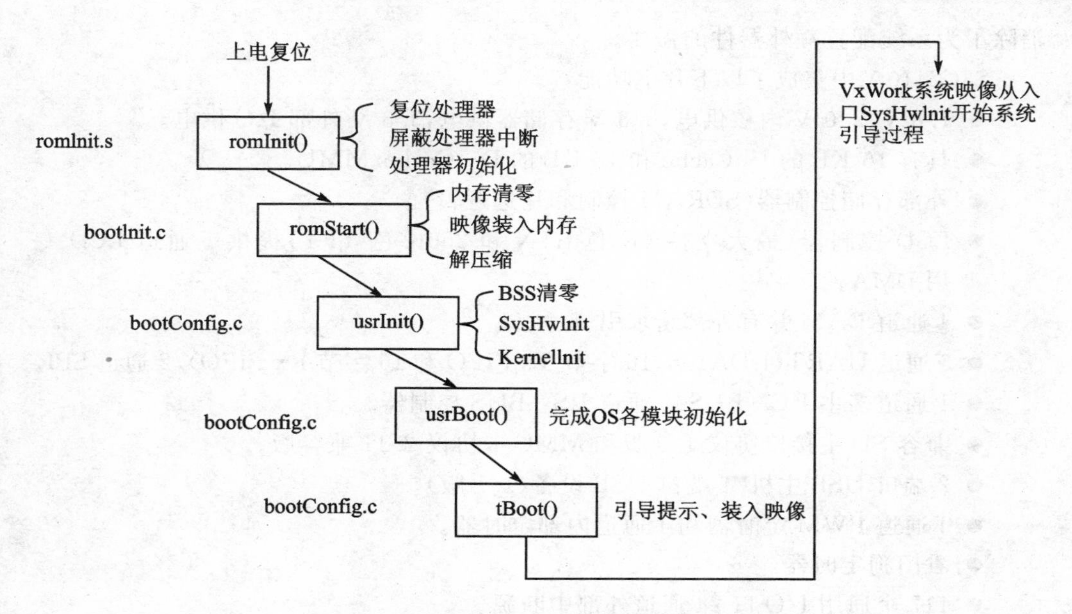

图 3.2　BootROM Image 初始化逻辑示意图

3.4　ARM9 S3C2410A 介绍

Samsung 公司推出的 16/32 位 RISC 处理器 S3C2410A，为手持设备和一般类型应用提供了低价格、低功耗、高性能小型微控制器的解决方案。为了降低整个系统的成本，S3C2410A 提供了以下丰富的内容设备：分开的 16 KB 指令 Cache 和 16 KB 数据 Cache，MMU 虚拟存储器管理，LCD 控制器（支持 STC&TFT），支持 NAND FLASH 系统引导，系统管理器（片选逻辑和 SDRAM 控制器），3 通道 UART，4 通道 DMA，4 通道 PWM 定时器，I/O 端口，RTC，8 通道 10 位 ADC 和触摸屏接口，I^2C 总线接口，USB 主机，USB 设备，SD 主卡接口，MMC 卡接口，2 通道的 SPI 以及内部 PLL 时钟倍频器。

S3C2410A 采用了 ARM920T 内核，0.18 μm 工艺的 CMOS 标准宏单元和存储单元。它的低功耗、精简和出色的全静态设计特别适用于对成本和功耗敏感的应用场合。同样，它采用了一种叫做 Advanced Microcontroller Bus Architecture（AMBA）的新型总线结构。

S3C2410A 的显著特性是它的 CPU 内核是一个由 Advanced RISC Machines（ARM）有限公司设计的 16/32 位 ARM920T RISC 处理器。ARM920T 实现了 MMU，AMBA BUS 和 Harvard 高速缓冲体系结构。这一结构具有独立的 16 KB 指令 Cache 和 16 KB 数据 Cache，每个都是由 8 字长的行(line)构成。

通过提供一系列完整的体系外围设备，S3C2410A 大大减少了整个体系的成本，消除了为系统配置额外器件的需要。

S3C2410A 中集成了以下片上功能：

- 1.8 V、2.0 V 内核供电，3.3 V 存储器供电，3.3 V 外部 I/O 供电。
- 具有 16 KB 的 I－Cache 和 16 KB 的 D－Cache/MMU。
- 外部存储控制器(SDRAM 控制和片选逻辑)。
- LCD 控制器(最大支持 4K 色 STN 和 256K 色 TFT)提供 1 通道 LCD 专用 DMA。
- 4 通道 DAM 并有外部请求引脚。
- 3 通道 UART(IrDA1.6，16 字节 Tx FIFO 和 16 字节 Rx FIFO)/2 通道 SPI。
- 1 通道多主 I^2C－BUS/1 通道 IIS－BUS 控制器。
- 兼容 SD 主接口协议 1.0 版和 MMC 卡协议 2.11 兼容版。
- 2 端口 USB 主机/1 端口 USB 设备(1.1 版)。
- 4 通道 PWM 定时器和 1 通道内部定时器。
- 看门狗定时器。
- 117 个通用 I/O 口 24 通道外部中断源。
- 空号控制模式：具有普通、慢速、空闲和掉电模式。
- 8 通道 10 比特 ADC 和触摸屏、接口。
- 具有日历功能的 RTC。
- 具有 PLL 片上时钟发生器。

S3C2410A 其他特征如下所示：

(1) 体系结构。

- 为手持设备和通用嵌入式应用提供片上集成系统解决方案。
- 16/32 位 RISC 体系结构和 ARM920T 内核强大指令集。
- 加强的 ARM 体系结构 MMU 用于支持 WinCE、EPOC32 和 Linux。
- 具备指令高速存储缓冲器(I-Cache)，数据高速存储缓冲器(D-Cache)。
- 采用 ARM920T CPU 内核支持 ARM 调试体系结构。
- 内部高级微控总线(ARMA)体系结构(ARMA2.0，AHB/APB)。

(2) 系统管理器。

- 支持大/小端方式。
- 寻址空间：每 bank 128 MB(总共 1 GB)。
- 支持可编程的每 bank 8/16/32 位数据总线带宽。
- 从 bank0 到 bank 6 都采用固定的 bank 起始寻址。
- bank 7 具有可编程起始地址和大小。
- 8 个存储器 bank：

—其中 6 个适用于 ROM，SRAM 和其他；

—另外 2 个适用于 ROM/SRAM 和同步 DRAM。

- 所有的存储器 bank 都具有可编程的操作周期。
- 支持掉电时的 SDRAM 自刷新模式。
- 支持各种型号的 ROM 引导(NOR/NAND Flash,EEPROM,或其他)。

(3) NAND FLASH 启动引导。

- 支持从 NAND FLASH 存储器的启动。
- 采用 4 KB 内部缓冲器进行启动引导。
- 支持启动之后 NAND 存储器仍然作为外部存储器使用。

(4) Cache 存储器。

- 64 项全相连模式,采用 I - Cache(16 KB)和 D - Cache(16 KB)。
- 每行 8 字长度,其中每行带有 1 个有效位和 2 个 dirty 位。
- 伪随机数或轮转循环替换算法。
- 采用写穿式(write - through)或写回式(write - back)cache 操作来更新主存储器。
- 写缓冲器可以保存 16 个字的数据和 4 个地址。

(5) 时钟和电源管理。

- 片上 MPLL 和 UPLL:

—采用 UPLL 产生操作 USB 主机/设备的时钟;

—MPLL 产生最大 266 MHz(在 2.0 V 内核的电压下)操作 MCU 所需要的时钟。

- 通过软件可以有选择性地为每个功能模块提供时钟。
- 电源模式:

—正常模式:正常运行模式;

—慢速模式:不加 PLL 的低时钟频率模式;

—空闲模式:只停止 CPU 的时钟;

—掉电模式:切断所有外设和内核的电源。

- 可以通过 EINT[15:0]或 RTC 报警中断从掉电模式中唤醒处理器。

(6) 中断控制器。

- 55 个中断源(1 个看门狗定时器,5 个定时器,9 个 UARTs,24 个外部中断,4 个 DMA,2 个 RTC,2 个 ADC,1 个 I^2C,2 个 SPI,1 个 SDI,2 个 USB,1 个 LCD 和 1 个电池故障)。
- 电平/边沿触发模式的外部中断源。
- 可编程的边沿/电平触发极性。
- 支持为紧急中断请求提供快速中断服务。

(7) 具有脉冲带宽调制功能的定时器。

- 4 通道 16 位具有 PWM 功能的定时器,1 通道 16 位内部定时器,可基于 DMA 或中断工作。

- 可编程的占空比周期、频率和极性。
- 能产生死区。
- 支持外部时钟源。

(8) RTC(实时时钟)。

- 全面的时钟特性:秒、分、时、日期、星期、月和年。
- 32.768 kHz 工作。
- 具有报警中断。
- 具有节拍中断。

(9) 通用 I/O 端口。

- 24 个外部中断端口。
- 多功能输入/输出端口。

(10) UART 端口。

- 3 通道 UART,可以基于 DMA 模式或中断模式工作。
- 支持 5 位、6 位、7 位或者 8 位串行数据发送/接收。
- 支持外部时钟作为 UART 的运行时钟(UEXTCLK)。
- 可编程的波特率。
- 支持 IrDA 1.0。
- 具有测试用的回环模式。
- 每个通道都具有内部 16 字节的发送 FIFO 和 16 字节的接收 FIFO。

(11) DMA 控制器。

- 4 通道的 DMA 控制器。
- 支持存储器到存储器、I/O 到存储器、存储器到 I/O 以及 I/O 到 I/O 的传输。
- 采用猝发传送模式加快传输速率。

(12) A/D 转换和触摸屏接口。

- 8 通道多路复用 ADC。
- 最大 500 ksps/10 位精度。

(13) LCD 控制器 STN LCD 显示特性。

- 支持 3 种类型的 STN LCD 显示屏:4 位双扫描、4 位单扫描、8 位单扫描显示类型。
- 支持单色模式、4 级、16 级灰度 STN LCD、256 色和 4096 色 STN LCD。
- 支持多种不同尺寸的液晶屏。
- LCD 的典型尺寸是:640×480、320×240、160×160 及其他。
- 最大虚拟屏幕大小是 4 MB。
- 256 色模式下支持的最大虚拟屏幕是:4 096×1 024、2 048×2 048、1 024×4 096等。

(14) TFT 彩色显示屏。

- 支持彩色 TFT 的 1、2、4 或 8 bbp 调色显示。
- 支持 16 bbp 无调色真彩显示。
- 在 24 bbp 模式下支持最大 16 M 色 TFT。
- 支持多种不同尺寸的液晶屏。
- 典型实屏尺寸：640×480、320×240、160×160 及其他。
- 最大虚拟屏幕大小 4 MB。
- 64K 色彩模式下最大的虚拟尺寸为 2 048×1 024 及其他。

(15) 看门狗定时器。

- 16 位看门狗定时器。
- 在定时器溢出时发生中断请求或系统复位。

(16) I^2C 总线接口。

- 1 通道多主 I^2C 总线。
- 支持串行 8 位数据位双向数据传输。
- 标准模式下数据传输速度可达 100 kbps，快速模式下可达到 400 kbps。

(17) IIS 总线接口。

- 1 通道音频 IIS 总线接口，可基于 DMA 方式工作。
- 串行，每通道 8/16 位数据传输。
- 发送和接收具备 128 字节(64 字节加 64 字节)FIFO。
- 支持 IIS 格式和 MSB－justified 数据格式。

(18) USB 主设备。

- 2 个 USB 主设备接口。
- 遵从 OHCI Rev1.0 标准。
- 兼容 USB ver1.1 标准。

(19) USB 从设备。

- 1 个 USB 从设备接口。
- 具备 5 个 Endpoints。
- 兼容 USB ver1.1 标准。

(20) SD 主机接口。

- 兼容 SD 存储卡协议 1.0 版。
- 兼容 SDIO 卡协议 1.0 版。
- 具备发送和接收 FIFO。
- 基于 DMA 或中断模式工作。
- 兼容 MMC 卡协议 2.11 版。

(21) SPI 接口。

- 兼容 2 通道 SPI 协议 2.11 版。
- 发送和接收有 2×8 位的移位寄存器。
- 可以基于 DMA 或中断模式工作。

(22) 工作电压。

- 内核:1.8 V 最高 200 MHz(S3C2410A－20);
  2.0 V 最高 266 MHz(S3C2410－26)。
- 存储器和 IO 口:3.3 V。

(23) 操作频率。

- 最高达到 266 MHz。

(24) 封装。

- 272 - FBGA。

3.5 MagicARM2410 实验箱介绍

本书以广州致远公司的 MagicARM2410 实验箱作为开发应用对象,进行相关的 BSP 开发和实验设计,MagicARM2410 实验箱功能如图 3.3 所示。

实验箱的资源如表 3.1 所列。

表 3.1 MagicARM 实验箱资源

序 号	名 称	描 述
1	处理器	ARM902T 处理器 S3C2410A,工作频率高达 203 MHz, 16 KB 指令 Cache,16 KB 数据 Cache,MMU 功能
2	SDRAM	64 MB
3	NAND Flash	64 MB
4	NOR Flash	2 MB
5	EEPROM	256 Byte
6	液晶屏	8 英寸 640×480 真彩 TFT 液晶屏
7	以太网接口	1 个 10/100M 以太网接口
8	USB 接口	4 个 USB 主机口,1 个 USB 设备口
9	音频接口	IIS 数字音频输入/输出接口(有 2 个扬声器 1 个咪头)
10	PCMCIA 接口	1 个 68 Pin PCMCIA 接口
11	CF 卡接口	1 个(PCMCIA 接口扩展)

续表 3.1

序 号	名 称	描 述
12	IDE 硬盘接口	1个(PCMCIA 接口扩展)
13	SD/MMC 卡接口	1个
14	RS-232	2路
15	IrDA	1路
16	RS-485	1路
17	CAN 接口	1路(CAN 控制器 SJA1000)
18	ADC	CPU 内置,2路直流电压测量
19	DAC	1路 PWM DAC 输出
20	直流电机	1个
21	步进电机	1个
22	RTC	CPU 内置,实验箱上有 RTC 后备电池
23	WDT	CPU 内置
24	数码管	8位动态数码管(ZLG7290 驱动)
25	键盘	16键小键盘(ZLG7290 驱动)
26	独立按键	1个(接到中断输入引脚)
27	蜂鸣器	1个,直流蜂鸣器
28	独立 LED	4个
29	GPRS PACK 接口	1个,用来扩展 GPRS 模块
30	VGA PACK	1个,用来扩展 VGA 输出接口
31	总线扩展接口	2个(1个16位总线的、1个32位总线的)
32	JTAG 接口	20Pin JTAG 调试接口
33	JTAG 仿真器	Wiggler JTAG 仿真器
34	GPRS 模块	选件
35	GPS 模块	选件
36	CS8900 PACK 板	10M 以太网。选件
37	触摸屏	4线电阻式。选件
38	VGA PACK 板	在 Linux 和 Win CE 系统下均不闪烁。选件

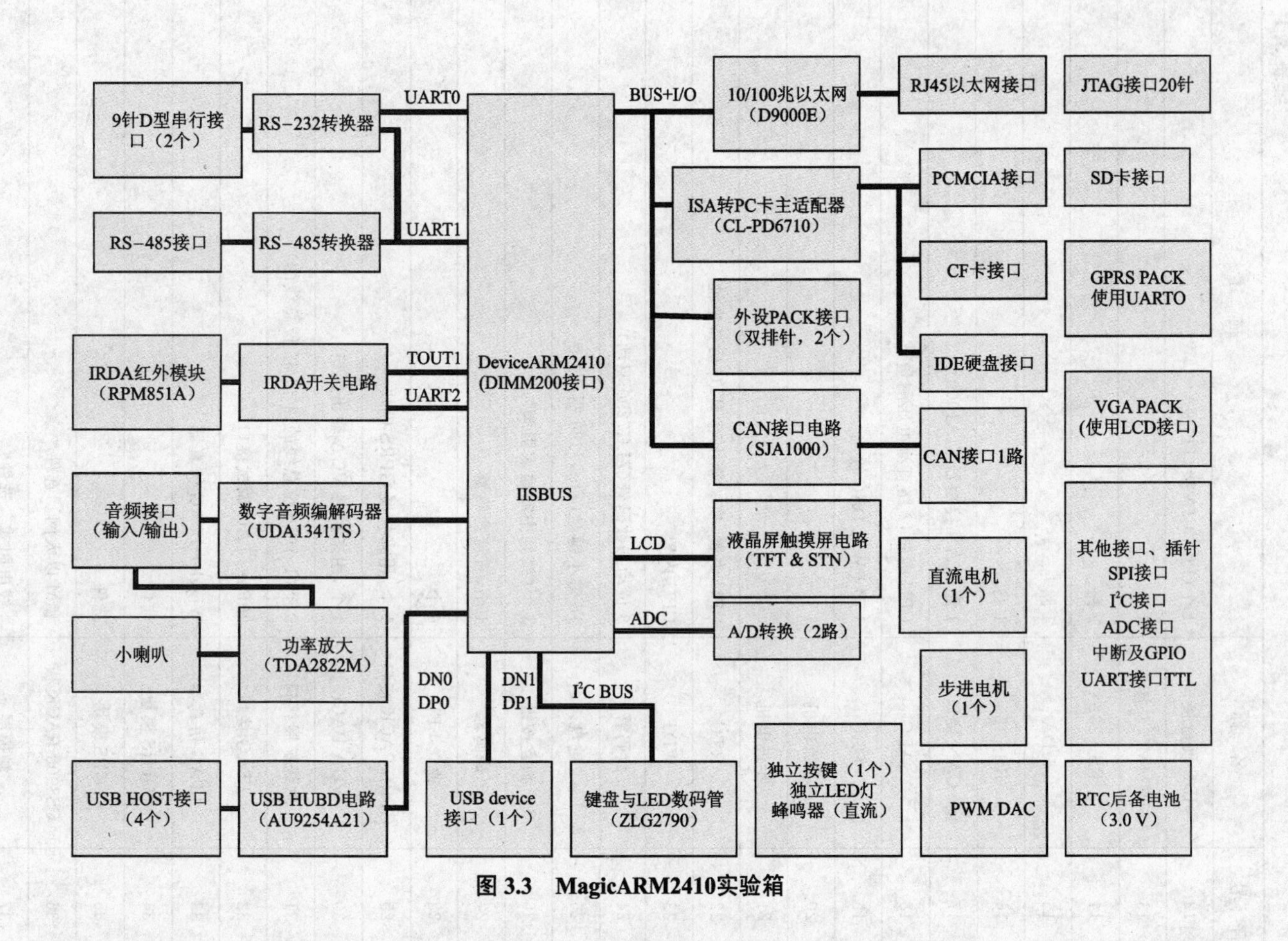

图 3.3　MagicARM2410实验箱

3.6　BSP 移植的基本流程

VxWorks 操作系统开发最重要的就是 BSP 的开发。BSP 是实时嵌入式操作系统 VxWorks 的重要组件，负责完成系统硬件的初始化以及驱动工作，提供 VxWorks 访问硬件设备的接口，用户在开发应用软件过程中能够最大限度地脱离硬件设备的制约。BSP 与内核、驱动程序以及应用程序的关系如图 3.4 所示。

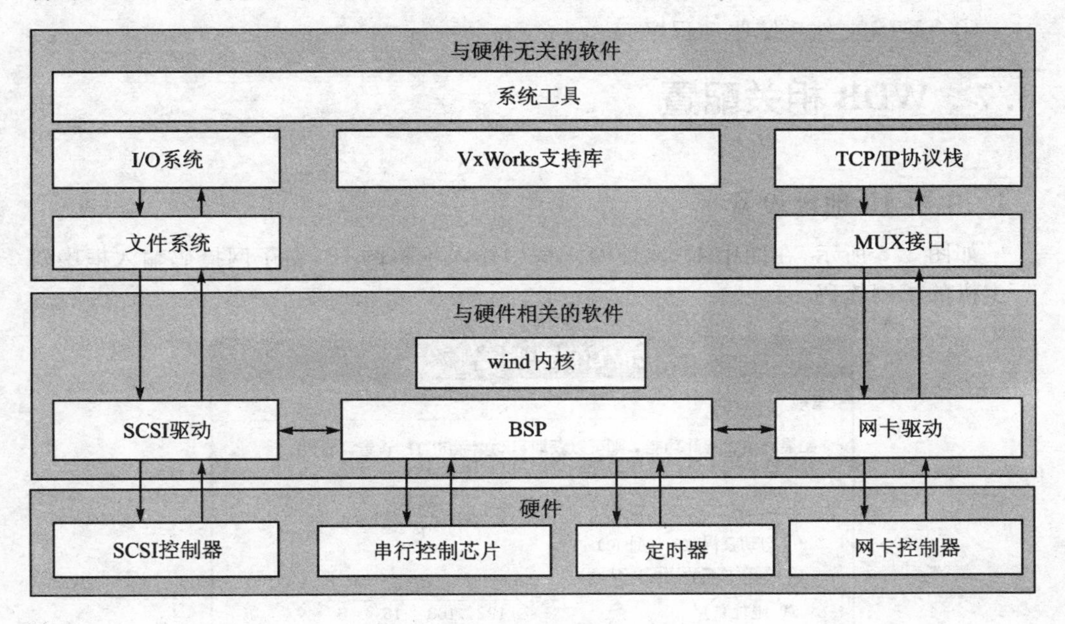

图 3.4　MagicARM2410 实验箱

如图 3.4 所示，BSP 的作用实际上就是在内核和硬件之间建立一个桥梁，以便进行数据交换。

BSP 的开发总体上分为 5 个阶段：

① 配置开发环境和选择调试工具；

② 编写 BSP 初始化代码；

③ 明确硬件的具体情况，配置一个最小化内核；

④ 启动并测试上一步创建的最小内核；

⑤ 编写其他驱动(如网络驱动、文件系统驱动)。

BSP 在初始化时，所需要完成的工作包括：

① CPU 寄存器的初始化；

② SDRAM 内存的初始化；

③ 实时时钟初始化；

④ 网卡初始化；

⑤ 串口初始化；

⑥ FLASH 初始化。

对于实验箱上的设备驱动程序主要包括：

① 串口驱动程序；

② 网卡驱动程序；

③ FLASH 芯片读写驱动程序；

④ TFFS 文件系统驱动程序。

3.7 WDB 相关配置

1. 主机 IP 地址设置

如图 3.5 所示，在图中 IP 地址输入框中输入主机的 IP，在子网掩码输入框中输入主机的子网掩码。

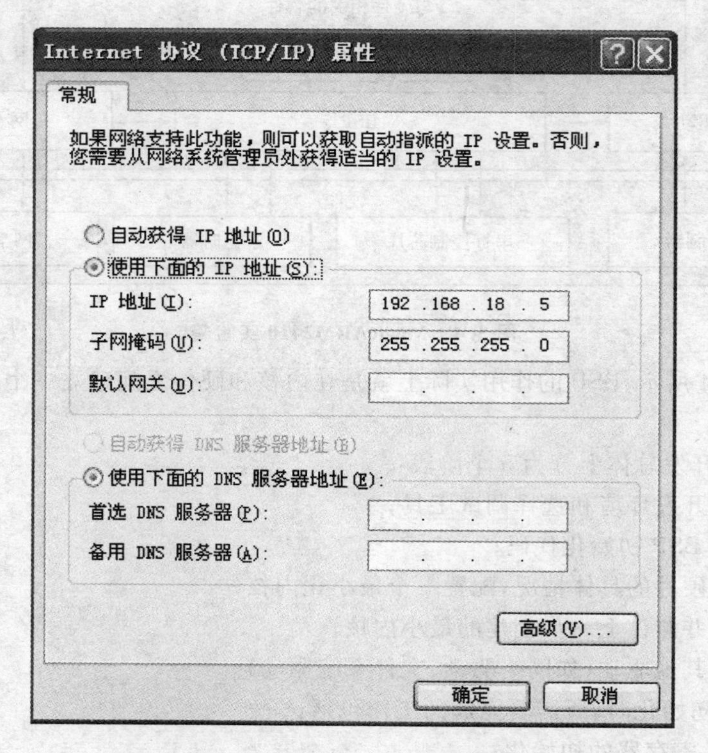

图 3.5 主机 IP 地址设置图

2. Tornado FTP Server 设置

如图 3.6 所示，在图中选择“Security”菜单，然后在弹出的下拉菜单中选择“Users/rights”项，如图 3.7 所示。

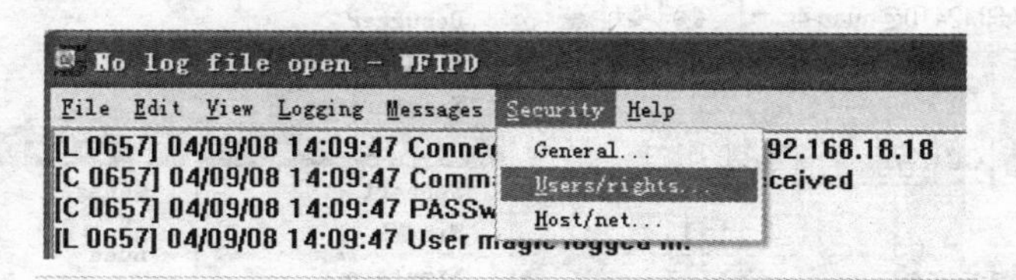

图 3.6 FTP Server 设置图

在图 3.7 中“User Name”中输入用户名“magic”，User Name 和密码要与 config.h 中“DEFAULT_BOOT_LINE”一样，Home Directory 为：“C:\MagicARM2410”，单击“Done”后路径变成大写。

然后，在 C 盘下新建一个名为“MagicARM2410”的文件夹，把 VxWorks 镜像复制进去。

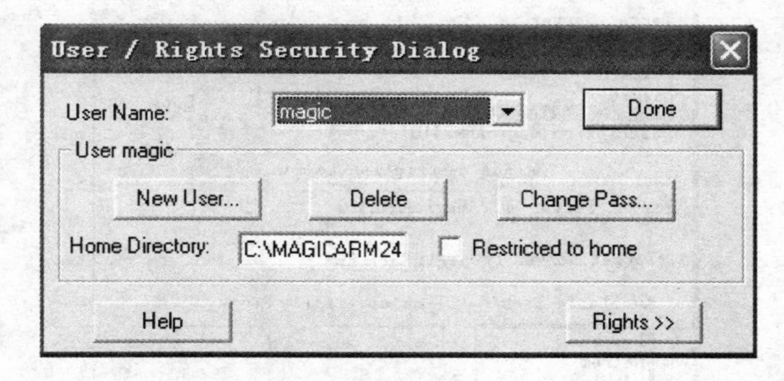

图 3.7 FTP Server User/Rights 设置图

在 C 盘下新建一个名为“MagicARM2410”的文件夹，把 VxWorks 镜像复制进去。

3. Target Server 设置

在 Tornado 开发环境下选择“Tools”菜单下的“Target Server”选项的“Configure”，如图 3.8 所示。

在弹出的配置页中新建一个名为“MagicARM410”的用户，如图 3.9 所示，并完成其他配置。

如图 3.10 所示，完成各个配置项的内容，选择“OK”。

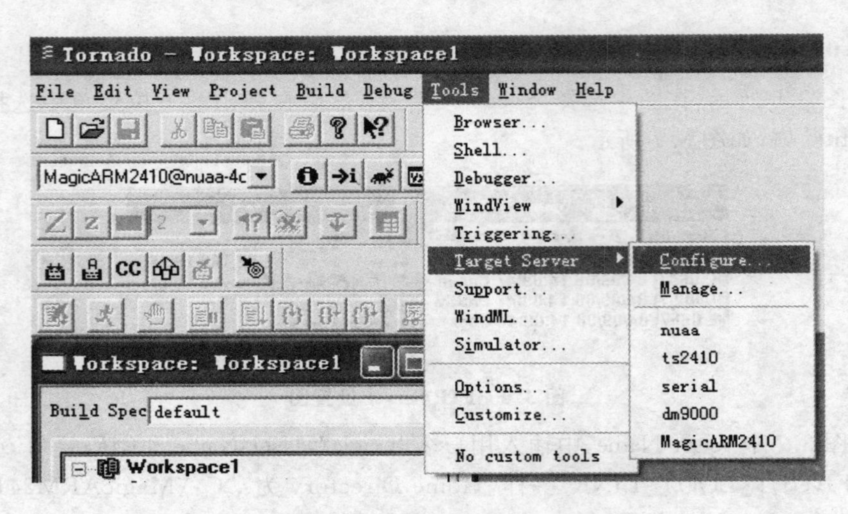

图 3.8　Tornado 开发环境 Configure 项

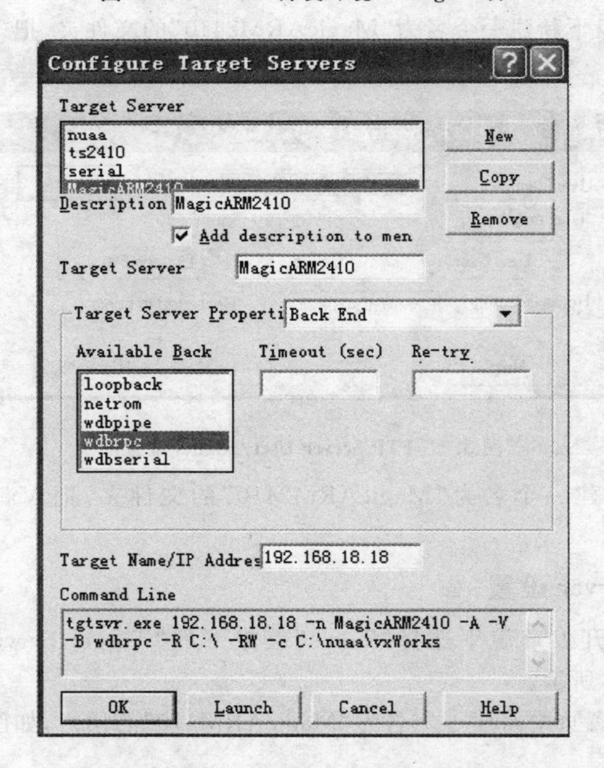

图 3.9　Tornado 开发环境 Configure 对话框

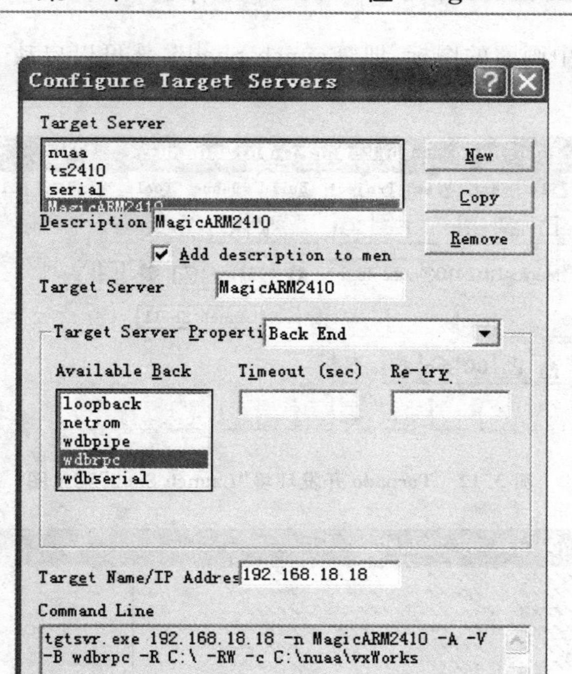

图 3.10　Tornado 开发环境 Configure 对话框选项图

配置完毕后，在“Target Server”菜单项目中可以看到刚才新建的用户“MagicARM2410”，如图 3.11 所示。此时，选择“MagicARM2410”项，在 PC 机桌面的右下角会出现图标。

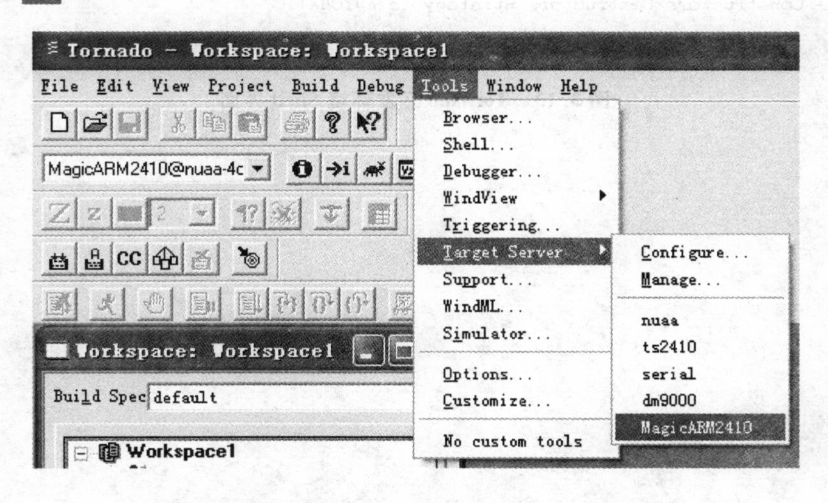

图 3.11　Tornado 开发环境 Target Server 菜单图

单击图 3.12 中画圈的图标，即“Launch Shell”，就可以启动 WShell，如图 3.13 所示。

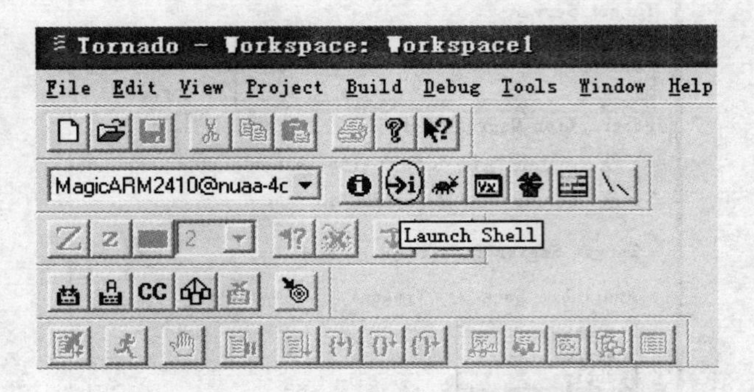

图 3.12　Tornado 开发环境“Launch Shell”菜单图

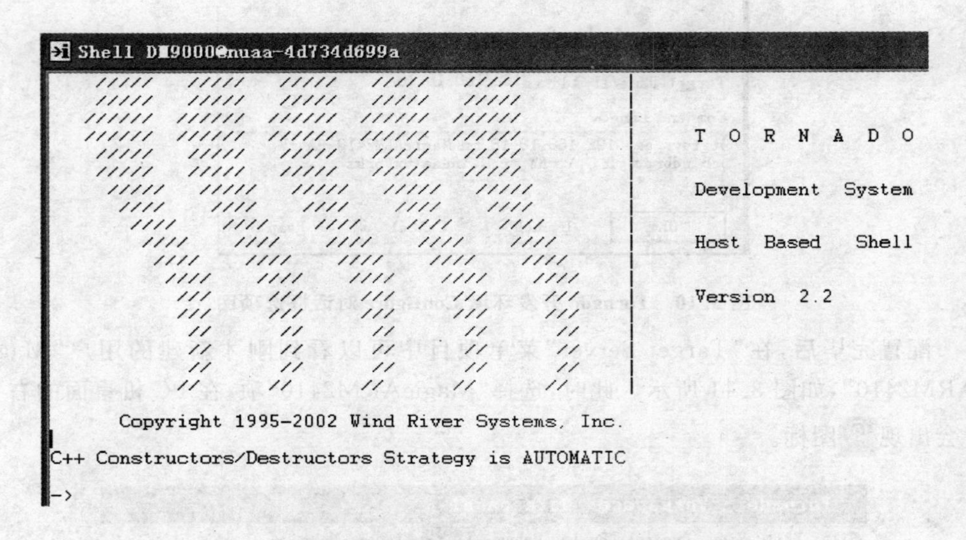

图 3.13　Tornado 开发环境 Shell 界面

第 4 章

TTY 设备驱动程序设计

4.1 TTY 设备驱动编写概述

终端是一种字符型设备,它有多种类型,通常使用 TTY 作为各种类型的终端设备的简称。TTY 是 Teletype 的缩写,是最早出现的一种终端设备,类似电传打字机,是由 Teletype 公司生产的。

串行端口终端(Serial Port Terminal)是指使用计算机串行端口连接的终端设备。计算机把每个串行端口都看作是一个字符设备。串行端口设备过去通常被称为终端设备,因为那时其最大用途就是用来连接终端。这些串行端口所对应的设备名称是/tyCo/0、/tyCo/1 等,分别对应于 DOS 系统下的 COM1、COM2 等。若要向一个端口发送数据,在命令行上把标准输出重定向到这些特殊文件名上即可。VxWorks 系统下最重要的终端特殊设备就是串行端口终端,它既支持 VxWorks 下的 I/O 系统,又支持目标机代理的接口;既可以采用中断方式工作,也可以采用轮询方式工作。除此之外,串行驱动程序还要实现一些与设备无关的功能,如图 4.1 所示为串行设备驱动程序的工作模式。

串口通信是由串行通信控制器(SCC,Serial Communication Controller)控制。1 个 SCC 芯片一般有 2~4 个通道(CHAN),1 个通道物理上对应 1 个串口,每个 SCC 通道有独立的通道缓冲区,能独立地进行配置。在控制上,软件为每个通道维护一组控制结构信息。

在 VxWorks 系统用户程序,采用串口终端(TTY)设备方式访问串口。从 VxWorks 5.3 开始,系统采用 3 层抽象软件结构:标准 I/O 库(ioLib)→TTY 库(ttyDrv/tyLib)→底层 SCC 驱动(xxDrv),如图 4.2 所示。

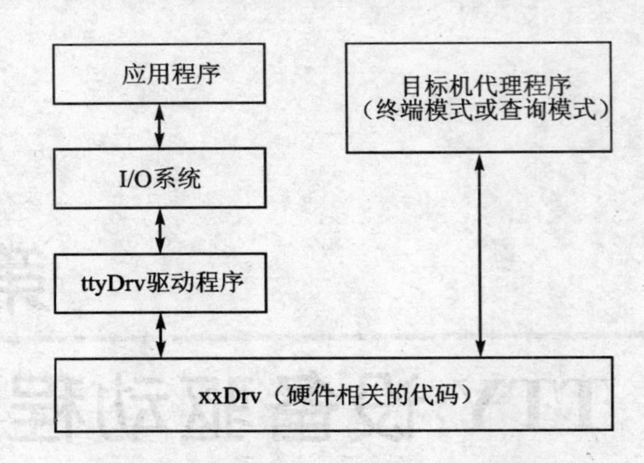

图 4.1　串行设备驱动程序的工作模式

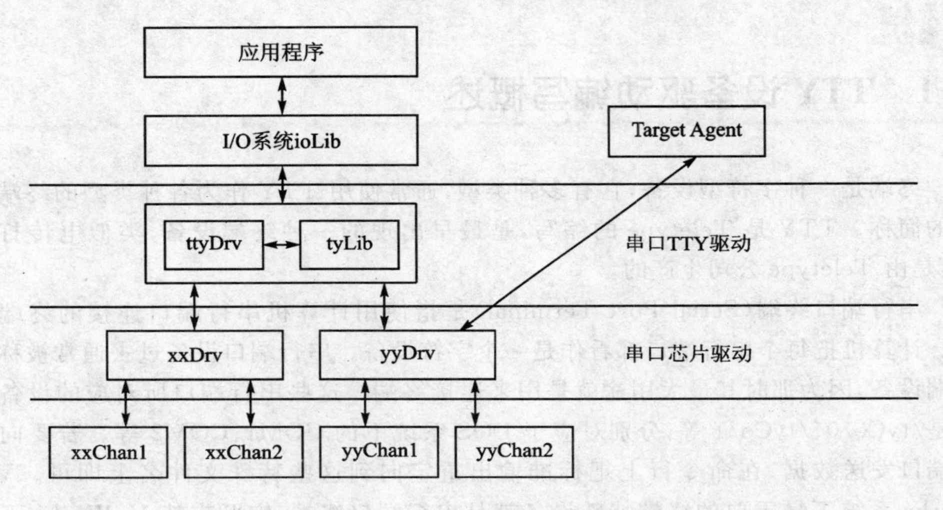

图 4.2　串口软件结构

在 Tornado“安装目录/target/src/drv/serial/”下有 tyCoDrv 驱动库的实现代码；“安装目录/target/src/drv/sio/”下是 SCC 驱动 xxDrv 的实现代码；TTY 库以目标码形式提供。串口 TTY 驱动（ttyDrv/tyLib）使 I/O 系统独立于具体 SCC 驱动，可以同时支持多个不同类型的 SCC 驱动。SCC 驱动（xxDrv/yyDrv）实际处理不同串口芯片的细节，对串口 TTY 驱动提供标准设备 I/O 接口和 target agent 两种接口。串口芯片驱动支持终端和查询两种工作模式。

4.1.1　TTY 驱动

用户通过基本的 I/O 实现串口输入输出，实现 TTY 的驱动，与其他驱动程序不同，TTY 驱动分别由两个库 ttyDrv 和 tyLib 实现，如图 4.3 所示。

● I/O 系统的驱动程序安装和设备安装。

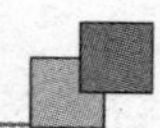

- 驱动程序函数处理 I/O 系统的基本 I/O 调用：open/read/write/ioctl 等。
- 支持 select。
- 维护输入输出函数的环形队列和互斥访问。

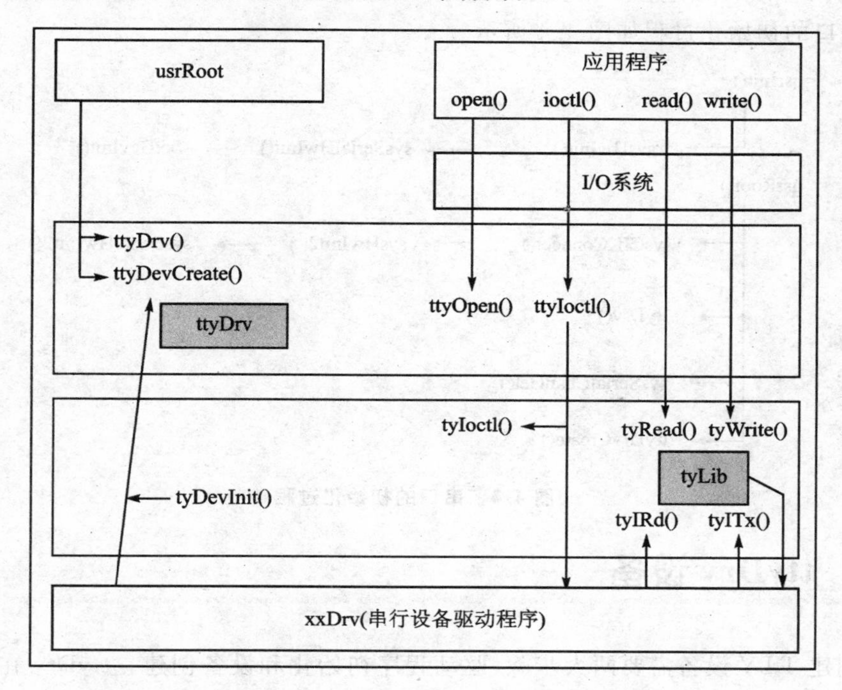

图 4.3　ttyDrv 与 ttyLib 之间的关系

4.1.2　SCC 驱动：xxDrv

SCC 驱动采用中断方式时，通过回调函数在设备和 I/O 系统之间传送数据。TTY 设备需要使用串口的中断方式，在创建 TTY 设备时，ttyDevCreate()负责设置回调函数。

4.2　串口启动和初始化过程

系统启动后，首先在 usrConfig.c 中，usrInit()函数调用 sysHwInit()函数，完成对系统硬件基本的初始化工作。接下来 sysLib.c 中的 sysHwInit()函数调用 sysSerial.c 中的 sysSerialHwInit()函数对 BSP 串行器件进行初始化，使其处于等候状态；sysSerialHwInit()再通过 xxDevInit()复位串行通道。在 usrInit()函数的最后，产生根任务 usrRoot()，usrRoot()调用 sysClkConnect()。sysHwInit2()主要安装系统中断，它调用 sysSerialHwInit2()连接串行中断。如果定义了 INCLUDE_TTY_

DEV，而没有定义 INCLUDE_TYCODRV_5_2，在 usrRoot()任务中调用 ttyDrv()来初始化串行设备驱动，并通过 ttyDevCreate()函数创建串行设备。串行驱动是在 VxWorks 系统开始过程中被初始化的。

串口的初始化过程如图 4.4 所示。

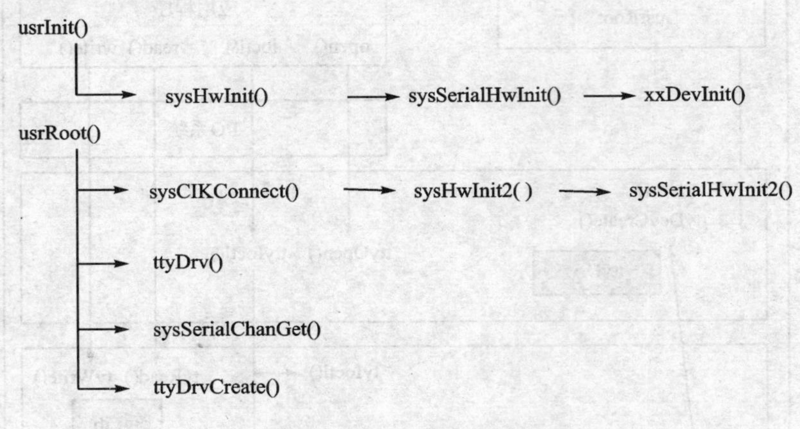

图 4.4　串口的初始化过程

4.3　ttyDrv 设备

创建 TTY 设备需要两大步骤：驱动程序初始化和设备创建。ttyDrv 在 I/O 系统与具体驱动之间，管理两者之间的通信。如图 4.5 所示，ttyDrv 函数将调用 iosDrvInstall()函数在系统的驱动程序表中安装 tty 驱动程序。所有的用户应用的串口 I/O 请求通过 ttDrv 和 tyLib 实现，ttyDrv()安装的函数分别来自于 ttyDrv 和 tyLib：

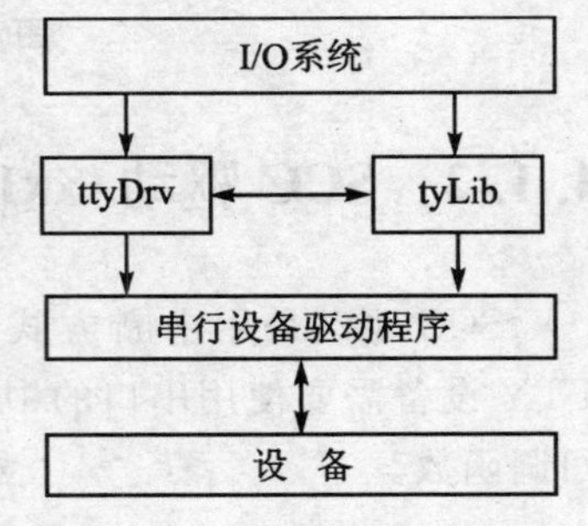

图 4.5　TTY 的功能

- ttyDrv 负责 open 和 ioctl，即 ttyOpen()函数和 ttyIoctl()函数。
- tyLib 负责 read 和 write，即 tyRead()函数和 tyWrite()函数。
- 设备创建在系统初始化时完成，creat()函数和 close()函数对应的驱动程序函数为 NULL。

ttyDrv 数据流向和系统层次非常清晰，其数据流图如图 4.6 所示。

ttyDrv 的相应函数被 I/O 系统调用来响应，同时完成缓冲区互斥管理工作，任务同步操作。I/O 系统所感受的是在与 ttyDrv 设备通信。接下来，ttyDrv 将 I/O 系统的请求包进行处理之后传递给驱动设备，完成实际的 I/O 操作。

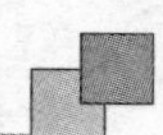

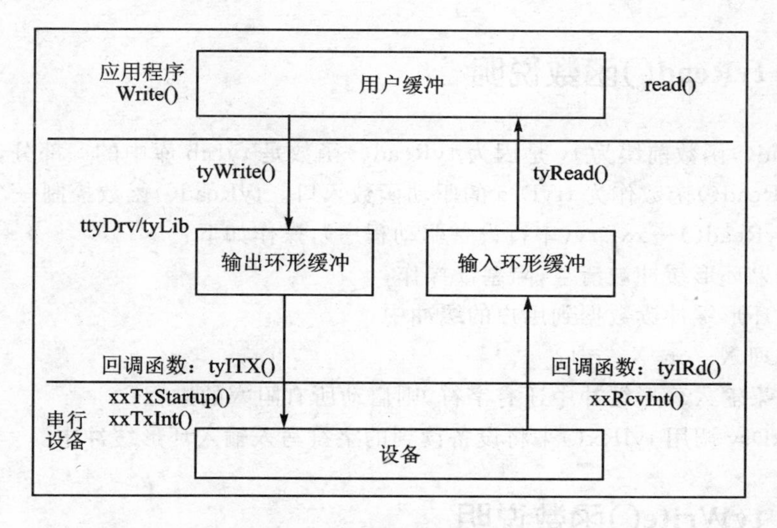

图 4.6 ttyDrv 的数据流图

4.3.1 ttyDrv()函数说明

作为字符型设备，ttyDrv 的实现方式与字符设备相同，ttyDrv()函数调用 iosDrvInstall()函数安装驱动程序，分别安装 ttyOpen()、ttyIoctl()、tyRead()、tyWrite()到系统驱动程序表中，供 I/O 系统调用。如果 BSP 中没有定义 INCLUDE_TYCODRV_5_2，系统则在 usrConfig. c 中的 usrRoot 任务中自动调用 ttyDrv()。

4.3.2 ttyDrvCreate()函数说明

创建设备函数：

STATUS ttyDrvCreate(devName，pSioChan，rdBufSize，wrtBufSize)

ttyDevCreate()函数是由 usrConfig. c 中的 usrRoot 任务调用的。根据 NUM_TTY 的值，usrRoot()函数调用 ttyDevCreate()函数创建相应数量的设备，该函数完成的具体工作如下：

- 分配并初始化设备描述结构空间；
- 进行 tyLib 库初始化，tyLib 提供初始化函数 tyDevInit()；selectLib 初始化，创建输入输出的环形队列，创建信号量；
- 调用 iosDevAdd()将设备加入设备列表；
- 安装 tyLib 提供的输入输出回调函数，底层 SCC 驱动负责调用；
- 以中断方式启动串口通道。

4.3.3 tyRead()函数说明

tyRead()函数前缀为 ty 是因为 tyRead()函数是 tyLib 库中的一部分，ttyDrv()函数将 tyRead()函数作为 ttyDrv 的驱动函数入口。tyRead()函数控制一个缓冲区，read()←tyRead()←xxDrv(串行设备驱动程序)，操作如下：

- 如果环形缓冲被清空，阻塞读操作；
- 从环形缓冲读数据到用户的缓冲中；
- 处理 X-on/X-off；
- 如果输入环形缓冲中还有字符，则启动所有阻塞的进程；
- xxDrv 调用 tyIRx()来将设备读到的字符写入输入环形缓冲中。

4.3.4 tyWrite()函数说明

如同 tyRead()函数，tyWrite()函数也是 tyLib 库中的一部分，控制一个环形缓冲区，调用时，tyWrite()函数会将数据从用户缓冲区复制到该缓冲区中，write()→tyWrite()→xxDrv(串行设备驱动程序)写过程，操作如下：

- 如果环形缓冲满，则阻塞写操作；
- 从用户的缓冲中读取数据到环形缓冲中；
- 在一定情况下激活 xxDrv 开始传输数据周期(一个传输周期中先传送数据到设备，然后等待中断，继续传输下一个数据)；
- xxDrv 通过调用 tyITx()从输出环形缓冲区读数据。

4.3.5 ttyIoctle()函数说明

ttyIoctl()函数对应 I/O 系统 ioctl()的调用。当调用 ttyIoctl()函数时，这个函数首先访问实际驱动程序所提供的 xxIoctl()函数。如果所需的功能在 xxIoctl()函数中未提供，ttyIoctl()函数还会继续调用 tyIoctl()函数；如果 tyIoctl()函数执行也失败，ttyIoctl()最终也会返回失败。I/O 系统串行设备控制过程如图 4.7 所示。

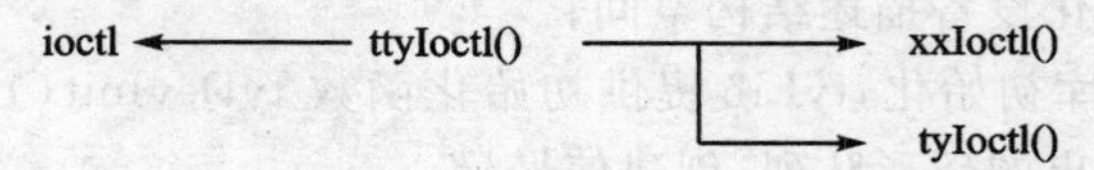

图 4.7 串行设备控制操作流程

下面列出了 ttyIoctl()所支持的功能：

- FIONREAD，获取缓冲区中可读取的字符数；
- FIONWRITE，获取缓冲区中将要写入的字符数；

- FIOFLUSH,清除所有缓冲区中所有字符;
- FIOWFLUSH,清除写缓冲区中所有字符;
- FIORFLUSH,清除读缓冲区中所有字符;
- FIOSETOPTIONS,设置串行设备相关选项;
- FIOGETOPTIONS,获取串行设备相关选项;
- FIOCANCEL,取消对设备的读写操作;
- FIOISATTY,是否是TTY设备;
- FIOSELECT,select功能所指定的命令;
- FIOUNSELECT,unselect功能所制定的命令。

4.4 S3C2410串口驱动设计

编写串行设备驱动主要有以下几项工作。

- 初始化:确定系统要支持的串行通道个数,初始化设备的描述符,编写设备的初始化代码。
- 编写入口点函数。
- 编写设备中断服务程序(ISR)。
- 修改sysSerial.c文件。

4.4.1 串口初始化过程

串口初始化顺序如图4.4所示,主要涉及到两大步骤。

1. 在usrInit()中进行设备的初始化

由函数sysHwInit()调用sysSerialHwInit()函数,进行设备资源的设置,填写S3C2410_CHAN结构,初始化硬件,并关闭串口的中断。

2. 在usrRoot中执行进一步操作

① 在函数usrRoot()中调用sysSerialHwInit2()函数,开启中断。

② 在函数usrIosCoreInit()中调用ttyDrv()函数,初始化ttyLib。

③ 在函数usrIosCoreInit()中调用usrSerialInit()函数,然后在usrSerialInit()函数中调用ttyDevCreate()函数,创建串口设备。其中创建时ttyDevCreate()的第二个参数SIO_CHAN需要通过sysSerialChanGet()获得。

编写驱动程序,首先要根据具体的串行设备定义设备描述符的结构(xx_DEV)。该结构是注册在设备列表中的,它与描述设备的设备头(DRV_HEAD)以及设备相关的成员构成一个“设备”,在文件C:\Tornado2.2\target\h\sioLib.h中定义如下:

```
typedef struct sio_chan                        /* 一个串行通道 */
    {
        SIO_DRV_FUNCS * pDrvFuncs;
    } SIO_CHAN;

struct sio_drv_funcs                           /* 驱动函数 */
    {
        int (* ioctl)(SIO_CHAN *    pSioChan, int cmd,    void * arg);

        int (* txStartup)(SIO_CHAN *    pSioChan);

        int (* callbackInstall)( SIO_CHAN *    pSioChan, int callbackType,
                                 STATUS( * callback)(void *, ...),
                                 void * callbackArg);

        int (* pollInput)(SIO_CHAN *    pSioChan, char *    inChar);

        int (* pollOutput)(SIO_CHAN *    pSioChan, char     outChar);
    };

typedef struct s3c2410x_CHAN
{
    /* must be first */
    SIO_CHAN    sio;                 /* 标准 SIO_CHAN 成员,并且必须为第一个; */

    /* callbacks */
    STATUS    ( * getTxChar) ();            /* 安装 Tx 回调函数 */
    STATUS    ( * putRcvChar)();            /* 安装 Rx 回调函数 */
    void *      getTxArg;                   /* Tx 回调函数的参数 */
    void *      putRcvArg;                  /* Rx 回调函数的参数 */

    UINT32 *    regs;                       /* s3c2410x 寄存器 */
    UINT8     intLevelRx;                   /* Rx 中断号 */
    UINT8     intLevelTx;                   /* Tx 中断号 */

    UINT32   channelMode;                   /* such as INT, POLL modes */
    int         baudRate;                   /* the current baud rate */
    UINT32   xtal;                          /* UART clock frequency */

    int    options;
}s3c2410x_CHAN;
```

本书案例定义了两个串口设备，在文件中 C:\Tornado2.2\target\config\MagicARM2410\sysSerial.c 定义了串口资源，如下：

```
typedef struct
    {
        UINT        vector;
        UINT32 *    baseAdrs;
        UINT        intLevel;
    }SYS_s3c2410x_CHAN_PARAS;

LOCAL SYS_s3c2410x_CHAN_PARAS devParas[] =
    {
        {INT_VEC_UART_0, (UINT32 *)UART_0_BASE_ADR, INT_LVL_UART_0},
        {INT_VEC_UART_1, (UINT32 *)UART_1_BASE_ADR, INT_LVL_UART_1}
    };

LOCAL s3c2410x_CHAN s3c2410xChan[N_s3c2410x_UART_CHANNELS];
```

其中 N_s3c2410x_UART_CHANNELS 定义为 2;

```
SIO_CHAN * sysSioChans [] =
    {
        &s3c2410xChan[0].sio,     /* /tyCo/0 */
        &s3c2410xChan[1].sio      /* /tyCo/1 */
    };
```

s3c2410Chan 数据结构的填写主要由 sysSerialHwInit()和 s3c2410SioDevInit()两个函数完成。其中，sysSerialHwInit()完成资源定义；s3c2410SioDevInit()则负责处理函数的指定，具体代码如下：

```
void sysSerialHwInit (void)
{
        int i;

        for(i = 0; i < N_s3c2410x_UART_CHANNELS; i++)
        {
            s3c2410xChan[i].regs = devParas[i].baseAdrs;
            s3c2410xChan[i].baudRate = CONSOLE_BAUD_RATE;
            s3c2410xChan[i].xtal = UART_XTAL_FREQ;

            s3c2410xChan[i].intLevelRx = devParas[i].intLevel;
            s3c2410xChan[i].intLevelTx = devParas[i].intLevel;

            /*
```

```
         * Initialise driver functions, getTxChar, putRcvChar and channelMode
         * and initialise UART
         */

         s3c2410xSioDevInit(&s3c2410xChan[i]);
      }
}

void sysSerialHwInit2 (void)
{
      int i;

      for(i = 0; i < N_s3c2410x_UART_CHANNELS; i++)
      {
         /*
          * Connect and enable the interrupt.
          * We would like to check the return value from this and log a message
          * if it failed. However, logLib has not been initialised yet, so we
          * cannot log a message, so there's little point in checking it.
          */
         (void)intConnect(
                              INUM_TO_IVEC(devParas[i].vector),
                              s3c2410xSioInt,
                              (int)&s3c2410xChan[i]
                           );
         s3c2410xIntLvlEnable(devParas[i].intLevel);
      }
}

SIO_CHAN * sysSerialChanGet(int channel)  /* 串口号 */
{
if(channel < 0 || channel >= (int)(NELEMENTS(sysSioChans)))
     {
         return (SIO_CHAN *)ERROR;
     }

     /* 根据串口号返回描述结构的指针 */
     return sysSioChans[channel];
}

void sysSerialReset (void)
{
```

```
    int i;

    for(i = 0; i < N_s3c2410x_UART_CHANNELS; i++)
    {
        /* disable serial interrupts */
        intDisable (devParas[i].intLevel);
    }

}
```

xxDevInit()函数是系统启动过程中首先调用的一个底层函数，被 sysSerail.c 中的 sysSerialHwInit()函数调用，用于对目标系统的串行设备描述符初始化之后，将操作函数安装到 SIO_DRV_FUNCS 结构中，具体代码如下：

```
LOCAL SIO_DRV_FUNCS s3c2410xSioDrvFuncs =
    {
        (int (*)())s3c2410xIoctl,
        s3c2410xTxStartup,
        (int (*)())s3c2410xCallbackInstall,
        s3c2410xPollInput,
        s3c2410xPollOutput
    };

void s3c2410xSioDevInit
    (
        s3c2410x_CHAN *     pChan       /* 串口描述指针结构 */
    )
{
        int oldlevel = intLock();

        /* 设置驱动程序的操作例程 */
        pChan->sio.pDrvFuncs = &s3c2410xSioDrvFuncs;

        pChan->getTxChar = s3c2410xDummyCallback;
        pChan->putRcvChar = s3c2410xDummyCallback;

        pChan->channelMode = SIO_MODE_POLL;    /* undefined */

        /* 硬件初始化 */
        s3c2410xInitChannel(pChan);

        intUnlock(oldlevel);
```

```
}

LOCAL void s3c2410xInitChannel
(
    s3c2410x_CHAN *     pChan      /* 串口描述指针结构 */
)
{
    UINT32     tempUINT32;

    /* 设置 UCLK, polling interrupt 寄存器 */
     s3c2410x_UART_REG_WRITE(pChan,OFFSET_UCON,
                        CLK_PCLK + TxMode_IntPoll + RxMode_IntPoll);

    /* 打开串口子中断 */
    s3c2410x_INT_REG_READ(s3c2410x_INT_CSR_INTSUBMSK,tempUINT32);
    switch((int)(pChan->regs))
    {
        case UART_1_BASE_ADR:
            tempUINT32 &= ~((1<<SUBINT_LVL_RXD1)|(1<<SUBINT_LVL_TXD1));
            break;
        case UART_0_BASE_ADR:
        default:
            tempUINT32 &= ~((1<<SUBINT_LVL_RXD0)|(1<<SUBINT_LVL_TXD0));
     }

    s3c2410x_INT_REG_WRITE(s3c2410x_INT_CSR_INTSUBMSK,tempUINT32);

    /* 本例设置波特率为 9 600 */
    s3c2410xIoctl((SIO_CHAN *)pChan,
                SIO_BAUD_SET,pChan->baudRate );

    /* 设置串口为 8 位,无校验位 N,1 个停止位,轮询方式 */
    s3c2410xIoctl((SIO_CHAN *)pChan, SIO_HW_OPTS_SET, CLOCAL + CS8);

    s3c2410xIoctl((SIO_CHAN *)pChan, SIO_MODE_SET, SIO_MODE_POLL);

    /* 关闭 FIFO */
    s3c2410x_UART_REG_WRITE(pChan, OFFSET_UFCON, FIFO_OFF);

    /* 使能 UART 的引脚 */
    s3c2410x_IO_READ(rGPHCON, tempUINT32);
```

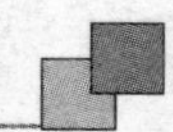

```
switch((int)(pChan->regs))
{
 case UART_1_BASE_ADR:
    tempUINT32| = (MASK_GPH4(2)
                 + MASK_GPH5(2) + MASK_GPH6(3) + MASK_GPH7(3));
        break;
    case UART_0_BASE_ADR:
    default:
        tempUINT32| = (MASK_GPH0(2)
                 + MASK_GPH1(2) + MASK_GPH2(2) + MASK_GPH3(2));
}
s3c2410x_IO_WRITE(rGPHCON,tempUINT32);

/* 清除接收缓冲 */
s3c2410x_UART_REG_READ(pChan, OFFSET_URXH, tempUINT32);
}
```

4.4.2 编写处理函数

如上文建立数据结构之后，则着手编写一些具体的通信函数，包括SIO_DRV_FUNCS结构中的函数、输入/输出ISR和一个辅助函数xxDevInit()。SIO_DRV_FUNCS中的主要函数如下：

- xxCallBackInstall，安装高层协议的处理例程；
- xxPollOutput，轮询方式输出例程；
- xxPollInput，轮询方式输入例程；
- xxIoctl，设备相关的一些控制码处理例程；
- xxTxStartup，启动一个传输过程；
- xxTxInt，发送中断处理例程；
- xxRxInt，接收中断处理例程。

1. 上层处理函数安装例程 xxCallBackInstall()

回调用于层间协作，上层应用将本层函数安装在下层，这个函数就是回调函数。而下层在一定条件下触发回调。例如作为一个驱动函数，本身是一个底层函数，它在收到一个驱动层的数据时，除了完成本层的处理工作外，还需要进行回调，将这个数据交给上层应用层来做进一步处理，这在分层的数据通信中很普遍。

```
LOCAL int s3c2410xCallbackInstall
(
    SIO_CHAN *      pSioChan,      /* 设备描述符指针 */
```

```
    int         callbackType,     /* 回调函数的类型  */
    STATUS     (*callback)(),   /* 回调函数指针  */
    void*      callbackArg      /* 回调函数的参数列表  */
)
{
    s3c2410x_CHAN * pChan = (s3c2410x_CHAN *)pSioChan;

    switch(callbackType)
    {
        case SIO_CALLBACK_GET_TX_CHAR:
            pChan->getTxChar      = callback;
            pChan->getTxArg        = callbackArg;
            return OK;

        case SIO_CALLBACK_PUT_RCV_CHAR:
            pChan->putRcvChar       = callback;
            pChan->putRcvArg      = callbackArg;
            return OK;

        default:
            return ENOSYS;
    }
}
```

2. 设备相关的一些控制码处理例程 xxIoctl()

指向驱动的标准 I/O 控制接口函数。此函数为任何驱动提供主要的控制接口。为了实现标准的 SIO 设备的 I/O 控制，用以下的定义：

- SIO_BAUD_SET，设置波特率；
- SIO_BAUD_GET，获取当前波特率；
- SIO_MODE_SET，设置波特率；
- SIO_MODE_GET，获取波特率；
- SIO_AVAIL_MODES_GET，获取可选模式；
- SIO_HW_OPTS_SET，设置硬件模式；
- SIO_HW_OPTS_GET，获取硬件模式；
- SIO_OPEN 打开一个通道；
- SIO_HUP 关闭一个通道。

函数原型如下：

```
LOCAL int s3c2410xIoctl
    (
```

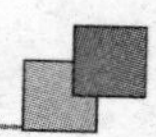

```
    SIO_CHAN *      pSioChan,           /* 通道描述符 */
    int             request,            /* 请求码 */
    int             arg                 /* 参数 */
)
{
    s3c2410x_CHAN * pChan = (s3c2410x_CHAN *) pSioChan;
    int oldlevel;                       /* 旧中断级别 */
    UINT32 tempUINT32 = 0;
    int lvl;

    switch (request)
    {
        case SIO_BAUD_SET:
            if(arg < s3c2410x_BAUD_MIN || arg > s3c2410x_BAUD_MAX)
                return(EIO);

            /* 关闭中断 */
            oldlevel = intLock ();
            /* 设置波特率 */
            s3c2410x_UART_REG_WRITE(pChan,OFFSET_UDIV,
                                    (((s3c2410x_PCLK/16)/arg) - 1));
            intUnlock(oldlevel);
            s3c2410x_UART_REG_READ(pChan,OFFSET_UDIV,tempUINT32);
            pChan->baudRate = ((s3c2410x_PCLK/16)/(tempUINT32 + 1));
            break;
        case SIO_BAUD_GET:
            /* 返回当前串口波特率 */
            *(int *)arg = pChan->baudRate;
            break;
        case SIO_MODE_SET:
            /* 设置当前串口工作模式:中断或者轮询 */
            switch(arg)
            {
                case SIO_MODE_INT:
                /* 中断模式下的设置 */

                    switch((int)(pChan->regs))
                    {
                    case UART_1_BASE_ADR:
                        s3c2410x_INT_REG_WRITE(s3c2410x_INT_CSR_SUBSRCPND,
                                               ((1<<SUBINT_LVL_TXD1)
                                               |(1<<SUBINT_LVL_RXD1)));
```

```
        break;
    case UART_0_BASE_ADR:
    default:
        s3c2410x_INT_REG_WRITE(s3c2410x_INT_CSR_SUBSRCPND,
                               ((1<<SUBINT_LVL_TXD0)
                               |(1<<SUBINT_LVL_RXD0)));
    }

    /* enable uart_int */
    intEnable(pChan->intLevelRx);

    /* enable subInterrupt for UART0. */
    s3c2410x_INT_REG_READ(s3c2410x_INT_CSR_INTSUBMSK,
                          tempUINT32);
    switch((int)(pChan->regs))
    {
        case UART_1_BASE_ADR:
            tempUINT32 &= ~((1<<SUBINT_LVL_RXD1)
                           |(1<<SUBINT_LVL_TXD1));
            break;
        case UART_0_BASE_ADR:
        default:
            tempUINT32 &= ~((1<<SUBINT_LVL_RXD0)
                           |(1<<SUBINT_LVL_TXD0));
    }

    s3c2410x_INT_REG_WRITE(s3c2410x_INT_CSR_INTSUBMSK,
                           tempUINT32);

    break;

case SIO_MODE_POLL:
/* 轮询模式下的设置 */

    /* disable uart_int */
    intDisable(pChan->intLevelRx);

    /* disable subInterrupt for UART0. */
    s3c2410x_INT_REG_READ(s3c2410x_INT_CSR_INTSUBMSK,
                          tempUINT32);
    switch((int)(pChan->regs))
    {
    case UART_1_BASE_ADR:
```

```
                    tempUINT32 |= ((1<<SUBINT_LVL_RXD1)
                                  |(1<<SUBINT_LVL_TXD1));
                    break;
                    case UART_0_BASE_ADR:
                    default:
                    tempUINT32 |= ((1<<SUBINT_LVL_RXD0)
                                  |(1<<SUBINT_LVL_TXD0));
                    }
                    s3c2410x_INT_REG_WRITE(s3c2410x_INT_CSR_INTSUBMSK,
                                           tempUINT32);
                    break;
                default:
                    return(EIO);
            }
        pChan->channelMode = arg;
        break;
    case SIO_MODE_GET:
        /* 获取当前的工作模式 */
        *(int *)arg = pChan->channelMode;
        break;
    case SIO_AVAIL_MODES_GET:
        /* 获取可选的工作模式 */
         *(int *)arg = SIO_MODE_INT | SIO_MODE_POLL;
         break;
    case SIO_HW_OPTS_SET:
        if(arg & 0xffffff00) return EIO;

        /* do nothing if options already set */
        if(pChan->options == arg) break;

        switch (arg & CSIZE)
        {
        case CS5:
            tempUINT32 = DATABIT_5; break;
        case CS6:
            tempUINT32 = DATABIT_6; break;
        case CS7:
            tempUINT32 = DATABIT_7; break;
        default:
        case CS8:
            tempUINT32 = DATABIT_8; break;
        }
```

```
if (arg & STOPB)
{
    tempUINT32 | = TWO_STOPBIT;
}
else
{
    /* tempUINT32 &= ~TWO_STOPBIT */;
}

switch (arg & (PARENB|PARODD))
{
    case PARENB|PARODD:
        tempUINT32 + = ODD_PARITY;
        break;
    case PARENB:
        tempUINT32 + = EVEN_PARITY;
        break;
    case 0:
    default:
        break;/* no parity */
}

lvl = intLock();
s3c2410x_UART_REG_WRITE(pChan,OFFSET_ULCON,tempUINT32);
intUnlock(lvl);

if (arg & CLOCAL)
{
    /* clocal disables hardware flow control */
    lvl = intLock();
    s3c2410x_UART_REG_WRITE(pChan,OFFSET_UMCON,AFC_OFF);
    intUnlock(lvl);
}
else
{
    lvl = intLock();
    s3c2410x_UART_REG_WRITE(pChan,OFFSET_UMCON,AFC_ON);
    intUnlock(lvl);
}
```

```
        pChan->options = arg;
        break;
    case SIO_HW_OPTS_GET:
        *(int*)arg = pChan->options;
        return (OK);
    case SIO_HUP:
        /* check if hupcl option is enabled */
        break;
    case SIO_OPEN:
        break; /* always open */
    default:
        return (ENOSYS);
    }
    return (OK);
}
```

3. 中断传输例程 xxTxStartup( )和 xxRxInt( )、xxRxInt( )

串口发送数据时需要系统启动串口，xxTxStartup( )就是用来启动串口数据发送，当发送成功后会触发串口中断，中断根据缓存中是否还有数据来决定是否继续发送，当缓存清空后串口将恢复初始状态，并等待下一次传送的启动。这部分包括 3 个底层函数编写：启动输出函数 xxTxStartup()、发送中断处理函数 xxTxInt()和接收中断处理函数 xxRxInt()。输出部分的处理流程如图 4.8 所示。

当向设备写入数据时，调用 tyWrite 将数据写入输出环形缓冲后调用 xxTxStartup()启动设备数据发送。

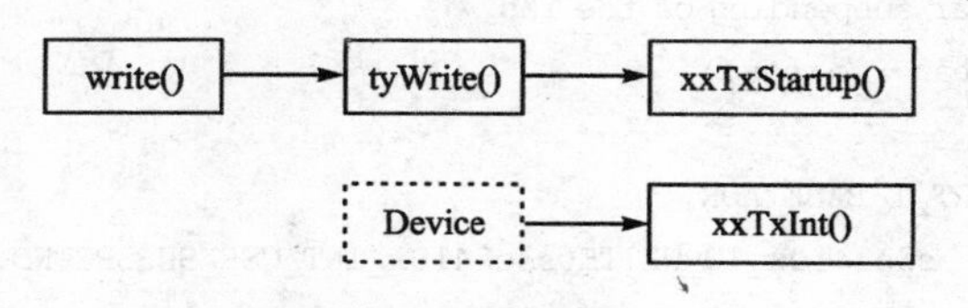

图 4.8 输出部分的处理流程

```
LOCAL int s3c2410xTxStartup
    (
        SIO_CHAN *     pSioChan    /* 串口描述符指针 */
    )
{
        s3c2410x_CHAN * pChan = (s3c2410x_CHAN *)pSioChan;

        if(pChan->channelMode == SIO_MODE_INT)
        {
            intEnable(pChan->intLevelTx);
```

```
                s3c2410xSioIntTx(pChan);
                return OK;
            }
    else
    {
                return ENOSYS;
            }
    }
```

用户通过 I/O 系统的 write()操作是调用 tyWrite()函数在驱动列表中的 ttyDrv 的写入口函数。tyWrite()把数据复制到环形缓冲中，并且调用 xxTxStartup()来初始化一个发送周期。当设备输出完毕后，设备就给 CPU 一个中断表示可以接收下一个字符，然后进入中断 xxIntTx()。函数 xxIntTx()调用回调函数 getTxChar 从高层协议取字节，把字节写入到设备。如果需要的话，在没有数据等待发送的时候，复位发送中断。在中断模式下，缓冲区中的数据发送完毕后，设备会产生一个中断发送下一个字符，这时就会调用中断处理程序 xxTxInt()来完成其余的数据发送工作。发送中断处理函数如下：

```
    void s3c2410xSioIntTx
        (
            s3c2410x_CHAN *     pChan      /* 串口描述符指针 */
        )
        {
            char outChar;

            /* clear subpending of the TXn */
    switch((int)(pChan->regs))
            {
            case UART_1_BASE_ADR:
                    s3c2410x_IO_WRITE(s3c2410x_INT_CSR_SUBSRCPND,
    (1<<SUBINT_LVL_TXD1));
                    break;
                case UART_0_BASE_ADR:
                default:
    s3c2410x_IO_WRITE(s3c2410x_INT_CSR_SUBSRCPND,
    (1<<SUBINT_LVL_TXD0) );
            }
            /* 用回调函数 getTxChar()从驱动程序环形缓冲中读取一个字符 */
            if(( * pChan->getTxChar) (pChan->getTxArg, &outChar) != ERROR)
            {
                /* 将待发送的字符写入到发送缓冲区 */
                s3c2410x_UART_REG_WRITE(pChan, OFFSET_UTXH,
```

```
(((UINT32)outChar)&0x000000ff));
        }
}
```

接收数据部分与发送部分有一点区别，这部分只是保留了输入中断处理函数，该中断处理函数在收到数据时，将寄存器中的数据通过回调函数写入到驱动程序的环形缓冲区中，输入函数之间的关系如图 4.9 所示。

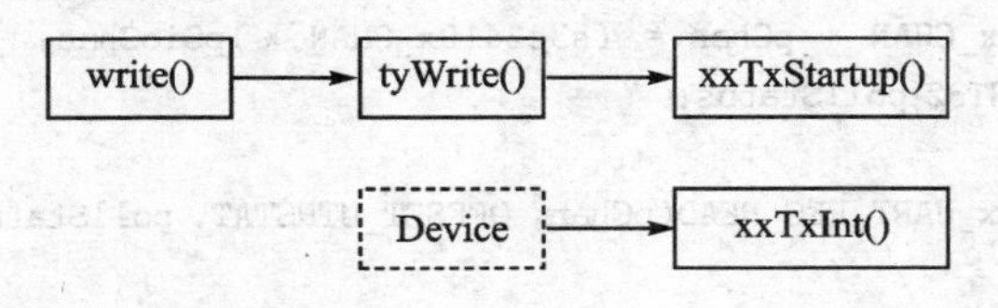

图 4.9 输入部分函数的调用关系

```
void s3c2410xSioIntRx
    (
        s3c2410x_CHAN *     pChan       /* 串口描述符指针 */
    )
{
        char inchar;

        /* clear subpending of the RXn */
        switch((int)(pChan->regs))
        {
case UART_1_BASE_ADR:
s3c2410x_IO_WRITE(s3c2410x_INT_CSR_SUBSRCPND,
                (1<<SUBINT_LVL_RXD1));
                break;
            case UART_0_BASE_ADR:
default:
s3c2410x_IO_WRITE(s3c2410x_INT_CSR_SUBSRCPND,
(1<<SUBINT_LVL_RXD0));
        }

        /* 从串口缓冲区中读取数据 */
        s3c2410x_UART_REG_READ(pChan, OFFSET_URXH, inchar);
    /* 用回调函数 putRcvChar()将读取的数据保存在驱动程序的环形缓冲区中 */
     (*pChan->putRcvChar) (pChan->putRcvArg, inchar);
}
```

4. 轮询方式下的函数编写

轮询方式下完成串口数据的接收和发送也是常见的方式，比如目标机代理。轮

询方式用到的函数如下：

```
LOCAL int s3c2410xPollOutput
    (
        SIO_CHAN * pSioChan,              /* 通道描述符指针 */
        char       outChar                /* 输出字符 */
    )
{
        s3c2410x_CHAN * pChan = (s3c2410x_CHAN *)pSioChan;
        FAST UINT32 pollStatus;

        s3c2410x_UART_REG_READ(pChan, OFFSET_UTRSTAT, pollStatus);

        /* 查询接收缓冲区的状态 */
        if(! (pollStatus & UTRSTAT_TRNSR_EM))
{
            return EAGAIN;
         }

         /* 发送数据 */
         s3c2410x_UART_REG_WRITE(pChan, OFFSET_UTXH, outChar);

         return OK;
}

LOCAL int s3c2410xPollInput
    (
        SIO_CHAN *      pSioChan,          /* 通道描述符指针 */
        char *         thisChar            /* 接收缓冲区指针 */
    )
{
        s3c2410x_CHAN * pChan = (s3c2410x_CHAN *)pSioChan;
        FAST UINT32 pollStatus;

s3c2410x_UART_REG_READ(pChan, OFFSET_UTRSTAT, pollStatus);

        if(! (pollStatus & UTRSTAT_RB_RDY))
        {
            return EAGAIN;
        }

        /* got a character */
        s3c2410x_UART_REG_READ(pChan, OFFSET_URXH, *thisChar);

        return OK;
}
```

第 5 章

VxWorks 块设备驱动程序设计

5.1 VxWorks 块设备简介

文件系统建立在块设备驱动之上，块设备是 VxWorks 系统中一种重要的设备。块设备驱动提供对设备上的块进行边界划分和读写；文件系统在此基础上建立文件结构提供应用程序访问，提供的块设备驱动有：

- ramDrv 虚拟 RAM 盘驱动；
- scsiLib SCSI 块驱动；
- tffsDrv 可选的 TrueFFS；
- ataDrv ATA/IDE 块设备驱动；
- cbioLib CBIO 驱动；
- 第三方块设备驱动。

VxWorks 支持的文件系统如表 5-1 所列。

表 5.1 VxWorks 支持的文件系统

dosFs	兼容 MS-DOS 的文件系统，兼顾实时性应用需求
rawFs	提供整个磁盘作为一个文件的访问方式
tapeFs	磁带文件系统
cdromFs	ISO-9660 标准格式的 CD-ROM 文件系统
TSFS	目标服务器的文件系统，通过调试代理和目标服务器之间的连接访问

伴随着 FLASH 技术的发展和普及，越来越多的嵌入式操作系统中使用 FLASH 作为主存介质。在嵌入式应用中，为了方便系统设计和文件管理，一般需要在 FLASH 上建立文件系统。文件系统建成后，用户就可以在 FLASH 上进行建立文件、删除文件、数据复制等工作，如同在普通的 PC 机系统下操作硬盘一样。

目前，NOR 和 NAND 是市场上两种最主要的非易失闪存技术。本书的开发对象 MagicARM2410 实验箱配备了 NOR FLASH 和 NAND FLASH 两种芯片，分别是 SST39VF1601 和 K9F1208UOB。本章节将以此两款 FLASH 芯片作为具体的开发对象，进行基于 TrueFFS 的块设备驱动开发工作。

5.2 TrueFFS 机制概述

5.2.1 TrueFFS 简介

Tornado 的 TrueFFS 是与 VxWorks 兼容的一种 M-Systems Flite 实现方式，它为种类繁多的 FLASH 存储设备提供了统一的块设备接口，并且具有可重入、线程安全的特点，支持大多数流行的 CPU 构架。

在 Tornado 系统中，为了均衡地使用 FLASH 的各个扇区，避免 FLASH 某些扇区过度使用而变成只读状态，可以使用 TrueFFS 文件系统来管理整个介质。TrueFFS 使用动态映射的方法来实现物理介质的均衡使用。当某块被修改、移动或收集垃圾时，这个映射会做出相应调整。TrueFFS 文件系统的磨损控制（wear-leveling）功能使所有可擦除块的擦除次数在使用周期内尽可能平衡，避免某些块因负担过重而损坏，从而在整体上提高了 FLASH 的寿命。

有了 Tornado 的 TrueFFS 机制，应用程序对 FLASH 存储设备的读写就像它们对拥有 MS-DOS 文件系统的磁碟设备的操作一样。这样一来，TrueFFS 就屏蔽了下层存储介质的差异，为开发者提供了统一的接口方式。

目前，嵌入式系统主要使用基于 NOR 和 NAND 这两种技术的非易失闪存技术的芯片作为存储介质。Intel 公司在 1988 年首先开发出 NOR FLASH 技术，彻底改变了过去 EPROM 和 EEPROM 一统天下的局面。NOR FLASH 的特点是芯片内执行（XIP，eXecute In Place），应用程序能够直接在 FLASH 上运行，而不再需要在 RAM 中执行代码，同时具有随机存取和对字节执行写（编程）操作的能力。NOR FLASH 的传输效率高，位交换和坏块的发生率很低，稳定性较高，尤其在 1～4 MB 的小容量 FLASH 应用时具有很高的成本效益。但是，NOR FLASH 的写入和擦除速度较低，适用于保存系统代码和相关配置参数。

东芝公司在 1989 年开发出了 NAND FLASH 结构，可以提供极高的单元密度，达到高存储密度，写（编程）和擦除操作的速率很快，支持速率超过 5 Mbps 的持续写操作，其区块擦除时间短至 2 ms，而 NOR FLASH 是 750 ms。但是，NAND FLASH 无法保证不产生坏块，需要 EDC/ECC 算法对坏块进行管理，自身的管理算法也较 NOR FLASH 复杂。基于系统稳定性的要求，适宜用于保存与系统稳定性无关的数据，如系统日志，跟踪信息。

在电路设计上，两者也各有特点。以 16 位的器件为例，NOR FLASH 器件大约需要 41 个 I/O 引脚；相对而言，NAND FLASH 器件约需 24 个引脚。NAND FLASH 器件能够复用指令、地址和数据总线，从而节省了引脚数量。

5.2.2 块读写均衡机制

FLASH 存储器擦写寿命是有限的，NOR FLASH 每个块的最大擦写次数是约十万次级别，NAND FLASH 每个块的最大擦写次数是一百万次级别，不能无限次重复使用。因此，随着使用次数的增多，部分块最终会变成只读状态而影响 FLASH 的整体寿命。为了尽可能延长器件的寿命，VxWorks 使用了 wear-leveling 技术，其目的就是平衡使用所有的存储单元，而不让某一单元过度使用。TrueFFS 使用一种基于一张动态维护表的 block-to-flash（块 FLASH 对应）传输系统来实现 wear-leveling 技术。当 FLASH 的块数据被修改、移动，或者碎片回收后，该维护表会自动调整。TrueFFS 把 FLASH 存储空间映射到一个特殊的连续存储块队列中，以便文件系统可以对它进行数据的读写。

5.2.3 碎片回收机制

任何存储设备使用后都会产生碎片，而且随着时间的拉长，碎片会逐渐增多，可使用的空间减少，造成剩余可用空间的读写次数大幅提升的恶性循环。如果没有有效的机制来回收这些碎片，FLASH 被频繁使用的那些块很快就会变成只读状态，导致 FLASH 器件的损坏。TrueFFS 使用一种被称为碎片回收（garbage collection）的机制来回收那些不再包含有效数据的块。TrueFFS 碎片回收算法会采用随机选择的处理方法，用来保证回收处理能够均匀地覆盖整个存储空间。

5.2.4 块分配和关联数据机制

为了提高数据的读取效率，TrueFFS 使用一种空间分配策略：尽量将有关联的数据（如由同一个文件的内容组成的多个块）集结到同一个单独擦除单元（erase unit）内的一段连续的区域中。为此，TrueFFS 尽量在同一个擦除单元（erase unit）内维持一个由多个物理上连续自由的块组成的存储池。如果这样连续的存储池无法实现，TrueFFS 会尽可能保证池中的所有块是在同一个擦除单元（erase unit）内。如果连这样的情况也不可能的话，TrueFFS 会尽可能把块池分配到一个拥有最多可用空间的擦除单元内。

运用集结关联数据的方法，具有几大优点：首先，如果 TrueFFS 必须从一个小的存储窗口（memory window）来访问 FLASH，那么集结了的关联数据可以减少调用映射物理块(physical blocks)到该窗口的次数,加快文件继续访问的速度;其次,这种策略可以减少碎片的产生。删除一个文件可以释放掉更容易回收的完整块,这意味着碎片回收会变得更快;除此之外,它可以使属于静态文件的多个块存放在同一地方,这样当 wear-leveling 算法决定移动静态区域时,转移这些块就变得更加容易了。

5.2.5 错误恢复机制

在实际操作时,系统对 FLASH 进行写操作的时候可能会出错,比如在响应文件系统写请求时,或者在碎片回收期间,甚至是在 TrueFFS 格式化或擦除 FLASH 时。TrueFFS 使用“先写后擦”(erase after write)的策略,当更新 FLASH 一个扇区的数据时,只有在更新操作完成并且新存储的数据校验成功后,先前的数据才会被允许擦掉。这样的结果是数据扇区始终保证写过程的完整性。以上操作成功的话,写入扇区的新数据有效;如果操作失败,则扇区的原始数据有效,这样有利于数据的稳定性。

5.2.6 引导映象和 TrueFFS 共享 FLASH 存储空间

缺省情况下,TrueFFS 格式化函数会把整个 FLASH 空间用于 TrueFFS,但也允许从一个偏移地址开始格式化 FLASH,这样可以在同一块 FLASH 芯片中前一段保存引导映象(BOOTROM. bin),后一段用于建立 TrueFFS 来保存系统映象(VxWorks. bin)。在 FLASH 的前面空间可以用于保存系统启动代码。随后,可以使用 TrueFFS 提供的函数来管理这个区域。使用中需要小心,避免写操作时超过 TrueFFS 这个区域,从而破坏掉 TrueFFS,出现导致系统无法启动的故障。

5.2.7 TrueFFS 构架解析

如图 5.1 所示,TrueFFS 由 1 个核心代码层和 3 个功能层组成,3 个功能层是:

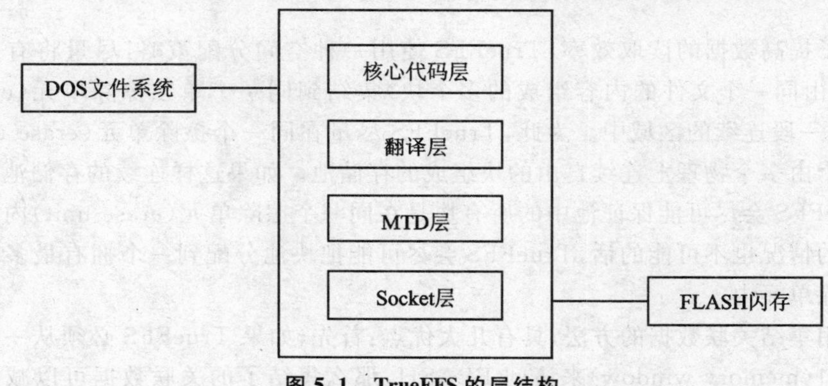

图 5.1 TrueFFS 的层结构

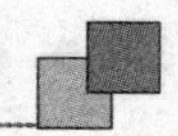

翻译层(translation layer)、MTD层(MTD layer)、Socket层(Socket layer)。

翻译层主要实现TrueFFS和dosFs之间的高级交互功能。它也包含了控制FLASH映射到块、wear-leveling、碎片回收和数据完整性所需的智能化处理功能。核心层将其他3层有机结合起来,处理全局问题,如信息量计时器、碎片回收和其他系统资源等。

5.3 Socket与MTD层

Socket层提供TrueFFS和板卡硬件(如FLASH卡)的接口服务,用来向系统注册Socket设备,检测设备拔插,硬件写保护等。MTD层(Memory Technology Drivers)主要功能是实现对具体的FLASH进行读、写、擦、ID识别等驱动,并设置与FLASH密切相关的一些参数。在FLASH设备驱动设计中,主要对MTD层进行修改,以便匹配不同的FLASH芯片。

MagicARM2410系统中使用了基于NOR技术的SST39VF1601和基于NAND技术的K9F1208UOB两种芯片,这两种芯片分别使用INCLUDE_TL_FTL和INCLUDE_TL_NFTL不同的翻译层。同时,这两种芯片都不在VxWorks自带支持的FLASH型号之内,所以需针对MagicARM2410目标系统自行编写两种芯片的MTD层驱动。另外,VxWorks 5.5.1并不支持基于NAND技术的翻译层,需要利用VxWorks的旧版本中的nftllite.c文件移植NAND技术的翻译层。

5.3.1 TrueFFS开发简介

为了在一个VxWorks映像中包含TrueFFS,需在config.h文件中定义INCLUDE_TFFS。以便VxWorks初始化代码调用tffsDrv()函数来创建管理TrueFFS所需的结构和全局变量,并为所有挂接了的FLASH设备注册socket组件驱动。在链接的时候,通过解析与tffsDrv()函数相关联的符号(symbols),可以将TrueFFS所必需的软件模块链接到VxWorks映象中。

为了支持TrueFFS,BSP中须包含sysTffs.c文件,它可将TFFS的Socket和MTD层编译链接到VxWorks镜像中。开发中需选取与系统对应的MTD和翻译层模块,同时如果目标系统包含了一个MMU单元,还需修改sysLib.c中的sysPhysMemDesc[]数组中对应的内存地址映射。重新编译VxWorks映象并重启目标系统后,即可以使用TrueFFS的各种功能函数。

5.3.2 配置TrueFFS

对于一个支持TrueFFS的BSP来说,所涉及到的文件必须放在target/config

下的 BSP 配置目录里。

映像启动后，自动运行函数 tffsDrv()。该函数自动为每一个 FLASH 设备注册一个 Socket 组件。此时，FLASH 设备还没有挂上块设备驱动。但是，Socket 组件驱动已经为调用 tffsDevFormat()函数提供了充足的条件。为了使用 TrueFFS，必须用这个函数来格式化 FLASH 媒体。为了在 Socket 组件的顶部创建一个 TrueFFS 块设备并将 mount dos 文件系统建立到这个块设备上，还需要调用 usrTffsConfig()函数。下面就具体讲讲这几个文件的修改：

① 修改 Makefile 文件。

为了加入 sysTffs.o 的编译，需要在 makefile 文件中加入如下的宏定义：

```
MACH_EXTRA = sysTffs.o
```

② 修改 config.h 文件。

对于大多数的 BSP 来说，包含 TrueFFS 需要在 config.h 文件中加入如下的两个宏定义：

```
#define INCLUDE_TFFS
#define INCLUDE_DOSFS
```

如果需要使用 tffsShow()函数和 tffsShowAll()函数来查看 Socket 信息，还要增加下面的宏定义：

```
#define INCLUDE_SHOW_ROUTINES
```

③ 修改 sysLib.c 文件。

如果用户的目标系统中包含了 MMU 模块，那么它的 BSP 在 sysLib.c 文件里面就定义了一个 sysPhysMemDesc[]表，需要在该表中增加对 FLASH 映射的描述。

④ 修改 sysTffs.c 文件。

该文件的主要功能就是定义一些 BSP 特殊的 Socket 代码，起到连接 FLASH 硬件和 VxWorks 的桥梁作用。缺省的 sysTffs.c 文件包含了所有的翻译层模块，所有的 MTD 层模块，以及 tffsBootImagePut()函数、tffsShow()函数、tffsShowAll()函数等工具函数。为了减小映像的大小，可通过编辑 sysTffs.c 文件来去除对目标系统无用的模块定义。

5.3.3 FLASH 的格式化函数

在使用 TrueFFS 之前，要先调用 tffsDevFormat()函数来格式化 FLASH。在格式化的过程中，该函数先擦除 FLASH，然后将 TrueFFS 数据管理结构（data-management structures）写到位于每个擦除单元（如扇区）起始处的位置。

函数 tffsDevFormat()需要两个输入参数：设备号（Socket 组件号）和一个指向

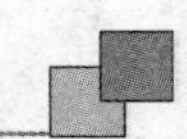

格式化参数(FormatParams)的指针。设备号由Socket组件在系统中注册的先后顺序决定。而FormatParams结构则是传递如何格式化FLASH的值。这个结构参数可以参见其头文件说明,默认的FormatParams结构定义如下:

```
#define STD_FORMAT_PARAMS {0, 99, 1, 0x10000l, NULL, {0,0,0,0}, NULL, 2, 0, NULL}
```

一般来说,可把这个指针赋为NULL(0),此时函数tffsDevFormat()使用在dosformt.h中定义的默认FormatParams结构。在这个默认的结构中定义的值能够满足绝大多数情况下的应用。但是,如果在FLASH上共用TrueFFS和引导映象(boot image)的话,就无法使用这个缺省结构了。此时,需要修改bootImageLen的值,以适应boot Image的大小,函数tffsDevFormat()在格式化的时候会保留出这一部分空间。

另外,VxWorks还提供了另一个格式化函数sysTffsFormat(),但是在它内部还是嵌套了tffsDevFormat()函数。该函数没有入参,使用自己的FormatParams参数来格式化FLASH。详细信息可以参阅BSP中sysTffs.c文件中的sysTffsFormat()函数。

关于使用这两种格式化函数的区别和利弊,本书后续章节将再做相关描述。

5.3.4 创建TrueFFS块设备

在创建一个逻辑TrueFFS块设备之前,首先需要运行tffsDrv()。正确配置了VxWorks后,系统在启动的时候自动加载。函数tffsDrv()为Tornado初始化TrueFFS,包括建立互斥信号量、全局变量和用来管理TrueFFS的数据结构,还包括为目标机上所有的FLASH设备注册Socket组件的驱动程序。

注册Socket组件的驱动程序从获取FLSocket中预先分配的5元素(5-element)TrueFFS内部数组开始。接下来是更新FLSocket结构以包含那些控制FLASH设备基本硬件接口的数据和函数指针。

当TrueFFS需要和具体的Socket硬件打交道时,需要使用设备号(0～4)作为索引来查找它的FLSocket结构,然后用相应结构中的函数来控制它的硬件接口需求。虽然这些Socket接口函数并没有提供完整的块设备接口,但是在使用函数tffsDevFormat()的接口上还是较为方便。这对于一个从未格式化的FLASH存储介质来说是非常重要的,通过这种方式在此阶段能够实现创建一个TrueFFS块设备。

在注册完一个Socket组件驱动后,就可以调用tffsDevCreate()函数来创建一个TrueFFS块设备。函数调用时,对应设备号(0～4)的输入参数需要进行指定,也就是进入FLSocket结构数组的索引,设备号对于dosFs是可见的。

在创建了TrueFFS块设备后,必须调用dosFsDevInit()函数将dos文件系统挂接(mount)到FLASH上,在这之后就可以实现如同在一个标准disk设备上读写

FLASH。为了方便使用，函数 usrTffsConfig()内部合成了 tffsDevCreate()和 dosFsDevInit()两个函数，并包含了创建 TrueFFS 块设备和挂接 dosFS 必要的函数，因此直接调用 usrTffsConfig()函数即可。以下是 bootCofig.c 中关于 TrueFFS 块设备的代码，当采用 TrueFFS 方式引导 VxWorks 内核时，tffsLoad 函数被调用。

```
#ifdef    INCLUDE_TFFS

#define    TFFS_MEM_DOSFS      0x200000
#include "../../src/config/usrTffs.c"
#include "sysTffs.c"      /* the BSP stub file, in the BSP directory */

/*****************************************************************
*
* tffsLoad - load a vxWorks image from a TFFS Flash disk
*
* RETURNS: OK, or ERROR if file can not be loaded.
*
* NOMANUAL
*/

LOCAL STATUS tffsLoad
(
    int     drive,              /* TrueFFS drive number (0 - (noOfDrives - 1)) */
    int     removable,          /* 0 - nonremovable flash media */
    char     * fileName,        /* file name to download */
    FUNCPTR * pEntry
)
{
    int fd;

    if (tffsDrv () != OK)
    {
        printErr ("Could not initialize.\n");
        return (ERROR);
    }

    printf ("Attaching to TFFS... ");

    dosFsInit (NUM_DOSFS_FILES);          /* initialize DOS - FS */

    if (usrTffsConfig (drive, removable, fileName) == ERROR)
    {
```

```
        printErr ("usrTffsConfig failed.\n");
        return (ERROR);
    }

    printErr ("done.\n");

    /* load the boot file */

    printErr ("Loading %s... ", fileName);

    if ((fd = open (fileName, O_RDONLY, 0)) == ERROR)
    {
        printErr ("\nCannot open \"%s\".\n", fileName);
        return (ERROR);
    }

    if (bootLoadModule (fd, pEntry) != OK)
        goto tffsLoadErr;

    close (fd);
    return (OK);

tffsLoadErr:

    close (fd);
    return (ERROR);
}

#endif      /* INCLUDE_TFFS */
```

5.3.5 TrueFFS 建立过程中的函数调用关系

TrueFFS 内部分配了一个包含 5 个 FLFlash 结构体的阵列，对应 5 个不同的 FLASH 设备。TrueFFS 使用这些 FLFlash 结构体来存储数据和回调函数的指针。比如，TrueFFS 使用 MTD 函数来处理对 FLASH 媒介基本的读写操作，而 FLFlash 结构体就包含这些 MTD 函数指针。当运行一个 MTD 识别程序时，系统就安装了这些函数指针。

FLFlash 结构体还包含一个指向 FLSocket 结构体的指针。TrueFFS 使用这些 FLSocket 结构体来存储数据和函数指针。与前者不同的是，FLSocket 中的回调函数是用于处理与 FLASH 设备硬件接口的，也就是 Socket 接口的相关函数调用。系

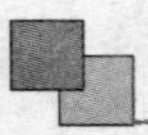

统在 sysTffsInit()函数中调用 rfaRegister()来注册这些函数。

在 VxWorks 中包含 TrueFFS 将会使 usrRoot()函数调用 tffsDrv()函数，从而引发一系列函数调用，如图 5.2 所示。

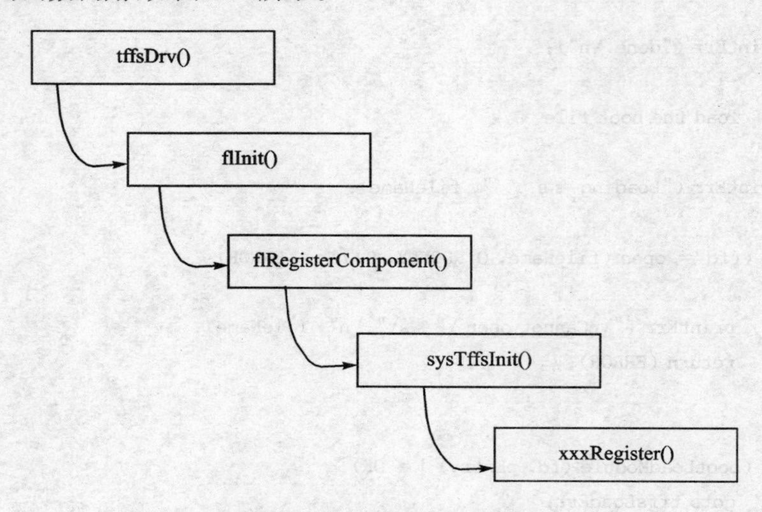

图 5.2　函数 tffsDrv 调用图

上述函数调用的目的之一就是用 TrueFFS 注册 Socket 驱动函数。一般情况下，注册工作都发生在 rfaRegister()函数中，该函数能更新 FLSocket 结构体。而此时，TrueFFS 已经对应 Socket 驱动中的服务程序给 FLSocket 结构体赋予了一个设备号，即卷标。TrueFFS 调用 FLSocket 结构体中引用的函数来处理与 FLASH 设备的硬件接口。

要创建一个 TrueFFS 块设备，需要调用 tffsDevCreate()函数。同样，这一调用也将发起一系列的函数调用，如图 5.3 所示。

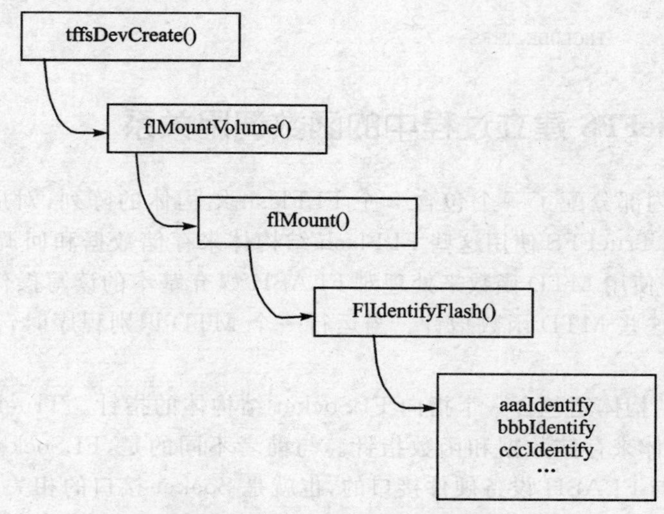

图 5.3　函数 tffsDevCreate 调用图

如图5.3所示，这些函数调用的目的之一是确认合适的MTD，该确认过程发生在flIdentifyFlash()函数中。函数flIdentifyFlash()通过逐个执行xxxIdentify()表中的程序来确定合适的MTD，相同的MTD可以存在多个不同的FLASH卷标。如果存在匹配的MTD，确认程序就会更新FLFlash结构体中的数据以及指向用于读、写、擦除、映射等操作的MTD程序指针。此外，确认过程还将完成在当前FLFlash结构体中涉及的FLSocket结构体的初始化。

5.4 MagicARM2410的NOR FLASH驱动设计

MTD是TrueFFS对FLASH编程时所需要的一个包括读、写、擦除、映射、识别等基本函数的软件模块。MagicARM2410系统中使用的是SST39VF1601 NOR FLASH芯片，MTD层在sst39vf1601mtd.c中实现。

TrueFFS驱动的移植工作主要集中MTD层上，并且和具体的FLASH存储器相关。如果目标系统中的FLASH芯片型号不在Tornado支持的范围内，则可以参考Tornado自带的相关FLASH驱动例程进行改动。

如图5.4所示，FLASH的驱动开发实现下述的功能模块。

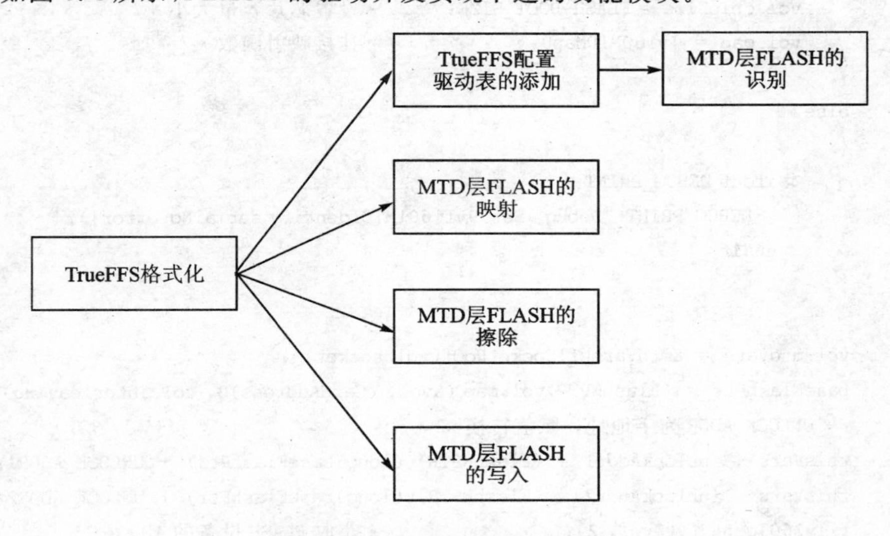

图5.4 FLASH驱动开发模块图

5.4.1 编写sst39vf1601MTDIdentify()函数

sst39vf1601MTDIdentify()函数是先于其他任何函数执行的，其函数原型定义如下：

```
FLStatus sst39vf1601MTDIdentify (FLFlash vol);
```

该函数主要完成 FLFlash 结构体部分参数的设置，如完成 FLASH 位宽、擦写速度、TrueFFS 暂用空间、擦写块大小、翻译层标识以及写、擦除、映射等回调函数的设置。

针对 MagicARM2410 系统，sst39vf1601MTDIdentify()函数原型如下：

```
FLStatus sst39vf1601MTDIdentify
(
    FLFlash vol
)
{
    FlashWPTR  baseFlashPtr;
    vol.interleaving = 1;
    flSetWindowBusWidth(vol.socket,16);      /* 设置 16 位的总线位宽 */
    flSetWindowSpeed(vol.socket,120);        /* 设置 120 ns 的窗速度 */

    if (vol.socket->serialNo == 0)
    {
        flSetWindowSize(vol.socket, FLASH_BOOT_SIZE>>12);
        vol.chipSize = FLASH_BOOT_SIZE;      /* 设置芯片容量大小 */
        vol.map = lv160MTDMap0;              /* 挂接映射函数 */
    }
    else
    {
        #ifdef DEBUG_PRINT
            DEBUG_PRINT("Debug: Sst39vf1601MTDIdentify serialNo error!);
        #endif
    }

    vol.mtdVars = &mtdVars[flSocketNoOf(vol.socket)];
    baseFlashPtr = (FlashWPTR)vol.map (&vol, (CardAddress)0, vol.interleaving);
    /* UNLOCK_ADDR 为字地址，赋值转换 x2 */
    thisVars->unlockAddr1 = (FlashWPTR)((long)baseFlashPtr) + UNLOCK_ADDR1;
    thisVars->unlockAddr2 = (FlashWPTR)((long)baseFlashPtr) + UNLOCK_ADDR2;
    fllv160Identify(&vol, 2);                /* 获取 FLASH 设备的 ID */

    if( vol.type != SST39VF1601_DEID)
    {
        #ifdef DEBUG_PRINT
            DEBUG_PRINT("Debug: Can not identify SST39VF1601.0x%x\n",vol.type);
        #endif
    return flUnknownMedia;
    }
```

```
    else
    {
        #ifdef DEBUG_PRINT
            DEBUG_PRINT("Debug: Identify SST39VF1601 media.0x%x\n",vol.type);
        #endif
    }
    vol.noOfChips = 0x1;                        /* 单芯片 */
    vol.erasableBlockSize = 0x1000;             /* 4 KB */
    vol.flags |= SUSPEND_FOR_WRITE;             /* 设置 NOR FLASH 标志位 */

    /* Register flash handlers */
    vol.write = lv160MTDWrite;                  /* 挂接 NOR FLASH 写函数 */
    vol.erase = lv160MTDErase;                  /* 挂接 NOR FLASH 擦除函数 */
    return flOK;
}
```

5.4.2 编写 sst39vf1601MTDMap()函数

SST39VF1601 FLASH 芯片带有通用的 SRAM 接口，可以轻松地挂接在 S3C2410 的地址和数据总线上。从编写程序的角度，如同有一个主机地址空间窗口沿着 FLASH 地址变化，系统的数据单元均匀地分布在 FLASH 的整个区域内。TrueFFS 寻址系统元数据(meta data)的时候，需要多次调用 MTD 的读函数来收集这些数据，从而使系统能够更有效地将整个 FLASH 映射到主机地址空间，并直接从存储窗口中读数。TrueFFS 用 FLFlash.map 指向默认的映射函数 flashMap()。该默认函数对那些能够直接映射到主机存储空间的 FLASH 设备来说是有效的，其内部调用一次 flMap()，原型如下：

```
void FAR0 * flMap ( FLSocket vol, CardAddress address );
```

该函数映射 FLASH 到一个特定的地址并返回指向该区域的指针，需要说明的是这个指针会有少许的偏移。在调用 flMap()函数后，FLASH 特定区域将被映射到 Socket 窗口成为本地存储空间的一部分。这样，就可以通过正常的寻址来访问它了。MagicARM2410 系统中整个 FLASH 在主机地址区都是透明的，sst39vf1601MTDMap()映射函数只要给 FLASH 存储器返回一个特定的地址指针即可，函数原型如下：

```
static void FAR0 *      sst39vf1601MTDMap
(
    FLFlash * vol,
    CardAddress addr,
```

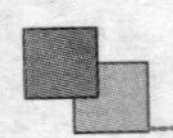


```
    int length
)
{
    UINT32 ulAddrMap;
    ulAddrMap  = FLASH_BOOT_ADRS + addr; /* 地址偏移 */
    return (void FAR0 *) ulAddrMap;
}
```

5.4.3 编写 sst39vf1601MTDErase()和 sst39vf1601MTDWrite()函数

擦除、写函数应该尽可能通用化，例如将 FLASH 操作的相关命令字序列定义成宏的形式，这样利于函数的复用，例如可将 SST39VF1601 的命令字序列定义如下：

```
#define SETUP_ERASE          0x80
#define SETUP_WRITE          0xa0
#define READ_ID              0x90
#define SECTOR_ERASE         0x30
#define BLOCK_ERASE          0x50
#define READ_ARRAY           0xf0
#define UNLOCK_1             0xaa
#define UNLOCK_2             0x55
#define UNLOCK_ADDR1         0x5555
#define UNLOCK_ADDR2         0x2aaa
```

sst39vf1601MTDErase() 和 sst39vf1601MTDWrite() 函数的编写参照 SST39VF1601 芯片的 Datasheet，按照芯片的时序和命令字编写即可，具体函数代码如下：

1. sst39vf1601MTDErase()函数

```
static FLStatus sst39vf1601MTDErase
(
    FLFlash vol,
    int firstErasableBlock,
    int numOfErasableBlocks
)
{
    FlashWPTR flashPtr;
    int iBlock;
    unsigned int offset;
    int level;
```

```
if(numOfErasableBlocks <= 0)
{
    return ERROR;
}
for (iBlock = 0; iBlock < numOfErasableBlocks; iBlock++)
{
    int i;
    offset = (firstErasableBlock + iBlock) * vol.erasableBlockSize;
    flashPtr = (FlashWPTR) vol.map(&vol, offset, vol.interleaving);
    *thisVars->unlockAddr1 = UNLOCK_1;
    *thisVars->unlockAddr2 = UNLOCK_2;
    *thisVars->unlockAddr1 = SETUP_ERASE;
    *thisVars->unlockAddr1 = UNLOCK_1;
    *thisVars->unlockAddr2 = UNLOCK_2;
    level = intLock();
    *flashPtr = SECTOR_ERASE;
    sst39vf1601OpOverDetect ((void *)flashPtr, 0x2000000);
    for(i = 0; i<vol.erasableBlockSize/2; i++,flashPtr++)
    {
        if(*flashPtr != 0xffff)
        {
            break;
        }
    }
    intUnlock(level);
    if(i < vol.erasableBlockSize/2)
    {
        #ifdef DEBUG_PRINT
            DEBUG_PRINT("Debug: sst39vf1601MTDErase fail.\n");
        #endif
            return flWriteFault;
    }
}
#ifdef DEBUG_PRINT
    DEBUG_PRINT("\Debug: sst39vf1601MTDErase OK ! \n");
#endif
    return flOK;
}
```

2. sst39vf1601MTDWrite()函数

```
static FLStatus sst39vf1601MTDWrite
```

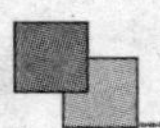


```
(
    FLFlash vol,
    CardAddress address,
    const void FAR1 * buffer,
    int length,
    FLBoolean overwrite
)
{
    int cLength;
    FlashWPTR flashPtr, flashTmp;
    volatile UINT16 * gBuffer;
    int level;
    flashTmp = flashPtr = (FlashWPTR) vol.map(&vol, address, length);
    if(length&1)
    {
        #ifdef DEBUG_PRINT
            DEBUG_PRINT("Debug: Warning! the data length can not divided by 2\n");
        #endif
    }
    cLength = length/2;
    gBuffer = (UINT16 *)buffer;
    while (cLength >= 1)
    {
        * thisVars->unlockAddr1 = UNLOCK_1;
        * thisVars->unlockAddr2 = UNLOCK_2;
        level = intLock();
        * thisVars->unlockAddr1 = SETUP_WRITE;
        * flashPtr = * gBuffer;
        if(sst39vf1601OpOverDetect ((void *)flashPtr, 0x1000000));
        if(* flashPtr != * gBuffer)
        {
            * flashPtr = READ_ARRAY;
            #ifdef DEBUG_PRINT
                DEBUG_PRINT("Debug: sst39vf1601MTDWrite timeout.\n");
            #endif

            return flWriteFault;
        }
        intUnlock(level);
        cLength--;
        flashPtr++;
        gBuffer++;
```

```
    }
    if (tffscmp((void FAR0 * ) flashTmp, buffer,length))
    {
        #ifdef DEBUG_PRINT
            DEBUG_PRINT("Debug: sst39vf1601MTDWrite fail.\n");
        #endif

        return flWriteFault;
    }
    return flOK;
}
```

5.4.4 编写 sst39vf1601OpOverDetect()函数

sst39vf1601OpOverDetect()函数用于检测 FLASH 设备是否处于写保护状态，TrueFFS 的擦除、写操作先需确认 FLASH 是否允许进行相关操作，而这之前是不能进行任何擦、写操作的。该函数具体代码如下：

```
static STATUS  sst39vf1601OpOverDetect
(
    void * ptr,
    int timeCounter
)
{
    FlashWPTR pFlash = ptr;
    INT16 buf1,buf2;
    buf1 = * pFlash & 0x40;
    while(1)
    {
        buf2  = * pFlash & 0x40;
        if(buf1 == buf2)
        {
            break;
        }
        else
        {
            buf1 = buf2;
        }
        if(timeCounter-- <= 0)
        {
            return ERROR;
```

```
        }
    }
    return OK;
}
```

5.4.5 注册 MTD

当 TrueFFS 试图将 MTD 与 FLASH 设备匹配时，需要执行 mtdTable[]表中的函数，该函数在/SRC/DRV/TFFS/tffsConfig. c 中有定义。在编写一个 MTD 后需要 TrueFFS 识别的话，就必须在此表中添加它的注册信息。针对 MagicARM2410 系统，需增加如下定义：

```
#ifdef      INCLUDE_MTD_SST39VF1601
sst39vf1601MTDIdentify,
#endif
```

当然，这之前需要一个条件包含声明，即在 config. h 中定义增加该宏。在同一片 NOR flash 上建立 BOOTROM 和 TrueFFS，按照下面的关键步骤进行。

1. sst39vf1601 FLASH 芯片容量划分

sst39vf1601 FLASH 芯片容量划分如表 5. 2 所列。

表 5. 2 sst39vf1601 FLASH 芯片容量表

地址范围	容　量	描　述
0x00000000～0x00080000	512 KB	用于存放系统的 BOOTROM
0x00080000～0x00200000	1 536 KB	用于建立 TrueFFS 存放 VxWorks 映像

据此定义如下宏：

```
#define ROM_BASE_ADRS           0x00000000  //FLASH 的起始地址
#define FLASH_BOOT_ADRS         0x00080000  //TrueFFS 的起始地址
#define FLASH_BOOT_SIZE         0x00180000  //TrueFFS 的容量大小
```

这样定义的好处在于，可以直接使用 tffsDevFormat()函数从 TrueFFS 的起始地址来格式化 FLASH，而不必使用 sysTffsFormat()函数中设置偏移量的方式来格式化 FLASH。不仅如此，TrueFFS 的相关操作被限制在 TrueFFS 设定的区域内，保证了 BOOTROM 区域不被改写，从而增强了系统的稳定性。

2. MMU 内存映射设置

为了使操作系统在使用 MMU 的情况下正确访问 FLASH，需在 sysLib. c 文件中的 sysPhysMemDesc 数组加入如下映射：

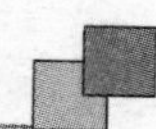

```
{
    (void * ) (ROM_BASE_ADRS),
    (void * ) (ROM_BASE_ADRS),
    ROUND_UP (ROM_SIZE_TOTAL, PAGE_SIZE),
    VM_STATE_MASK_VALID | VM_STATE_MASK_WRITABLE | VM_STATE_MASK_CACHEABLE,
    VM_STATE_VALID      | VM_STATE_WRITABLE      | VM_STATE_CACHEABLE_NOT
}
```

这其中 ROM_SIZE_TOTAL 为 0x00200000(2 MB)，即 sst39vf1601 FLASH 的实际大小。另外，cache 不能覆盖 FLASH 存储器空间，因为 FLASH 的操作指令序列是连续的，cache 的中断发生会使操作失败，所以参数设置为 VM_STATE_CACHEABLE_NOT。

3. 中断屏蔽

由于系统中只有一片 NOR FLASH，同时 VxWorks 的异常入口位于 NOR FLASH 存储器的开始处，异常发生时不能得到正常的入口指令，会导致系统程序跑飞。这种情况下需要在 sst39vf1601MTDErase()函数和 sst39vf1601MTDWrite()函数中加入屏蔽中断，因为 FLASH 在擦除、写和读 ID 状态时，不能正常读取 FLASH 中数据。

中断屏蔽函数为 intLock()和 intUnlock()，注意使用时需包含 intLib. h 头文件。

4. 低级格式化函数

为了防止文件系统的意外崩溃，需在 MTD 层中编写 NOR FLASH 存储器低级格式化程序，用于文件系统崩溃时格式化整个 NOR FLASH，以避免重新烧写 BOOTROM 来建立文件系统的繁琐步骤。

5.5 MagicARM2410 的 NAND FLASH 驱动程序设计

与 NOR FLASH 的线性结构不同，NAND FLASH 属于块状结构，有页块之区分，所以需要首先对 NAND FLASH 的结构进行分析，在了解芯片结构的基础上再进行针对性编程。

5.5.1 NAND FLASH 结构解读

如图 5.5 所示，NAND FLASH 主要以页(page)为单位进行读写，以块(block)为单位进行擦除。FLASH 页的大小和块的大小因不同类型块结构而不同，块结构有两种：小块结构和大块结构。NAND FLASH 在小块结构下每个块包含 32 个页，

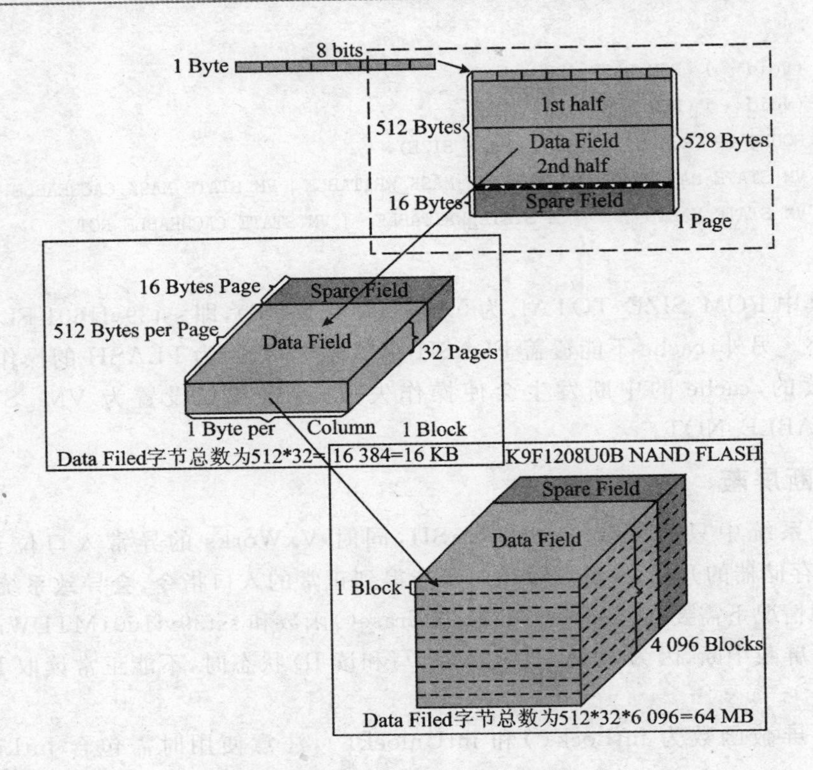

图 5.5　NAND FLASH 块结构

每页 512＋16 字节；在大块结构下每个块包含 64 个页，每页 2 048＋64 字节。MagicARM2410 开发箱使用的是 K9F1208U0B，属于小块结构的 NAND FLASH。

对于小块结构，每个页的 512 字节用于存放数据，其余 16 字节用于存放其他信息（包括：块好坏的标记、块的逻辑地址、页内数据的 ECC 校验和等）。NAND FLASH 的随机读取的效率很低，一般以页为单位进行读操作。系统在每读一页后会计算其校验和，并和存储在页内的冗余 16 字节位置的校验和做比较，以此来判断读出的数据是否正确。

大块结构和小块结构的 NAND FLASH 都有与页大小相同的页寄存器，用于数据缓存。当读数据时，先从 NAND FLASH 内存单元把数据读到页寄存器，外部通过访问 NAND FLASH I/O 端口获得页寄存器中数据（地址自动累加）；当写数据时，外部通过 NAND FLASH I/O 端口输入的数据首先缓存在页寄存器，写命令发出后才写入到内存单元中。

5.5.2　MagicARM2410 的 NAND FLASH 接口电路分析

S3C2410A 处理器拥有专门针对 NAND FLASH 的接口，可以很方便地和

NAND FLASH 对接，如图5.6所示。虽然 NAND FLASH 的接口比较简单，容易接到系统总线上，但 S3C2410A 处理器针对 NAND FLASH 还集成了硬件 ECC 校验，这将大大提高 NAND FLASH 的读写效率。如果芯片不支持硬件 ECC 校验功能，就需要软件完成 ECC 校验功能，这将消耗大量的 CPU 资源，使读写速度下降。

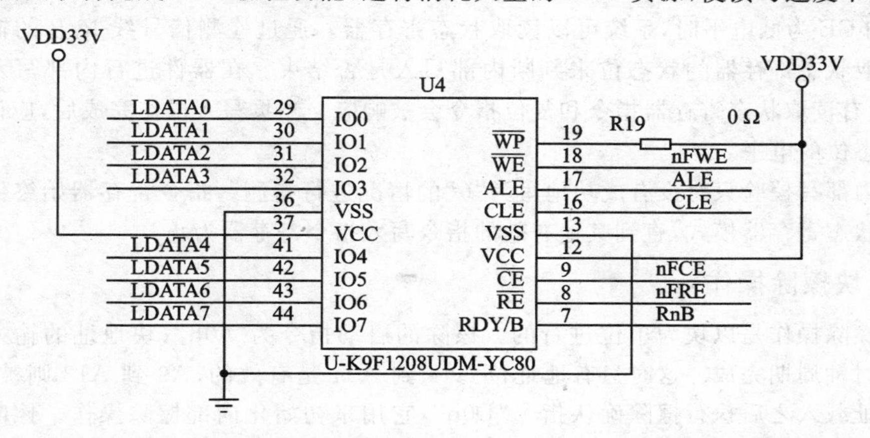

图5.6 MagicARM2410 的 NAND FLASH 接口电路原理图

5.5.3 NAND FLASH 编程说明

1. 页读操作

在初次上电时，器件进入缺省的“读方式1”模式。在这一模式下，页读操作通过将“00h”指令写入指令寄存器，接着写入3个地址(1个列地址、2个行地址)来启动。一旦页读指令被器件锁存，下面的页读操作就不需要再重复写入指令了。

写入指令和地址后，处理器可以通过对信号线 R/B 的分析来判断该操作是否完成。如果该信号为低电平，表示器件正“忙”；如果为高电平，说明器件内部操作完成，要读取的数据已经被送入了数据寄存器。外部控制器可以在以50 ns为周期的连续 RE 脉冲信号的控制下，从 I/O 口依次读出数据。连续页读操作中，输出的数据是从指定的列地址开始，直到该页的最后1个列地址的数据为止。

2. 页写操作

K9F1208UOB 的写入操作也以页为单位。写入操作必须在擦除之后，否则写入时将报错。页写入周期依次分3个步骤：①写入串行数据输入指令“80h”；②写入3个字节的地址信息；③串行写入数据。一次性串行写入数据量最大为528字节，它们首先被写入到器件内的页寄存器，接着器件进入一个内部写入过程，将数据从页寄存器写入存储宏单元。

串行数据写入完成后，需要写入“页写入确认”指令“10h”，这条指令将初始化器件的内部写入操作。如果单独写入“10h”而没有前面的步骤，则“10h”不起作用。在“10h”写入之后，K9F1208UOB 的内部写控制器将自动执行内部写入和校验工作，此时系统控制器就被释放出来了。

内部写入操作开始后，器件自动进入“读状态寄存器”模式。在这一模式下，当 RE 和 CE 为低电平时，系统可以读取状态寄存器。通过检测信号线 R/B 的输出或者读取状态寄存器的状态位来判断内部写入是否结束。在器件进行内部写入操作时，只有读取状态寄存器指令和复位指令会被响应。当页写入操作完成后，应该检测写状态位的电平。

内部写校验只对没有成功地写为“0”的情况进行检测。指令寄存器始终保持着读取状态寄存器模式，直到其他有效的指令写入指令寄存器为止。

3. 块擦除操作

擦除操作是以块为单位进行的。擦除的启动指令为“60h”，块地址的输入需要两个时钟周期完成。这时只有地址位 A14 到 A24 是有效的，A9 到 A13 则被忽略。块地址载入之后执行擦除确认指令“D0h”，它用来初始化内部擦除操作。擦除确认命令还用来防止外部干扰产生擦除操作的意外情况。器件检测到擦除确认命令输入后，在 WE 的上升沿启动内部写控制器，开始执行擦除和擦除校验。内部擦除操作完成后，检测写状态位状态，从而确定擦除操作是否有错误发生。

4. 读状态寄存器

K9F1208UOB 包含一个状态寄存器，该寄存器可以反映写入和擦除操作是否完成，或写入和擦除操作是否无错。写入“70h”指令后，开始读状态寄存器。状态寄存器的内容将在 CE 或 RE 下降沿时送至 I/O 端口。

器件一旦接收到读状态寄存器的指令，它就置状态寄存器为在读状态，直到有其他的指令输入。因此，如果在任意读操作过程中进行了状态寄存器的读操作，在连续页读的过程中就必须重发“00h”或“50h”指令。

5. 读器件 ID 操作

K9F1208UOB 器件具有一个产品鉴定识别码(ID)，系统控制器可以读出这个 ID，从而起到识别器件的作用。读 ID 的步骤是：写入“90h”指令，然后写入一个地址“00h”。在两个读周期下，厂商代码和器件代码将被连续输出至 I/O 口。

同样，一旦进入这种命令模式，器件将保持这种命令状态，直到接收到其他的指令为止。

6. 复位操作

器件提供一个复位(RESET)指令，通过向指令寄存器写入“FFh”来完成对器件的复位操作。当器件处于任意读模式、写入或擦除模式的“忙”状态时，发送复位指令可以使器件中止当前的操作，正在被修改的存储器宏单元的内容不再有效，指令寄存

器被清 0 并等待下一条指令的到来。当 WP 为高时，状态寄存器被清为“C0h”。

5.5.4　VxWorks 下的 NAND FLASH 驱动程序

VxWorks 下的 NAND FLASH 的驱动可以借鉴 Linux 下的相关驱动或者其他单片机下的驱动进行移植，芯片的读写过程差异较小。移植主要对原有驱动改造，使之与 VxWorks 的驱动框架进行匹配，需要修改和编写 f1208MTDIdentify() 函数，nandMTDWrite() 函数，nandMTDErase() 函数和编写 nandMTDRead() 函数。

其中，nftllite.c 文件包含了 NAND 闪存文件转换层的核心代码，该文件主要的对外接口函数如表 5.3 所列。

表 5.3　NFTL 主要对外接口函数

序　号	函数名
1	static FLStatus mountNFTL(unsigned volNo, TL * tl, FLFlash * flash, FLFlash * * volForCallback)
2	static FLStatus formatNFTL(unsigned volNo, TLFormatParams * formatParams, FLFlash * flash)
3	static const void FAR0 * mapSector(Anand vol, SectorNo sectorNo, CardAddress * physAddress)
4	static FLStatus writeSector(Anand vol, SectorNo sectorNo, void FAR1 * fromAddress)
5	static FLStatus deleteSector(Anand vol, SectorNo sectorNo, SectorNo noOfSectors)
6	static FLStatus defragment(Anand vol, long FAR2 * sectorsNeeded)
7	static FLStatus checkVolume(Anand vol)
8	static SectorNo sectorsInVolume(Anand vol)
9	static FLStatus NFTLInfo(Anand vol, TLInfo * tlInfo)
10	static FLStatus tlSetBusy(Anand vol, FLBoolean state)
11	static FLStatus readBBT(Anand vol, CardAddress FAR1 * buf, long FAR2 * mediaSize, unsigned FAR2 * noOfBB)

如表 5.3 所列，前两个函数，即函数 mountNFTL() 和函数 formatNFTL() 仅仅是在初始化时使用的，当我们使用 tffsDevCreate() 函数或者 tffsDevFormat() 函数时，就会最终调用它们。而后面的 mapSector() 函数等则是 NFTL 向系统提供的，诸如将来需要读写文件，查看当前介质的信息，如有多少扇区空闲，被占用多少等，是通过后面的函数来完成的。函数 defragment() 是一个碎片整理函数。如果没有这个函数，则不能进行碎片整理。nftllite.c 文件的内容很多，这里只介绍几个主要函数的调用流程。

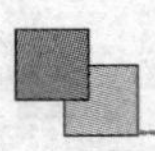

1. 函数 formatNFTL()流程

如流程图 5.7 所示，整个格式化的过程不会将出厂时坏块信息擦除掉，而是会在擦除之前将原有坏块信息读出，然后存入介质，供下一次格式化或者安装文件系统时使用。考虑到 NAND FLASH 的坏块，几乎每一重要数据都做了双保险。

格式化时首先判断起始单元是否合法，要看能否读到有效的起始单元。新出厂的介质，闪存上所有数据是十六进制“F”，不是有效的起始单元，因此不合法。格式化每一块的操作其实很简单，即将擦除次数及擦除标记等写入闪存的 OOB 区。

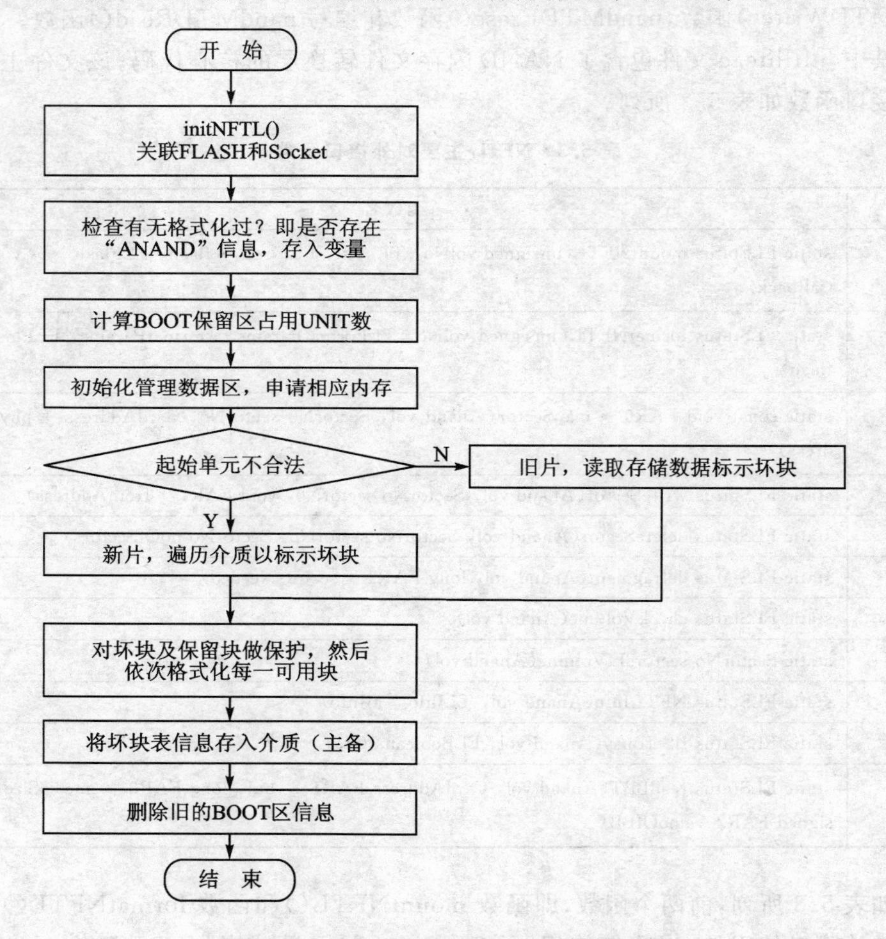

图 5.7 NAND FLASH 格式化主要流程

2. 函数 mountNFTL()流程

如图 5.8 所示，函数 mountNFTL()流程主要包括读取主区 BOOT 信息，读取备区 BOOT 信息，判断参数是否有效，安装 UNIT，对占用过多的虚拟单元链进行回收等关键步骤。

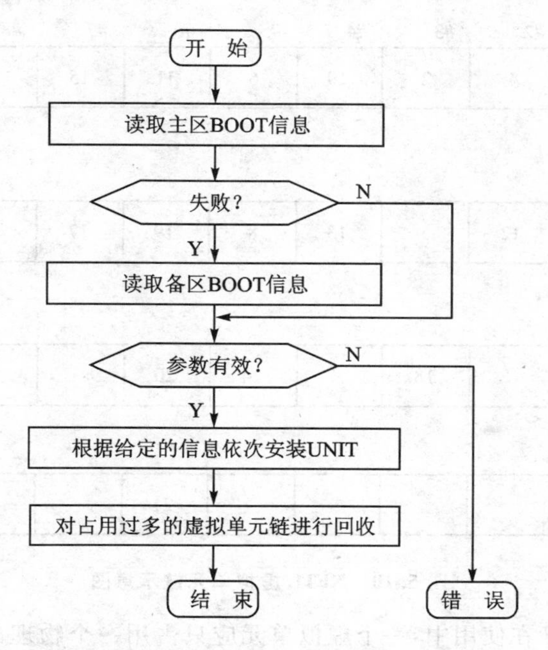

图 5.8 NAND FLASH MOUNT 主要流程

其中，基本参数合法性判断包括：介质上读出来的可用单元个数是否超过介质所有的单元个数，虚拟扇区个数是否与可用的单元个数相冲突等。

3. 函数 writeSector( )流程

单扇区写操作流程较为简单，如图 5.9 所示。复杂的地方在于执行磨损均衡算法及扇区分配两个步骤的处理。

NFTL 层存在一个虚拟的单元链，这个单元链的主要作用是为了减少擦除次数，如图 5.10 所示。

图 5.10 中，阿拉伯数字序号表示写入扇区的操作顺序，对应的位置就是对应的扇区号。如图 5.10，一共进行了 21 次扇区写操作。如果不借用虚拟扇区链，在操作时将不得不对第一个物理单元进行 14(＝1＋3＋1＋2＋1＋2＋3＋1)次擦除，因为每当有一个已经存在的扇区数据区时，都必须执行一次擦除操作才能写入。而在使用 4 个物理单元组成一个虚拟单元链后，将大大减少擦除次数。

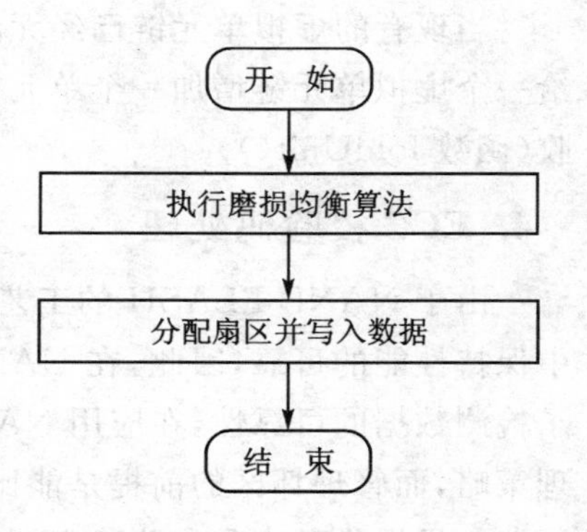

图 5.9 扇区写操作流程

通过将被置换的扇区打上删除标记，就能让系统软件自动到下一块的相应区去找到合法的数据，而不必在每次执行覆盖写操作时将一个物理块擦除，将有效减少介质的总擦除次数。

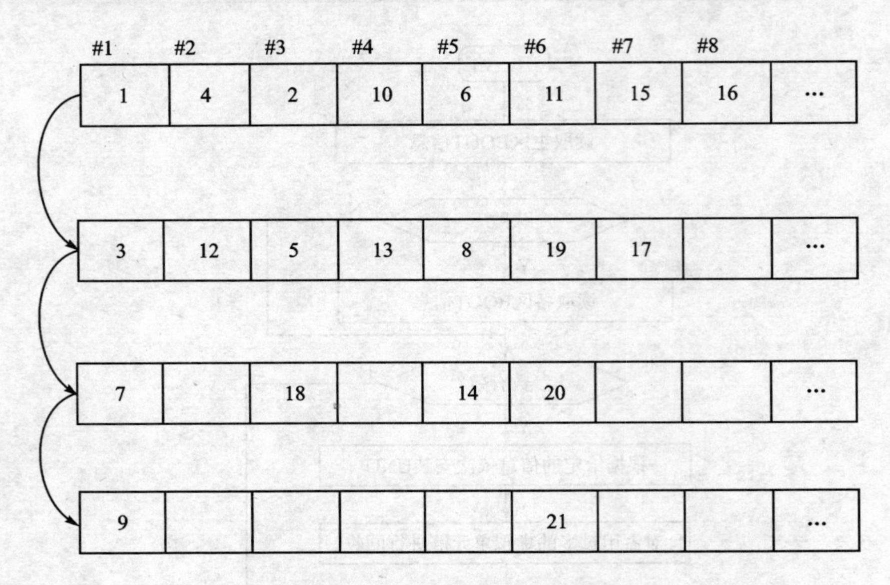

图 5.10　NFTL 虚拟单元链示意图

另一方面，由于在使用中，一个虚拟单元应只占用一个物理单元，所以回收操作常常会被不定期执行。即将一个虚拟单元链中的数据复制到一个独立的物理单元中去，这样将节省占用空间，完成重复利用。当满足如下条件时，将会执行虚拟单元链的回收操作：

- 可用单元数小于临界值(2)；
- 虚拟单元链的长度已经达到最大值(32，可调)；
- 虚拟单元数据已经不可再次写入时(数据已经写满)。

当现有的虚拟单元链已经无法完成数据写入而链长度还没有超出门限值时，会给一个虚拟单元链增加一个单元；如果不能分配(可用单元数太少)，则先执行单元回收(函数 foldUnit())。

4. ECC 校验码处理

由于 NAND FLASH 的工艺不能保证 NAND 的 Memory Array 在其生命周期中保持性能的可靠，因此，在 NAND FLASH 生产中及使用过程中会产生坏块。为了检测数据的可靠性，在应用 NAND FLASH 的系统中一般都会采用一定的坏区管理策略，而管理坏区的前提是能比较可靠的进行坏区检测。

如果操作时序和电路稳定性不存在问题的话，NAND FLASH 出错的时候一般不会造成整个 Block 或是 Page 不能读取或是全部出错，而是整个 Page(例如 512 B)中只有一个或几个位出错。

对数据的校验常用的有奇偶校验、CRC 校验等，而在 NAND FLASH 处理中，一般使用 ECC 校验。ECC 能纠正单位错误和检测双位错误，而且计算速度很快，但对

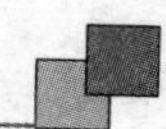

1 位以上的错误无法纠正，对 2 位以上的错误不保证能检测。

ECC 校验的代码实现参照 Linux 进行移植，需要在 MTD 层的代码中将 NFTL 层传进来的参数作一转换，比如将 SECTOR FLAGS 移到高地址上去，而将 UNIT 相关的 FLAG 移到低地址上去，读写时都做一层转换。移植过程较为复杂，详情可参见 MagicARM2410 开发箱提供的 BSP 中包含的 NAND FLASH 驱动部分。

5.6 TrueFFS 文件系统实验设计

5.6.1 实验目的

掌握 TrueFFS 文件系统开发过程。

5.6.2 实验设备

(1) 硬件：

- PC 机一台；
- MagicARM2410 教学实验开发平台一套。

(2) 软件：

- Windows98/XP/2000 系统；
- Tornado 交叉开发环境。

5.6.3 实验内容

NOR FLASH 的格式化，TrueFFS 文件系统的建立，通过 FTP 软件在 NOR FLASH 上，建立、上传、下载文件实验。

5.6.4 实验预习要求

FLASH 设备驱动程序设计。

5.6.5 实验原理

掌握如表 5.4 所列的 SST39VF1601 的命令字序列和编程步骤。

表 5.4　SST39VF1601 的命令字序列表

Command Sequence	1st Bus Write Cycle		2nd Bus Write Cycle		3rd Bus Write Cycle		4th Bus Write Cycle		5th Bus Write Cycle		6th Bus Write Cycle	
	Addr1	Data2	Addr1	Data2	Addr1	Data2	Addr1	Data2	Addr1	Data2	Addr1	Data2
Word - Program	5555H	AAH	2AAAH	55H	5555H	A0H	WA 3	Data				
Sector - Erase	5555H	AAH	2AAAH	55H	5555H	80H	5555H	AAH	2AAAH	55H	SAx4	30H
Block - Erase	5555H	AAH	2AAAH	55H	5555H	80H	5555H	AAH	2AAAH	55H	BAx4	50H
Chip - Erase	5555H	AAH	2AAAH	55H	5555H	80H	5555H	AAH	2AAAH	55H	5555H	10H
Erase - Suspend	XXXXH	B0H										
Erase - Resume	XXXXH	30H										
Query Sec ID5	5555H	AAH	2AAAH	55H	5555H	88H						
User Security ID Word - Program	5555H	AAH	2AAAH	55H	5555H	A5H	WA 6	Data				
User Security ID Program Lock - Out	5555H	AAH	2AAAH	55H	5555H	85H	XXH6	0000H				
Software ID Entry7,8	5555H	AAH	2AAAH	55H	5555H	90H						
CFI Query Entry	5555H	AAH	2AAAH	55H	5555H	98H						
Software ID Exit9,10/CFI Exit/Sec ID Exit	5555H	AAH	2AAAH	55H	5555H	F0H						
Software ID Exit/CFI Exit/Sec ID Exit	XXH	F0H										

编程步骤参见 SST39VF1601 的 datasheet。

5.6.6　实验步骤

(1) 在镜像下载后，在 WShell 中输入：tffsDevFormat(0,0)(该函数用于格式化 NOR FLASH)，在格式化的过程中，串口会有大量的 debug 输出信息，由此可见串口驱动的正确性。

(2) 在格式化完成之后，执行函数 usrTffsConfig(0,0,"/tffs0")，系统在每次重新启动后需要执行该命令，用于挂载 NOR FLASH 文件系统，如图 5.11 所示。

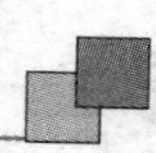

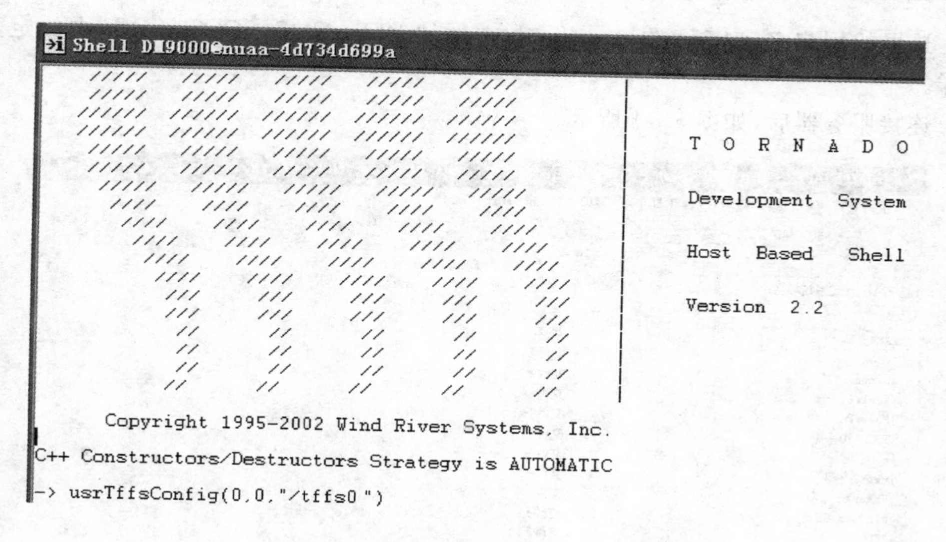

图 5.11 函数 usrTffsConfig( )函数执行图

(3) 挂载完成后，在 WShell 中输入“devs”，可以看到系统已经加载了 NorFlash，即“3 /tffs0/”如图 5.12 所示。函数“devs”用于查看系统设备。

(4) 此后通过 chdir("/tffs0")命令将当前目录设置为“/tffs0/”。每次重新启动后都要执行该命令，用于挂载。当然也可以放到系统初始化当中去执行。如果不设置当前目录，目标板的 ftp server 的工作目录为“host”，则无法上传文件。

(5) 通过 FTP 上传文件。

首先设置 FTP 软件，以 CuteFTP 工具为例，如图 5.13 所示。

```
-> devs
drv name
  0 /null
  1 /tyCo/0
  1 /tyCo/1
  5 host:
  6 /vio
  3 /tffs0/
value = 0 = 0x0
->
```

图 5.12 指令 devs 执行结果图

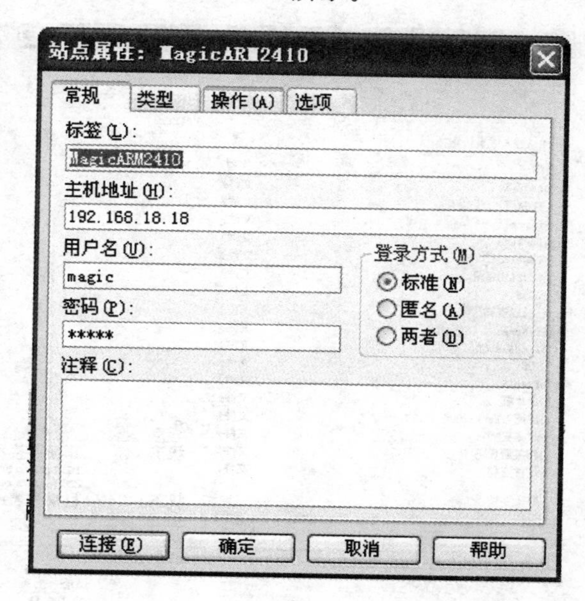

图 5.13 FTP 配置图

这里的用户名和密码要和 config.h 文件的 DEFAULT_BOOT_LINE 中的定义一样。

连接服务器后，如图 5.14 所示。

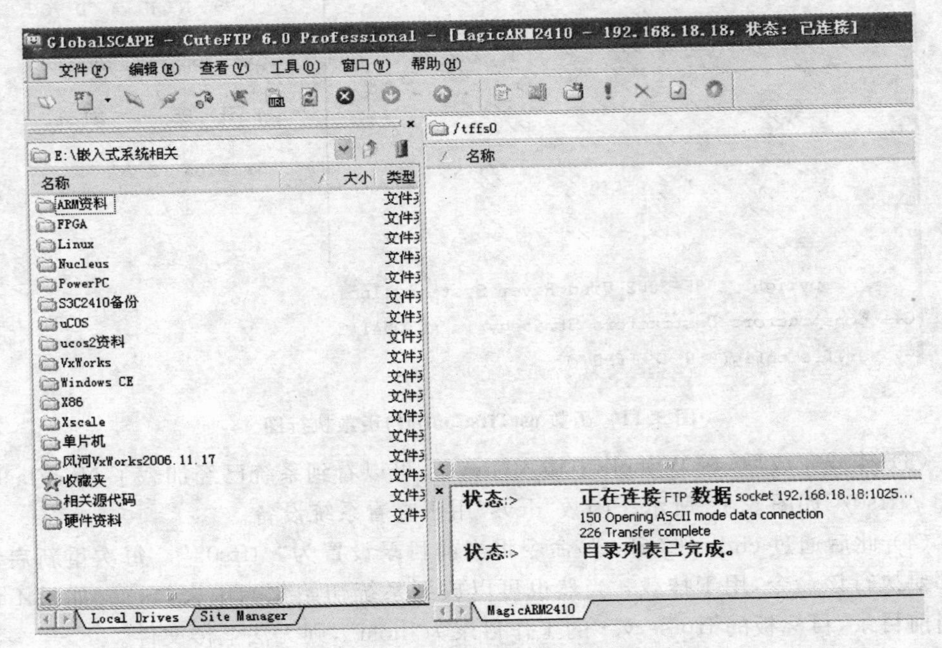

图 5.14　CuteFTP 工具连接图

上传文件后可以看到 NorFLASH 中的文件，如图 5.15 所示。

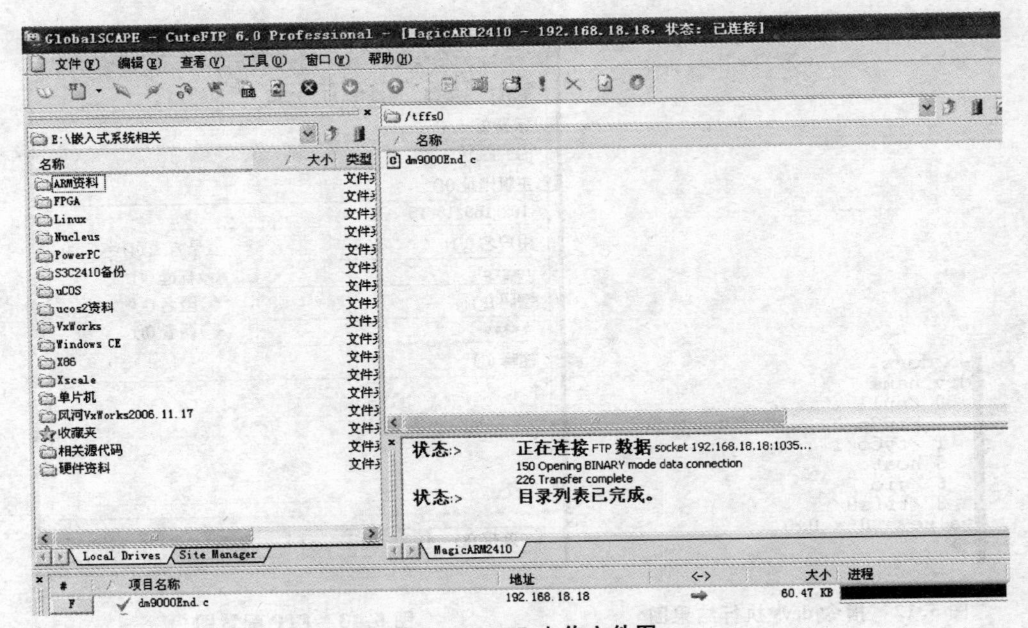

图 5.15　FTP 上传文件图

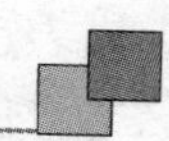

由于 NOR FLASH 中 BOOTROM 占用了 512 KB，TrueFFS 文件系统就只剩下 1.5 MB，加上 TrueFFS 自身的损耗，所以上传的文件尽量小于 1.4 MB。

通过 CuteFTP 软件可以在 NOR FLASH 上创建和删除文件。

使用提醒：

① 由于 S3C2410 自身的特性，一片 NOR FLASH 上使用 BOOT ROM 和 TrueFFS 文件系统的操作只能在 WShell 中执行，在 TShell 中无法运行。

② CuteFTP 在使用中因为版本问题可能导致无法上传和下载。

5.6.7 DOS 下实验方法

在 DOS 下给出实验范例，按照此步骤进行实验。

(1) 登陆 ftp，在 DOS 提示符下输入指令"ftp 192.168.18.18"，然后回车，PC 给出下面提示，按照 config 的配置，分别输入用户名和密码，登录成功。

Connected to 192.168.18.18.

220 VxWorks FTP server (VxWorks VxWorks5.5) ready.

User (192.168.18.18:(none)): s2410－dm9000

331 Password required

Password:

230 User logged in

(2) 观察当前目标机目录下是否有文件，在"ftp"提示符下，输入"dir"指令回车，则可以看到该目录下的文件，如下所示：

200 Port set okay

150 Opening BINARY mode data connection

226 Transfer complete

(3) 上传文件到目标机，将 PC 机上的 C 盘下的 readme.txt 文件上传给目标机，在"ftp"提示符输入"put c:\readme.txt"指令，然后回车，该文件则将上传给目标机，有如下提示：

200 Port set okay

150 Opening BINARY mode data connection

226 Transfer complete

ftp:发送 5 519 字节，用时 0.00Seconds 5 519 000.00Kbytes/sec

(4) 此时，再次查看目标机目录文件，按照步骤 2 进行操作，可以看到步骤 3 上传的文件"readme.txt"，提示如下。

200 Port set okay

150 Opening BINARY mode data connection

－rwx－－－－－－ 1 user group 5519 Jan 1 1980

README. TXT

226 Transfer complete

ftp:收到 65 字节,用时 0. 00Seconds 65 000. 00Kbytes/sec

(5) 定位主机的当前目录为 D 盘,在“ftp”提示符输入“lcd d:\”指令,然后回车,提示如下:

Local directory now D:\

(6) 下载目标机的“readme. txt”文件到主机的 D 盘,在“ftp”提示符输入“get readme. txt”指令,然后回车,提示如下:

200 Port set okay

150 Opening BINARY mode data connection

226 Transfer complete

ftp:收到 5 519 字节,用时 0. 13Seconds 44. 15Kbytes/sec

(7) 再次查询目标机目录文件,执行步骤 2,提示如下,说明目标机的文件没有受到影响。

200 Port set okay

150 Opening BINARY mode data connection

— rwx — — — — — — —   1 user   group   5519 Jan 1 1980

README. TXT

226 Transfer complete

ftp:收到 65 字节,用时 0. 00Seconds 65 000. 00Kbytes/sec.

(8) 删除目标机上的“readme. txt”文件,在“ftp”提示符输入“delete readme. txt”指令,然后回车,提示如下:

200 File deleted successfully

(9) 重新执行“dir”指令,看到目标机文件已经被删除了,如下:

200 Port set okay

150 Opening BINARY mode data connection

226 Transfer complete

(10) 离开 ftp 服务,在“ftp”提示符输入“bye”指令,然后回车,结束 ftp 服务,提示如下:

221 Bye... see you later

C:\Documents and Settings\space>

5.6.8 程序清单

参见 BSP 中的 lv160mtd. c、sysTffs. c、tffsConfig. c、usrTffs. c 等文件。

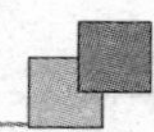

1. sysTffs.c 文件

```
/* sysTffs.c - MagicARM2410 system-dependent TrueFFS library */

/* Copyright 1984-1997 Wind River Systems, Inc. */
#include "copyright_wrs.h"

/* FAT-FTL Lite Software Development Kit
 * Copyright (C) M-Systems Ltd. 1995-1996      */


/*
DESCRIPTION
This library provides board-specific hardware access routines for TrueFFS.
In effect, these routines comprise the socket component driver (or drivers)
for your flash device hardware.  At socket registration time, TrueFFS stores
pointers to the functions of this socket component driver in an 'FLSocket'
structure.  When TrueFFS needs to access the flash device, it uses these
functions.

Because this file is, for the most part, a device driver that exports its
functionality by registering function pointers with TrueFFS, very few of the
functions defined here are externally callable.  For the record, these
external functions are flFitInSocketWindow() and flDelayLoop().  You should
never have any need to call these functions.

However, one of the most import functions defined in this file is neither
referenced in an 'FLSocket' structure, nor is it externally callable.  This
function is sysTffsInit().  TrueFFS calls this function at initialization
time to register socket component drivers for all the flash devices attached
to your target.  It is this call to sysTffs() that results in assigning
drive numbers to the flash devices on your target hardware.  Drive numbers
are assigned by the order in which the socket component drivers are registered.
The first to be registered is drive 0, the second is drive 1, and so on up to
4.  As shipped, TrueFFS supports up to five flash drives.

After registering socket component drivers for a flash device, you may
format the flash medium even though there is not yet a block device driver
associated with the flash (see the reference entry for the tffsDevCreate()
routine).  To format the flash medium for use with TrueFFS,
call tffsDevFormat() or, for some BSPs, sysTffsFormat().
```

```
The sysTffsFormat() routine is an optional but BSP-specific externally
callable helper function.  Internally, it calls tffsDevFormat() with a
pointer to a 'FormatParams' structure initialized to values that leave a
space on the flash device for a boot image. This space is outside the
region managed by TrueFFS.  This special region is necessary for boot
images because the normal translation and wear-leveling services of TrueFFS
are incompatible with the needs of the boot program and the boot image it
relies upon.  To write a boot image (or any other data) into this area,
use tffsBootImagePut().

Finally, this file also contains define statements for symbolic constants
that determine which MTDs, translation layer modules, and other utilities
are ultimately included in TrueFFS.  These defines are as follows:

.IP "INCLUDE_TL_NFTL"
To include the NAND-based translation layer module.
.IP "INCLUDE_TL_FTL"
To include the NOR-based translation layer module.
.IP "INCLUDE_TL_SSFDC"
To include the SSFDC-appropriate translation layer module.
.IP "INCLUDE_MTD_I28F016"
For Intel 28f016 flash devices.
.IP "INCLUDE_MTD_I28F008"
For Intel 28f008 flash devices.
.IP "INCLUDE_MTD_I28F008_BAJA"
For Intel 28f008 flash devices on the Heurikon Baja 4700.
.IP "INCLUDE_MTD_AMD"
For AMD, Fujitsu: 29F0{40,80,16} 8-bit flash devices.
.IP "INCLUDE_MTD_CDSN"
For Toshiba, Samsung: NAND CDSN flash devices.
.IP "INCLUDE_MTD_DOC2"
For Toshiba, Samsung: NAND DOC flash devices.
.IP "INCLUDE_MTD_CFISCS"
For CFI/SCS flash devices.
.IP "INCLUDE_MTD_WAMD"
For AMD, Fujitsu 29F0{40,80,16} 16-bit flash devices.
.IP "INCLUDE_TFFS_BOOT_IMAGE"
To include tffsBootImagePut() in TrueFFS for Tornado.
.LP
To exclude any of the modules mentioned above, edit sysTffs.c and undefine
its associated symbolic constant.
```

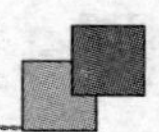

```
INCLUDE FILES: flsocket.h, tffsDrv.h

SEE ALSO : tffsDrv tffsConfig
*/

#include "vxWorks.h"
#include "config.h"
#include "tffs/flsocket.h"
#include "tffs/tffsDrv.h"


/* defines */
#define    INCLUDE_MTD_LV160
#undef     INCLUDE_MTD_VF160
#undef    INCLUDE_MTD_VF160
#undef     INCLUDE_MTD_I28F016          /* Intel: 28f016 */
#undef     INCLUDE_MTD_I28F008          /* Intel: 28f008 */
#undef     INCLUDE_MTD_AMD              /* AMD, Fujitsu: 29f0{40,80,16} 8bit */
#undef     INCLUDE_MTD_CDSN             /* Toshiba, Samsung: NAND, CDSN */
#undef     INCLUDE_MTD_DOC2             /* Toshiba, Samsung: NAND, DOC */
#undef     INCLUDE_MTD_CFISCS           /* CFI/SCS */
#undef     INCLUDE_MTD_WAMD             /* AMD, Fujitsu: 29f0{40,80,16} 16bit */
#undef     INCLUDE_TL_NFTL              /* NFTL translation layer */
#define    INCLUDE_TL_FTL               /* FTL translation layer */
#undef     INCLUDE_TL_SSFDC             /* SSFDC translation layer */
/* #define INCLUDE_TFFS_BOOT_IMAGE */ /* include tffsBootImagePut() */
#undef INCLUDE_TFFS_BOOT_IMAGE
#define FLASH_BOOT_ADRS                (0x00080000)
#define    FLASH_BOOT_SIZE           (0x00180000)

/* locals */


/* forward declarations */

LOCAL void          rfaWriteProtect (void);
LOCAL void          rfaWriteEnable (void);
LOCAL FLBoolean          rfaCardDetected
(
    FLSocket vol
);
LOCAL void          rfaVccOn
```

```
(
    FLSocket vol
);
LOCAL void          rfaVccOff
(
    FLSocket vol
);
#ifdef    SOCKET_12_VOLTS
LOCAL FLStatus         rfaVppOn
(
    FLSocket vol
);
LOCAL void          rfaVppOff
(
    FLSocket vol
);
#endif      /* SOCKET_12_VOLTS */
LOCAL FLBoolean        rfaGetAndClearCardChangeIndicator
(
    FLSocket vol
);
LOCAL FLBoolean        rfaWriteProtected
(
    FLSocket vol
);
LOCAL void          rfaSetWindow
(
    FLSocket vol
);
LOCAL void          rfaSetMappingContext
(
    FLSocket vol,
    unsigned page
);
LOCAL FLStatus         rfaSocketInit
(
    FLSocket vol
);
LOCAL FLStatus         rfaRegister (void);

#ifndef DOC
```

```
#ifdef INCLUDE_MTD_VF160
#include "vf160mtd.c"
#endif

#ifdef     INCLUDE_MTD_CDSN
#include "k9s1208mtd.c"
#endif

#include "tffsConfig.c"
#include "tffsDrv.c"
#endif /* DOC */

/*****************************************************************
*
* sysTffsInit - board-level initialization for TrueFFS
*
* This routine calls the socket registration routines for the socket component
* drivers that will be used with this BSP. The order of registration determines
* the logical drive number given to the drive associated with the socket.
*
* RETURNS: N/A
*/

LOCAL void sysTffsInit (void)
{

    rfaRegister ();
}

/*****************************************************************
*
* rfaRegister - registration routine for the RFA on MVME177
*
* This routine populates the 'vol' structure for a logical drive with the
* socket component driver routines for the RFA on the MVME177 board. All
* socket routines are referanced through the 'vol' structure and never
* from here directly
*
* RETURNS: flOK, or flTooManyComponents if there're too many drives
*/
```

```
LOCAL FLStatus rfaRegister (void)
{
    FLSocket vol = flSocketOf (noOfDrives);

    if (noOfDrives >= DRIVES)
        return (flTooManyComponents);

    tffsSocket[noOfDrives] = "RFA";

    vol.serialNo = noOfDrives;
    if  (noOfDrives == 0)
    vol.window.baseAddress = FLASH_BOOT_ADRS>> 12 ;
    noOfDrives++;

    /* fill in function pointers */

    vol.cardDetected          = rfaCardDetected;
    vol.VccOn                 = rfaVccOn;
    vol.VccOff                = rfaVccOff;
#ifdef SOCKET_12_VOLTS
    vol.VppOn                 = rfaVppOn;
    vol.VppOff                = rfaVppOff;
#endif
    vol.initSocket            = rfaSocketInit;
    vol.setWindow             = rfaSetWindow;
    vol.setMappingContext = rfaSetMappingContext;
    vol.getAndClearCardChangeIndicator =
                            rfaGetAndClearCardChangeIndicator;
    vol.writeProtected        = rfaWriteProtected;

    return (flOK);
}

/*******************************************************************
 *
 * rfaCardDetected - detect if a card is present (inserted)
 *
 * This routine detects if a card is present (inserted).
 *
 * RETURNS: TRUE, or FALSE if the card is not present.
 */
```

```
LOCAL FLBoolean rfaCardDetected
(
    FLSocket vol
)
{
    return (TRUE);
}

/*********************************************************************
 *
 * rfaVccOn - turn on Vcc (3.3/5 Volts)
 *
 * This routine turns on Vcc (3.3/5 Volts).  Vcc must be known to be good
 * on exit.
 *
 * RETURNS: N/A
 */

LOCAL void rfaVccOn
(
    FLSocket vol
)
{
    rfaWriteEnable ();
}

/*********************************************************************
 *
 * rfaVccOff - turn off Vcc (3.3/5 Volts)
 *
 * This routine turns off Vcc (3.3/5 Volts).
 *
 * RETURNS: N/A
 */

LOCAL void rfaVccOff
(
    FLSocket vol
)
{
    rfaWriteProtect ();
}
```

```
#ifdef SOCKET_12_VOLTS

/*******************************************************************
*
* rfaVppOn - turns on Vpp (12 Volts)
*
* This routine turns on Vpp (12 Volts). Vpp must be known to be good on exit.
*
* RETURNS: flOK always
*/

LOCAL FLStatus rfaVppOn
(
    FLSocket vol          /* pointer identifying drive */
)
{
    return (flOK);
}

/*******************************************************************
*
* rfaVppOff - turns off Vpp (12 Volts)
*
* This routine turns off Vpp (12 Volts).
*
* RETURNS: N/A
*/

LOCAL void rfaVppOff
(
    FLSocket vol          /* pointer identifying drive */
)
{
}

#endif    /* SOCKET_12_VOLTS */

/*******************************************************************
*
* rfaSocketInit - perform all necessary initializations of the socket
*
```

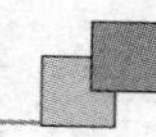

```
 * This routine performs all necessary initializations of the socket.
 *
 * RETURNS: flOK always
 */

LOCAL FLStatus rfaSocketInit
(
    FLSocket vol            /* pointer identifying drive */
)
{
    rfaWriteEnable ();

    vol.cardChanged = FALSE;

    /* enable memory window and map it at address 0 */
    rfaSetWindow (&vol);

    return (flOK);
}

/*******************************************************************
 *
 * rfaSetWindow - set current window attributes, Base address, size, etc
 *
 * This routine sets current window hardware attributes: Base address, size,
 * speed and bus width.  The requested settings are given in the 'vol.window'
 * structure.  If it is not possible to set the window size requested in
 * 'vol.window.size', the window size should be set to a larger value,
 * if possible. In any case, 'vol.window.size' should contain the
 * actual window size (in 4 KB units) on exit.
 *
 * RETURNS: N/A
 */

LOCAL void rfaSetWindow
(
    FLSocket vol            /* pointer identifying drive */
)
{
    /* Physical base as a 4K page */
    if(vol.serialNo == 0)
    {
```

```
        vol.window.baseAddress = FLASH_BOOT_ADRS >> 12;
        flSetWindowSize (&vol, FLASH_BOOT_SIZE >> 12);
    }
}

/****************************************************************
*
* rfaSetMappingContext - sets the window mapping register to a card address
*
* This routine sets the window mapping register to a card address.
* The window should be set to the value of 'vol.window.currentPage',
* which is the card address divided by 4 KB. An address over 128 MB,
* (page over 32 KB) specifies an attribute-space address. On entry to this
* routine vol.window.currentPage is the page already mapped into the window.
* (In otherwords the page that was mapped by the last call to this routine.)
*
* The page to map is guaranteed to be on a full window-size boundary.
*
* RETURNS: N/A
*/

LOCAL void rfaSetMappingContext
(
    FLSocket vol,           /* pointer identifying drive */
    unsigned page           /* page to be mapped */
)
{
}

/****************************************************************
*
* rfaGetAndClearCardChangeIndicator - return the hardware card-change indicator
*
* This routine returns the hardware card-change indicator and clears it if set.
*
* RETURNS: FALSE, or TRUE if the card has been changed
*/

LOCAL FLBoolean rfaGetAndClearCardChangeIndicator
(
    FLSocket vol            /* pointer identifying drive */
)
```

```
    {
        return (FALSE);
    }

/*****************************************************************
 *
 * rfaWriteProtected - return the write-protect state of the media
 *
 * This routine returns the write-protect state of the media
 *
 * RETURNS: FALSE, or TRUE if the card is write-protected
 */

LOCAL FLBoolean rfaWriteProtected
(
    FLSocket vol            /* pointer identifying drive */
)
{
    return (FALSE);
}

/*****************************************************************
 *
 * rfaWriteProtect - disable write access to the RFA
 *
 * This routine disables write access to the RFA.
 *
 * RETURNS: N/A
 */

LOCAL void rfaWriteProtect (void)
{
    /* clear GPOEN1 bit (#17), make sure GPIO1 bit (#13) is clear  */
    /* * VMECHIP2_IOCR = ( * VMECHIP2_IOCR) & ((~IOCR_GPOEN1) & (~IOCR_GPIO01_
HIGH)); */
}

/*****************************************************************
 *
 * rfaWriteEnable - enable write access to the RFA
 *
 * This routine enables write access to the RFA.
```

```
 *
 * RETURNS: N/A
 */

LOCAL void  rfaWriteEnable (void)
{
    /* set GPOEN1 bit (#17), make sure GPI01 bit (#13) is clear */
    /* *VMECHIP2_IOCR = ((*VMECHIP2_IOCR) | IOCR_GPOEN1) & (~IOCR_GPIO01_
HIGH); */
}

/******************************************************************
 *
 * flFitInSocketWindow - check whether the flash array fits in the socket window
 *
 * This routine checks whether the flash array fits in the socket window.
 *
 * RETURNS: A chip size guaranteed to fit in the socket window.
 */

long int flFitInSocketWindow
(
    long int chipSize,          /* size of single physical chip in bytes */
    int      interleaving,      /* flash chip interleaving (1,2,4 etc) */
    long int windowSize         /* socket window size in bytes */
)
{
    if (chipSize * interleaving > windowSize) /* doesn't fit in socket window */
    {
        int  roundedSizeBits;

        /* fit chip in the socket window */
        chipSize = windowSize / interleaving;

        /* round chip size at powers of 2 */
        for (roundedSizeBits = 0; (0x1L << roundedSizeBits) <= chipSize;
            roundedSizeBits++)
        ;

        chipSize = (0x1L << (roundedSizeBits - 1));
    }
```

```
    return (chipSize);
}

#if     FALSE
/*******************************************************************
 *
 * sysTffsCpy - copy memory from one location to another
 *
 * This routine copies <size> characters from the object pointed
 * to by <source> into the object pointed to by <destination>. If copying
 * takes place between objects that overlap, the behavior is undefined.
 *
 * INCLUDE FILES: string.h
 *
 * RETURNS: A pointer to <destination>.
 *
 * NOMANUAL
 */

void * sysTffsCpy
(
    void *          destination,    /* destination of copy */
    const void *    source,         /* source of copy */
    size_t          size            /* size of memory to copy */
)
{
    bcopy ((char *) source, (char *) destination, (size_t) size);
    return (destination);
}

/*******************************************************************
 *
 * sysTffsSet - set a block of memory
 *
 * This routine stores <c> converted to an `unsigned char' in each of the
 * elements of the array of `unsigned char' beginning at <m>, with size <size>.
 *
 * INCLUDE FILES: string.h
 *
 * RETURNS: A pointer to <m>.
 *
 * NOMANUAL
 */
```

```
void * sysTffsSet
(
    void * m,                        /* block of memory */
    int    c,                        /* character to store */
    size_t size                      /* size of memory */
)
{
    bfill ((char *) m, (int) size, c);
    return (m);
}
#endif      /* FALSE */

/*******************************************************************
*
* flDelayLoop - consume the specified time
*
* This routine consumes the specified time.
*
* RETURNS: N/A
*/

void flDelayLoop
(
    int cycles                  /* loop count to be consumed */
)
{
    while (--cycles)
    ;
}

/*******************************************************************
*
* sysTffsFormat - format the flash memory above an offset
*
* This routine formats the flash memory.  Because this function defines
* the symbolic constant, HALF_FORMAT, the lower half of the specified flash
* memory is left unformatted.  If the lower half of the flash memory was
* previously formated by TrueFFS, and you are trying to format the upper half,
* you need to erase the lower half of the flash memory before you format the
* upper half.  To do this, you could use:
```

```
 * .CS
 * tffsRawio(0, 3, 0, 8)
 * .CE
 * The first argument in the tffsRawio() command shown above is the TrueFFS
 * drive number, 0.  The second argument, 3, is the function number (also
 * known as TFFS_PHYSICAL_ERASE).  The third argument, 0, specifies the unit
 * number of the first erase unit you want to erase.  The fourth argument, 8,
 * specifies how many erase units you want to erase.
 *
 * RETURNS: OK, or ERROR if it fails.
 */

STATUS sysTffsFormat (void)
{
    STATUS status;
    tffsDevFormatParams params =
    {
#define     HALF_FORMAT     /* lower 0.5MB for bootimage, upper 1.5MB for TrueFFS */
#ifdef     HALF_FORMAT
    {0x000800001, 99, 1, 0x100001, NULL, {0,0,0,0}, NULL, 2, 0, NULL},
#else
    {0x000000001, 99, 1, 0x100001, NULL, {0,0,0,0}, NULL, 2, 0, NULL},
#endif     /* HALF_FORMAT */
    FTL_FORMAT_IF_NEEDED
    };

    /* we assume that the drive number 0 is RFA */

    status = tffsDevFormat (0, (int)&params);
    return (status);
}
```

2. lv160mtd.c 文件

```
#include "tffs/flflash.h"
#include "tffs/backgrnd.h"
#include "stdio.h"
#include "config.h"
#include "intLib.h"

/*需与 sysTffs.c 中相同定义一致*/
#define FLASH_BOOT_ADRS                (0x00080000)
```

```
#define FLASH_BOOT_SIZE                (0x00180000)


typedef struct
{
    FlashWPTR     unlockAddr1;
    FlashWPTR       unlockAddr2;
} Vars;

static Vars mtdVars[DRIVES];

#define thisVars     ((Vars *) vol.mtdVars)

#define SETUP_ERASE            0x80
#define SETUP_WRITE            0xa0
#define READ_ID                0x90
#define SECTOR_ERASE        0x30
#define BLOCK_ERASE            0x50              /* 块擦除 */
#define READ_ARRAY           0xf0               /* 软件 ID 读模式退出 */

#define UNLOCK_1            0xaa                 /* 解锁写入值 */
#define     UNLOCK_2            0x55

#define UNLOCK_ADDR1        0x5555               /* 解锁偏移地址[字] */
#define     UNLOCK_ADDR2      0x2aaa


/* JEDEC ids for this MTD */
#define AM29LV160_DEID 0x2249                    /* 设备 ID */
#define SST39VF1601_DEID 0x234b

#define DEBUG_PRINT              printErr
#undef   DEBUG_PRINT

static STATUS   lv160OpOverDetect
(
     void * ptr,   int timeCounter
);


/*-----------------------------------------------------------------------
 Procedure:      lv160MTDMap ID:1
```

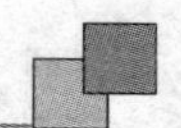

```
 Purpose:         映射 Flash 片内地址为 CPU 全局地址
 Input:           addr - 相对地址
 Output:
 Errors:
-----------------------------------------------------------------------*/
static void FAR0 *      lv160MTDMap0
(
    FLFlash * vol,
    CardAddress addr,
    int length
)
{
    UINT32 ret;
    ret = FLASH_BOOT_ADRS + addr;
    return (void FAR0 * )ret;
}

/*-----------------------------------------------------------------------
 Procedure:       lv160OpOverDetect ID:1
 Purpose:         探测 write,erase 操作是否结束,超时错误
 Input:
 Output:
 Errors:
-----------------------------------------------------------------------*/
static STATUS  lv160OpOverDetect(void * ptr,  int timeCounter)
{
    FlashWPTR pFlash = ptr;
    INT16 buf1,buf2;

    buf1 =  * pFlash & 0x40;
    while(1)
    {
            buf2  =  * pFlash & 0x40;
            if(buf1 == buf2)
                break;
            else
                buf1 = buf2;
            if(timeCounter-- <= 0)
            {
                return ERROR;
            }
    }
```

```
    return OK;
}
/*-------------------------------------------------------------------
 Procedure:     lv160MTDWrite ID:1
 Purpose:       MTD 写 Flash 函数
 Input:
 Output:
 Errors:
-------------------------------------------------------------------*/
static FLStatus lv160MTDWrite
(
     FLFlash vol,
     CardAddress address,
     const void FAR1 *buffer,
     int length,
     FLBoolean overwrite
)
{
    int cLength;
    FlashWPTR flashPtr, flashTmp;
    volatile UINT16 *gBuffer;
    int level;

    flashTmp = flashPtr = (FlashWPTR) vol.map(&vol, address, length);
    if(length&1)
    {
        printf("warning! the data length can not divided by 2.");
    }
    cLength = length/2;

    gBuffer = (UINT16 *)buffer;

    while (cLength >= 1)
    {
         *thisVars->unlockAddr1 = UNLOCK_1;
         *thisVars->unlockAddr2 = UNLOCK_2;
        level = intLock();
         *thisVars->unlockAddr1 = SETUP_WRITE;

         *flashPtr = *gBuffer;

        if(lv160OpOverDetect((void *)flashPtr, 0x1000000));
```

```
        if( * flashPtr != * gBuffer)
        {
            * flashPtr = READ_ARRAY;
            #ifdef DEBUG_PRINT
            DEBUG_PRINT("Debug: lv160MTDWrite timeout.\n");
            #endif
            return flWriteFault;
        }
            intUnlock(level);
            cLength--;
            flashPtr++;
            gBuffer++;
    }

    if (tffscmp((void FAR0 * ) flashTmp, buffer,length))
    {
        /* verify the data */
        #ifdef DEBUG_PRINT
          DEBUG_PRINT("Debug: lv160MTDWrite fail.\n");
        #endif
        return flWriteFault;
    }

    return flOK;
}
/*-----------------------------------------------------------------
 Procedure:      lv160MTDErase ID:1
 Purpose:        MTD 擦除 FLASH 函数
 Input:
 Output:
 Errors:
-----------------------------------------------------------------*/
static FLStatus lv160MTDErase
(
    FLFlash vol,
    int firstErasableBlock,
    int numOfErasableBlocks
)
{
    int iBlock;                 FlashWPTR flashPtr;
    unsigned int offset;
    int level;
```

```
    if(numOfErasableBlocks <= 0) return ERROR;

    for (iBlock = 0; iBlock < numOfErasableBlocks; iBlock++)
    {
        int i;
        offset = (firstErasableBlock + iBlock) * vol.erasableBlockSize;
        flashPtr = (FlashWPTR) vol.map(&vol, offset, vol.interleaving);

        *thisVars->unlockAddr1 = UNLOCK_1;
        *thisVars->unlockAddr2 = UNLOCK_2;
        *thisVars->unlockAddr1 = SETUP_ERASE;
        *thisVars->unlockAddr1 = UNLOCK_1;
        *thisVars->unlockAddr2 = UNLOCK_2;
        level = intLock();
        *flashPtr = SECTOR_ERASE;

        lv160OpOverDetect((void *)flashPtr, 0x2000000);
        for(i=0; i<vol.erasableBlockSize/2; i++,flashPtr++)
        {
            if(*flashPtr != 0xffff)  break;
        }
        intUnlock(level);
        if(i < vol.erasableBlockSize/2)
        {
            #ifdef DEBUG_PRINT
                DEBUG_PRINT("Debug: lv160MTDErase fail.\n");
            #endif
            return flWriteFault;
        }
    }
    printf("\Erase ok\n");
    return flOK;
}
/*-----------------------------------------------------------------------
 Procedure:      fllv160Identify ID:1
 Purpose:        MTD 读 FLASH 标识
 Input:
 Output:
 Errors:
-----------------------------------------------------------------------*/
static FLStatus fllv160Identify
(
```

```
    FLFlash vol,
    UINT32 offset
)
{

    FlashWPTR flashPtr;

    /* 这里调用map函数 */
    flashPtr = (FlashWPTR) vol.map(&vol, 0, vol.interleaving);

    *thisVars->unlockAddr1 = UNLOCK_1;
    *thisVars->unlockAddr2 = UNLOCK_2;
    *thisVars->unlockAddr1 = READ_ID;
    flashPtr = (FlashWPTR) vol.map(&vol, offset, vol.interleaving);
    vol.type = *flashPtr;
    flashPtr = (FlashWPTR)vol.map(&vol, 0, vol.interleaving) ;
    *flashPtr = READ_ARRAY;     /* 回到读状态 */

    return flOK;
}

/*-----------------------------------------------------------------
 Procedure:      lv160MTDIdentify ID:1
 Purpose:        MTD读FLASH标识[extern]
 Input:
 Output:
 Errors:
-----------------------------------------------------------------*/
FLStatus lv160MTDIdentify
(
    FLFlash vol
)
{
    FlashWPTR  baseFlashPtr;

    vol.interleaving = 1;

    flSetWindowBusWidth(vol.socket,16);/* use 16-bits */
    flSetWindowSpeed(vol.socket,120);  /* 120 ns */
    if (vol.socket->serialNo == 0)
    {
        flSetWindowSize(vol.socket, FLASH_BOOT_SIZE>>12);
```

```
        vol.chipSize = FLASH_BOOT_SIZE;
        vol.map = lv160MTDMap0;
    }
    else
    {
        printf("DD\n");
    }
    vol.mtdVars = &mtdVars[flSocketNoOf(vol.socket)];
    baseFlashPtr = (FlashWPTR)vol.map (&vol, (CardAddress)0, vol.interleaving);
    /* UNLOCK_ADDR 为字地址，赋值转换 x2 */
    thisVars->unlockAddr1 = (FlashWPTR)((long)baseFlashPtr) + UNLOCK_ADDR1;
    thisVars->unlockAddr2 = (FlashWPTR)((long)baseFlashPtr) + UNLOCK_ADDR2;

    fllv160Identify(&vol, 2);                /* get flash device ID. */
    if(vol.type! = SST39VF1601_DEID/* && (vol.type != SST39VF160_DEID) && (vol.
    type      != SST39VF3201_DEID) */)
    {
        #ifdef DEBUG_PRINT
            DEBUG_PRINT("Debug: can not identify AM29LV160 media.0x%x\n",
                        vol.type);
        #endif
        return flUnknownMedia;
    }
    else printf("Identify SST39VF1601 media.0x%x\n",vol.type);
    vol.noOfChips = 0x1;                         /* one chip. */
    vol.erasableBlockSize = 0x10000;           /* 64k bytes. */
    vol.flags |= SUSPEND_FOR_WRITE;
    /* Register our flash handlers */
    vol.write = lv160MTDWrite;
    vol.erase = lv160MTDErase;

    return flOK;
}
```

第6章

网络设备驱动程序设计

6.1 网卡设备驱动设计概述

网卡设备驱动设计在 VxWorks 开发中是极其重要的。在开发机——目标机模式中,通过网线进行下载调试是 Tornado 调试器最常用和最方便的方式。VxWorks 优秀的实时性能使其在网络通信当中具有极广泛的应用,如网络交换机、路由器的软件系统等。

6.1.1 数据交换

如图 6.1 所示,VxWorks 网络模块指出了 VxWorks 应用与网络设备驱动程序之间的各层协议。网络系统各层之间需要进行频繁的数据交换,各层遵循事先约定的协议,诸如数据交换格式等内容。数据发送过程中,网络传输协议层通过应用程序接口(如套接字)获得来自应用程序层的数据,并将其传送给网络协议层(如 IP 协议等)。网络接口层利用物理层(硬件)将数据发送到网络上,数据接收过程则是发送过程的反过程,数据的流向与发送相反。

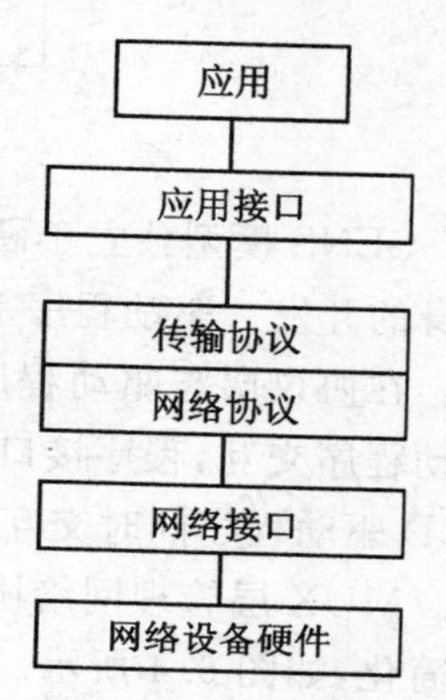

图 6.1 VxWorks 网络模块

6.1.2 网络接口驱动程序

网络数据链路层通过网络接口驱动程序,负责在两个相邻节点之间的线路上无差错地进行数据的发送和接收。传输的数据以帧为单位,每一帧数据包括一定数量的数据和一些必要的控制信息。

在数据发送时，如果接收方检测到接收数据有错误，则通知对方重发，直到该帧数据接收正确为止。每一帧的控制信息中，包括同步信息，地址信息，差错控制以及流量控制信息等。

VxWorks 支持两种类型网络设备驱动程序，VxWorks BSD 4.3 网络驱动程序和 VxWorks 可裁减的增强型网络堆栈（SENS，Scalable Enhanced NetWorks Stack）。VxWorks BSD 4.3 协议栈符合 BSD 4.3 标准，提供了网络设备驱动程序与 IP 协议的紧密结合；VxWorks SENS 协议栈提供了可替换的网络设备驱动程序，Wind River 把专为使用这个网络堆栈而编写的网络接口设备驱动程序称为增强的网络驱动程序（END，Enhanced NetWorks Drivers）。

两种不同模式的网络设备驱动程序模型的结构也不相同。网卡驱动程序在整个 VxWorks 网络接口中的角色可以从图 6.2 中看出。SENS 模型包括 3 大部分：协议驱动程序，多元接口 MUX 层，增强的网络驱动程序（END driver）。

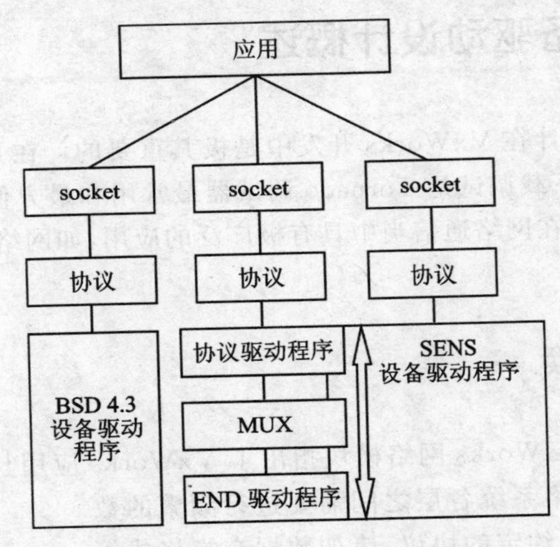

图 6.2　VxWorks 的网络系统

SENS 模型独立于硬件设备接口，开发者可以专注于驱动程序（END 驱动程序）本身的开发。驱动程序上与上层协议（如 TCP/IP）之间，SENS 模型提供了协议驱动层。在协议层与驱动程序之间，SENS 模型提供了 MUX 层。MUX 层直接与 END 驱动程序交互，程序接口提供了独立于网络协议的驱动程序接口，能够与多个独立的 END 驱动程序同时交互，如图 6.3 所示。

MUX 层管理网络协议接口和底层驱动程序之间的通信，使发送、接收数据的过程简化，如图 6.4 所示。

MUX 层作为独立的一个网络层有其自己的功能函数，但这些功能函数只是其上下两层通信的接口。网络协议层和网络驱动与 MUX 接口的调用关系如图 6.4 所示。网络协议提供下面的接口功能函数：

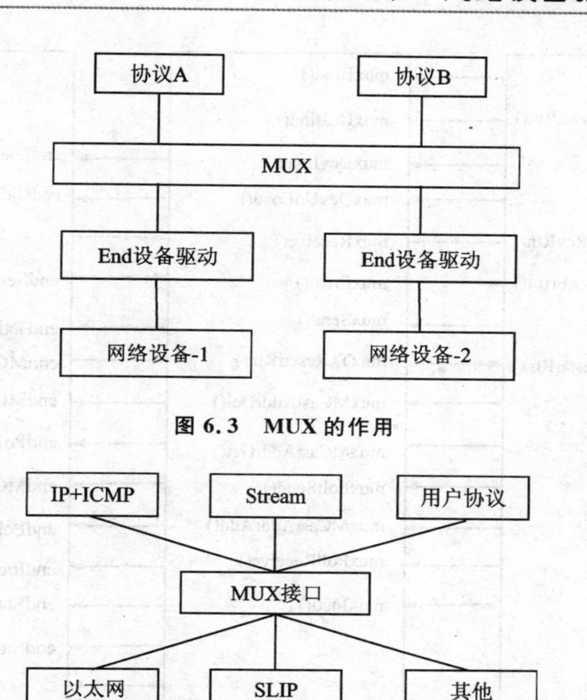

图 6.3　MUX 的作用

图 6.4　VxWork 网络协议与 MUX 接口

- stackShutdownRtn()；
- stackError()；
- stackRcvRtn()；
- stackTxRestartRtn()。

当 MUX 接口层需要与协议层相互通信时，就调用以上的功能函数。想要使网络协议层能够使用 MUX，必须至少实现以上 4 个功能函数。MUX 则实现 muxBind()、muxUnBind()、muxDevload()等等。网络协议层和网络驱动接口都要根据各自的需要使用 MUX 接入点，如图 6.5 所示。由于 MUX 是由系统提供的，不需要在应用时再进行额外的编码工作。只要在使用时，填入正确的参数即可。

例如在 VxWorks 中，muxDevLoad 是这样定义的：

```
END_OBJ * muxDevLoad
(
    int unit,  /* 设备号码 */
    END_OBJ * ( * endLoad)(char * ,void * ),  /* 调用设备函数 */
    char * pInitString,  /* 初始化字符串 */
    BOOL loaning,  /* 存储标识 */
    void * pBSP  /* 调用 BSP 功能的函数 */
)
```

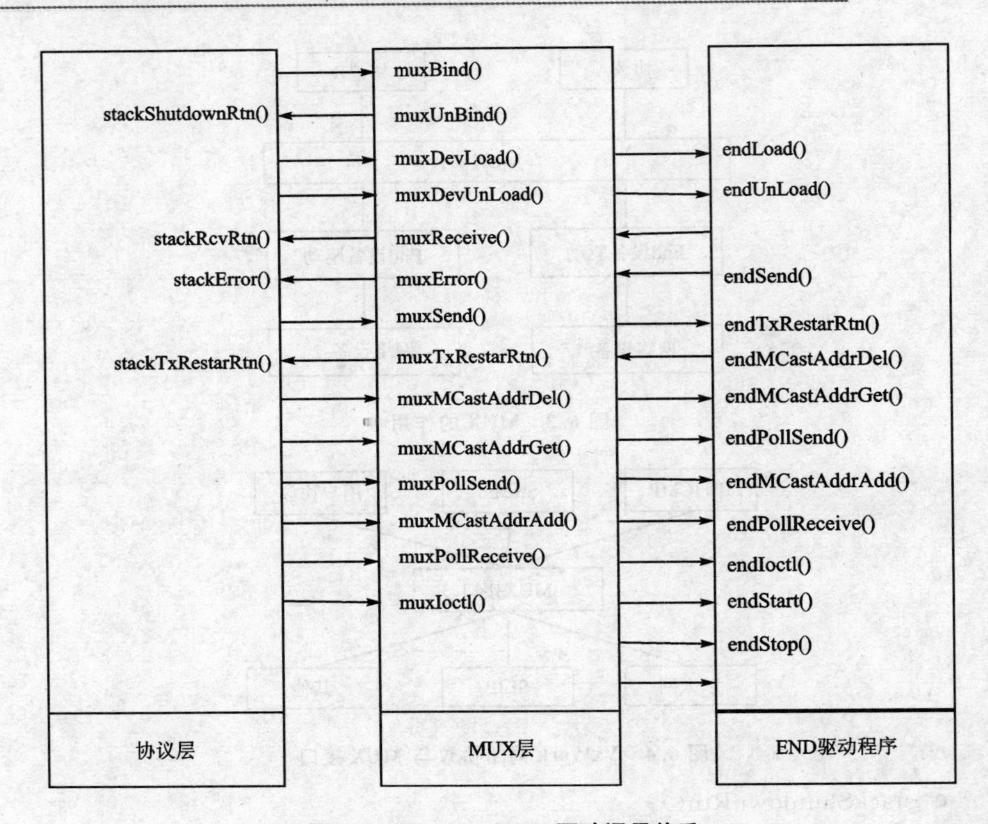

图 6.5 VxWorks MUX 驱动调用关系

其他功能函数在 muxLib. h 文件中有详细定义。网络接口的驱动程序要完成 endLoad()、endUnload()、endSend()等功能函数。MUX 使用这些功能函数来与网络驱动程序通信。当编写或加载一个使用 MUX 的网络驱动程序时，必须实现图 6.5 中 END 的所有功能。这些功能函数都是针对具体的网络接口，即每一个网络驱动程序中都要有这些功能函数。

6.2 END 设备驱动程序装载过程

END 设备驱动程序装载过程通常分 3 个步骤完成，即指定 END 设备、装载 END 设备和启动 END 设备。

6.2.1 系统 END 设备选定

网络设备驱动程序安装时，要修改相应的 BSP 程序。BSP 中的 END 设备驱动入口表 END_TBL_ENTRY 的结构如下：

```
typedef struct end_tbl_entry
{
    int unit;                                          /* 设备单元 */
    END_OBJ * ( * endLoadFunc)  (char * , void * )    /* 设置装载函数 */
    char *  endLoadString;                             /* 装载字符串,作为装载函数的输入参数 */
    BOOL endLoad;                                      /* 是否借出缓冲区 */
    void *  pBsp;                                      /* BSP 私有信息 */
    BOOL processed;                                    /* 此设备单元是否已经做处理 */
} END_TBL_ENTRY;
```

在 configNeth 文件中定义的 END_TBL_ENTRY 结构的数组 endDevTbl[],原型结构如下:

```
#define LOAD_FUNC_0      sysEtherEndLoad
#define LOAD_STRING_0      "0x18000000:0x0"
END_TBL_ENTRY endDevTbl [] =
{
   { 0, LOAD_FUNC_0, LOAD_STRING_0, 1, BSP_0, FALSE},
   { 0, END_TBL_END, NULL, 0, NULL, FALSE},
};
```

上面的数组描述了系统中网络设备的装载函数入口点及其相关参数。在网络设备表 endDevTbl[]添加了网络设备程序的装载函数入口点,如上面程序所示,入口函数为 sysEtherEndLoad(),还有 string 的参数等。然后在 config. h 中加入"#define INCLUDE_END",系统将初始化 MUX,通过 MUX 装载网络设备驱动程序。

通常为了保证 VxWorks 网络驱动程序的编程规范和系统的移植性、兼容性、可裁剪性等要求,在 configNet. h 中定义了系统所支持的网络设备,文件如下:

```
#ifndef INCconfigNeth
#define INCconfigNeth

#include "vxWorks.h"
#include "end.h"

IMPORT END_OBJ * sysEtherEndLoad(char * ,void * );

#define LOAD_FUNC_0       sysEtherEndLoad            /* 网卡驱动程序装载函数 */
#define LOAD_STRING_0       "0x18000000:0x0"          /* 驱动程序装载参数 */
#define BSP_0            NULL

END_TBL_ENTRY endDevTbl [] =
{
   { 0, LOAD_FUNC_0, LOAD_STRING_0, 1, BSP_0, FALSE},
```

```
    { 0, END_TBL_END, NULL, 0, NULL, FALSE},
};
#endif /* INCconfigNeth */
```

6.2.2 装载及启动 END 设备

在进行网络设备的开发和测试时，需要了解的相关内容如下：

- 实现装载和启动驱动程序的任务；
- 处理驱动程序注册中断程序的任务；
- 调用驱动程序处理数据包的任务。

1. 装载及启动 END 设备的系统函数

实现本功能，需要以下系统函数提供相应的功能：

(1) muxDevLoad()装载函数

该函数是实现装载指定设备的驱动程序。要装载 END 设备，系统一定要调用该函数，函数原型如下：

```
END_OBJ * muxDevLoad()
{
    …
    bzero((char *) devName, END_NAME_MAX);
    if (endLoad((char *) devName, NULL) != 0)
    goto muxLoadErr;
    if (endFindByName((char *) devName, unit) != NULL)
        goto muxLoadErr;
        …
    sprintf((char *) initString, "%d:%s", unit, pInitSpring);
    pNew = (END_OBJ *) endLoad((char *) initString, pBSP);
    if (pNew == NULL) goto muxLoadErr;
        …
    return (pNew);
    muxLoadErr;
        …
    return (NULL);
}
```

(2) muxDevStart()启动函数

该函数为启动 END 设备函数，系统调用示例如下：

```
for (count = 0, pDevTbl = endDevTbl; pDevTbl->endLoadFunc != END_TBL_END;
    pDevTbl++, count++)
```

```
{
    /* 查询 WDB 是否安装该设备 */
    if (! pDevTbl->processed)
    {
        /* 安装设备驱动程序 */
        pCookie = muxDevLoad( pDevTbl->unit,
                              pDevTbl->endLoadFunc,
                              pDevTbl->endLoadString,
                              pDevTbl->endLoad,
                              pDevTbl->pBsp);
        if (pCookie == NULL)
        {
            printf("muxDevLoad failed for device entry %d! \n", count);
        }
        else
        {
            pDevTbl->processed = TRUE;
            /* 启动 END 设备 */
            if (muxDevStart(pCookie) == ERROR)
            {
                printf("muxDevStart failed for device entry %d! \n", count);
            }
        }
    }
}
```

(3) muxBind()绑定函数

该函数实现将协议绑定到指定的 END 设备上。系统调用 ipAttach()函数，该函数调用 muxBind()函数，绑定协议堆栈到 MUX 上的一个指定的网络接口。当一个网络接口被关闭时，ipDetach()函数将释放网络接口所关联的 TCP/IP 堆栈模块。

2. 装载及启动 END 设备驱动程序过程

下面将介绍 END 驱动程序的执行过程。系统启动时，VxWorks 创建一个系统任务 tUsrRoot，该任务会执行下列操作：

- 初始化网络任务的工作队列；
- 创建任务 tNetTask 来处理网络任务工作队列中的项目；
- 调用 muxDevLoad()加载网络驱动程序；
- 调用 muxDevStart()开始驱动程序。

tUsrRoot 任务通过查找 endTbl 并将表中的相应参数传给 muxDevLoad()函数，muxDevLoad()函数根据传入的参数调用驱动程序入口点 endLoad()函数，end-

Load()函数必须完成硬件初始化，同时设置 END_OBJ 的大部分成员。系统在执行完 muxDevLoad()函数后，就完成了 END 设备加载。tUsrRoot 接下来通过调用 muxDevStart()函数来启动驱动程序，首先调用 sysIntConnect()函数，将 ISR 挂接在某个中断向量上。至此，MUX 的初始化以及 END 驱动程序初始化工作已经完成。要想使用网络设备，还需要将协议绑定到指定的 END 设备上，需要调用 muxBind()函数来实现。

网络设备驱动程序的安装过程如图 6.6 所示。

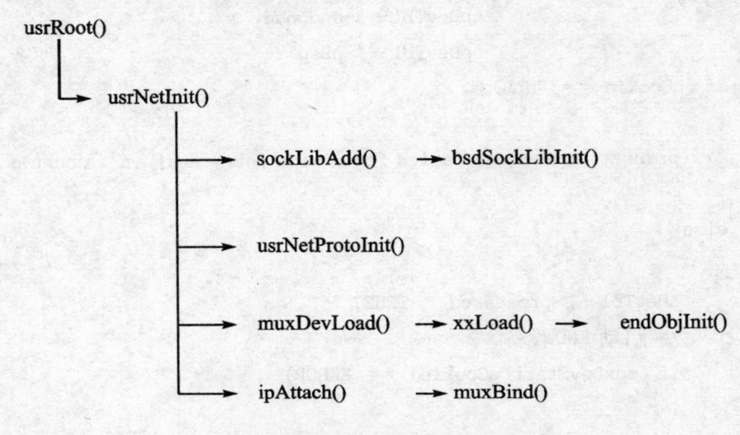

图 6.6　网络初始化顺序

系统通过在 usrRoot()函数中调用 usrNetInit()函数完成 MUX 的初始化、装载网络设备表 endDevTbl[]中描述的所有设备并将 IP 协议绑定到网络引导设备等工作。

6.3　DM9000 网络芯片

本书所编写的网络驱动以 MagicARM2410 平台上的 DM9000 网络芯片为对象进行开发和设计。

DM9000 是 Davicom 公司设计生产的一款高度集成的低功耗的快速以太网 MAC 控制器，带有一个通用处理器接口、一个 10/100M PHY 和 4K 双字节 SRAM 缓存数据区。DM9000 芯片为低功耗处理器而设计，工作电压 3.3 V，最高可输入 5 V的端口电压。DM9000 提供了一个 MII 接口，用于连接 HPNA 设备或其他支持 MII 接口的收发器。DM9000 物理协议层接口完全支持 10 Mbps 下使用 3 类、4 类、5 类非屏蔽双绞线，以及 100 Mbps 下使用 5 类非屏蔽双绞线，完全符合 IEEE 802.3u规格。DM9000 具有自动协调功能，可以自动完成配置，最大限度地适合线路带宽，支持 IEEE 802.3x 全双工流量控制。DM9000 编程简单，用户可以较为容易地移植不同系统下的端口驱动程序。

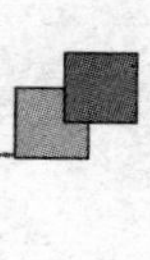

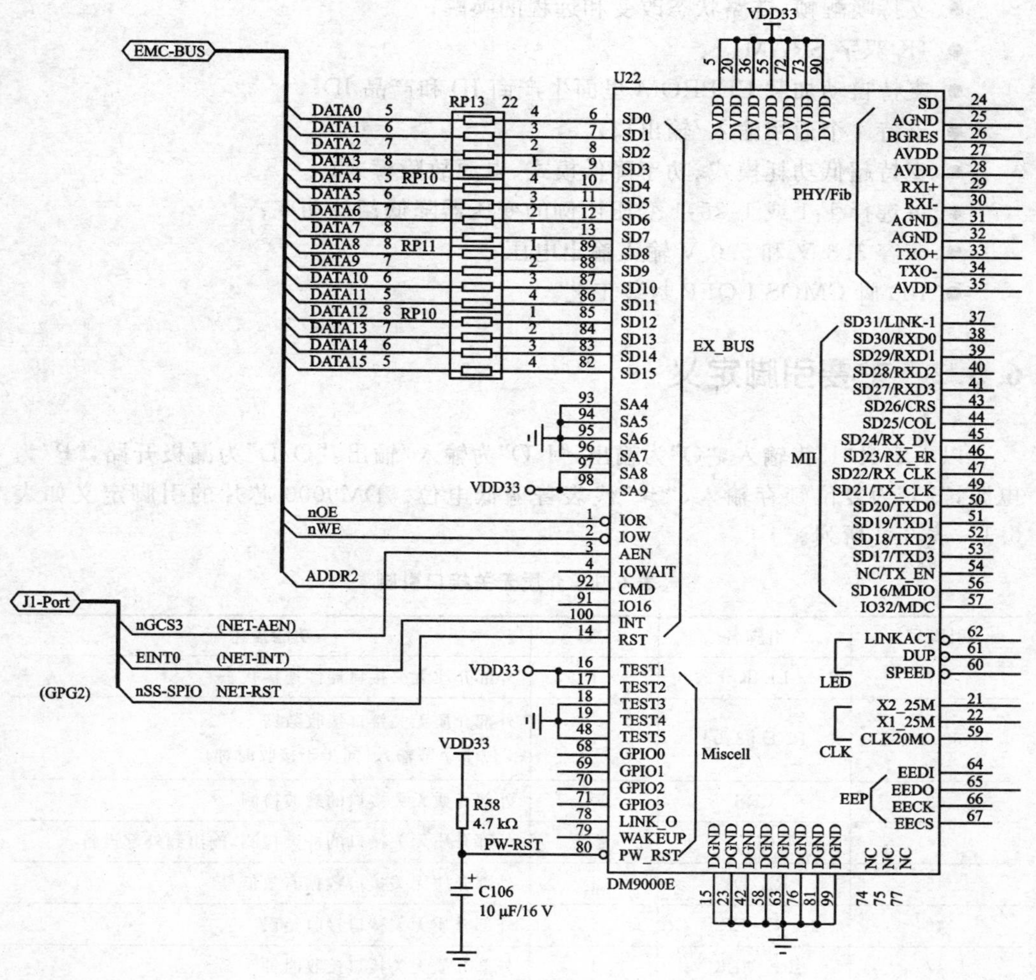

图 6.7　MagicARM2410 实验箱 DM9000 电路原理图

6.3.1　DM9000 主要性能

DM9000 芯片的性能如下：

- 支持处理器读写内部存储器的数据操作，命令包括 BYTE/WORD/DOUBLE WORD 3 种长度；
- 集成 10/100M 自适应收发器；
- 支持介质无关接口；
- 支持背压模式半双工流量控制模式；
- IEEE 802.3x 流量控制的全双工模式；

- 支持唤醒帧，链路状态改变和远程的唤醒；
- 4K 双字 SRAM；
- 支持自动加载 EEPROM 里面生产商 ID 和产品 ID；
- 支持 4 个通用输入/输出口；
- 支持超低功耗模式、功率降低模式、电源故障模式；
- 可选择 1:1 或 1.25:1 变压比例的变压器降低额外功率；
- 兼容 3.3 V 和 5.0 V 输入输出电压；
- 100 脚 CMOS LQFP 封装工艺。

6.3.2 主要引脚定义

以下定义“I”为输入，“O”为输出，“I/O”为输入/输出，“O/D”为漏极开路，“P”为电源，“LI”为复位锁存输入，“#”代表普遍低电位。DM9000 芯片的引脚定义如表 6.1～表 6.8 所列。

表 6.1 介质无关接口引脚

引脚号	引脚名	I/O	功能描述
37	LINK_I	I	外部介质无关接口器件连接状态
38、39、40、41	RXD [3:0]	I	外部介质无关接口接收数据 4 位半字节输入(同步于接收时钟)
43	CRS	I/O	外部介质无关接口的载波检测
44	COL	I/O	外部介质无关接口的冲突检测，输出到外部设备
45	RX_DV	I	外部介质无关接口数据有效信号
46	RX_ER	I	外部介质无关接口接收错误
47	RX_CLK	I	外部介质无关接口接收时钟
49	TX_CLK	I/O	外部介质无关接口发送时钟
50～53	TXD[3:0]	O	外部介质无关接口发送数据低 4 位输出 TXD[2:0]决定内部存储空间基址：(TXD [2:0]) * 10H + 300H
54	MDIO	I/O	外部介质无关接口串行数据通信
57	MDC	O	外部介质无关串行数据通信口时钟，且与中断引脚有关 该引脚高电平时候，中断引脚低电平有效；否则高有效

注意：以上介质无关端口内部都自带 60 kΩ 的下拉电阻。

表 6.2　处理器接口引脚

引脚号	引脚名	I/O	功能描述
1	IOR#	I	处理器读命令 低电平有效,极性能够被 EEPROM 修改,详细请参考对 EEPROM 内容的描述
2	IOW#	I	处理器写命令 低电平有效,同样能修改极性
3	AEN#	I	芯片选择,低电平有效
4	IOWAIT	O	处理器命令就绪 当上一指令没有结束,该引脚电平拉低表示当前指令需要等待
14	RST	I	硬件复位信号,高电平有效复位
1～6 82～89	SD0～15	I/O	015 位的数据地址复用总线,由 CMD 引脚决定当期访问类型
93～98	SA4～9	I	地址线 4～9;仅作芯片选择信号 (SA4～9:TXD0～2,011)被选中
92	CMD	I	访问类型 高电平是访问数据端口;低电平是访问地址端口
91	IO16	O	字命令标志,默认低电平有效 当访问外部数据存储器是字或双字宽度时,被置位
100	INT	O	中断请求信号 高电平有效,极性能被修改
37～53 56	SD31～16	I/O	双字模式,高 16 位数据引脚
57	IO32	O	双字命令标志,默认低电平有效

注意:以上引脚除了 SD8、SD9 和 IO16 3 个,其他内部都自带 60 kΩ 的下拉电阻。

表 6.3　EEPROM 引脚

引脚号	引脚名	I/O	功能描述
64	EEDI	I	数据输入引脚
65	EEDO	I/O	EEPROM 数据引脚 与 WAKEUP 引脚一起定义访问数据存储器的总线宽度 WAKEUP　EEDO　总线宽度 0　0　16 位 0　1　32 位 1　0　8 位 1　1　未定义

续表 6.3

引脚号	引脚名	I/O	功能描述
66	EECK	I	时钟信号
67	EECS	I/O	片选 也做 LED 模式选择引脚 高电平时,LED 模式 1,否则模式 0

注意:EECS、EECK、EEDO 引脚内部自带 60 kΩ 下拉电阻。

表 6.4 时钟引脚

引脚号	引脚名	I/O	功能描述
21	X2_25M	O	25 MHz 晶振输出
22	X1_25M	I	25 MHz 晶振输入
59	CLK20MO	O	20 MHz 晶振再生输出给外部介质无关设备,自带 60 kΩ 下拉电阻

表 6.5 LED 引脚

引脚号	引脚名	I/O	功能描述
60	SPEED100#	O	低电平指示 100M 带宽指示,高电平指示 10M 带宽
61	DUP#	O	全双工指示 LED LED 模式 0 时,低电平显示工作在 10M 带宽,或在 100M 带宽浮动
62	LINK&ACT#	O	连接 LED,在模式 0 时,只作物理层的载波监听检测连接状态

表 6.6 10/100 物理层与光纤接口

引脚号	引脚名	I/O	功能描述
24	SD	I	光纤信号检测 PECL 电平信号,显示光纤接收是否有效
25	DGGND	P	带隙地信号线
26	BGRES	I/O	带隙引脚
27	AVDD	P	带隙与电源保护环
28	AVDD	P	接收端口电源
29	RXI+	I	物理层接收端的正极
30	RXI−	I	物理层接收端的负极
31	AGND	P	接收端口地
32	AGND	P	发送端口地
33	TXO+	O	物理层发送端口正极

续表 6.6

引脚号	引脚名	I/O	功能描述
34	TXO−	O	发送端口负极
35	AVDD	P	物理层发送端口负极

表 6.7 各种其他功能引脚

引脚号	引脚名	I/O	功能描述
16～19	TEST1～4	I	工作模式 Test1～4(1,1,0,0)正常工作状态
48	TEST5	I	必须接地
68～69	GPIO0～3	I/O	通用 I/O 端口 通用端口控制寄存器和通用端口寄存器能编程该系列引脚 GPIO0 默认输出为高来关闭物理层和其他外部介质无关器件 GPIO1～3 默认为输入引脚
78	LINK_O	O	电缆连接状态显示输出,高电平有效
79	WAKEUP	O	流出一个唤醒信号当唤醒事件发生 内置 60 kΩ 的下拉电阻
80	PW_RST#	I	上电复位 低电平激活 DM9000 的重新初始化,当该引脚测试到电平变化 5 μs 后初始化
74,75,77	NC		无用

表 6.8 电源引脚

引脚号	引脚名	I/O	功能描述
5,20,36,55,72,90,73	DVDD	P	数字电源
15,23,42,58,63,81,99,76	DGND	P	数字地

6.3.3 DM9000 主要寄存器

DM9000 的主要寄存器包括数据读写端口寄存器、DM9000 状态寄存器和 DM9000 状态控制寄存器。其中,数据读写端口寄存器主要是用于发送和接收数据包时,对 DM9000 芯片内部的发送数据缓冲区和接收数据缓冲区进行读写时使用。DM9000 控制寄存器用来控制 DM9000 芯片来完成相应的动作,其中有中断控制等寄存器。这些寄存器用来完成所有有关 DM9000 芯片的操作。DM9000 状态寄存器用来反映 DM9000 的连接状态，以及任务执行过程中所产生的异常信息。表 6.9 中列出了 DM9000 的相关寄存器。

表 6.9 DM9000 主要寄存器列表

寄存器名	功能描述	片内地址	默认值
NCR	网络控制寄存器	00H	00H
NSR	网络状态寄存器	01H	00H
TCR	TX 控制寄存器	02H	00H
TSR I	TX 状态寄存器 I	03H	00H
TSR II	TX 状态寄存器 II	04H	00H
RCR	RX 控制寄存器	05H	00H
RSR	RX 状态寄存器	06H	00H
ROCR	接收溢出计数器	07H	00H
FCTR	流量控制寄存器	09H	38H
FCR	RX 流量控制寄存器	0AH	00H
EPCR	EEPROM & PHY 控制寄存器	0BH	00H
EPAR	EEPROM & PHY 地址寄存器	0CH	40H
EPDRL	EEPROM & PHY 低字节寄存器	0DH	XXH
EPDRH	EEPROM & PHY 高字节寄存器	0EH	XXH
WCR	唤醒寄存器	0FH	00H
PAR	IP 地址寄存器	10H～15H	由 EE 数据决定
MAR	多播地址寄存器	16H～1DH	XXH
GPCR	通用控制寄存器	1EH	01H
GPR	通用寄存器	1FH	XXH
TRPAL	TX SRAM 读取当前地址低字节	22H	00H
TRPAH	TX SRAM 读取当前地址高字节	23H	00H
RWPAL	RX SRAM 写地址低字节	24H	04H
RWPAH	RX SRAM 写地址高字节	25H	0CH
VID	设备 ID	28H～29H	0A46H
PID	产品 ID	2AH～2BH	9000H
CHIPR	芯片版本	2CH	00H
SMCR	特殊模式控制寄存器	2FH	00H
MRCMDX	数据读取寄存器(指针不增)	F0H	XXH
MRCMD	数据读取寄存器(指针自增)	F2H	XXH
MRRL	读取地址寄存器(低字节)	F4H	00H

续表 9.9

寄存器名	功能描述	片内地址	默认值
MRRH	读取地址寄存器(高字节)	F5H	00H
MWCMDX	缓冲内存写入(地址不自增)	F6H	XXH
MWCMD	缓冲内存写入(地址自增)	F8H	XXH
MWRL	写操作地址低字节	FAH	00H
MWRH	写操作地址高字节	FBH	00H
TXPLL	TX 包长度低字节寄存器	FCH	XXH
TXPLH	TX 包长度高字节寄存器	FDH	XXH
ISR	中断状态寄存器	FEH	00H
IMR	中断使能寄存器	FFH	00H

6.3.4 DM9000 芯片复位和初始化

复位与初始化功能是芯片必须要实现的一个基本功能，在系统启用了网络芯片之后，就要对网络芯片进行基本的配置，设置网络芯片的寄存器，使网络芯片处于工作状态。如果系统检测到网络芯片处于非正常状态时，需要复位网络芯片重新进行配置，以便使网络芯片从未知状态恢复到正常工作状态，相应的设计代码如下：

```
static void dmfe_reset_dm9000( END_DEVICE * dev )
{
    UCHAR tmp;

    /* 开启 DM9000 芯片，设置为寄存器 GPR 寄存器为 0 来打开芯片 */
    DM9000_OUT_REG(0x1f,0x00);   /* GPR */
    uDelay(20);

    /* soft reset */
    DM9000_OUT_REG( 0x00, 0x03 );
    uDelay(20);
    DM9000_OUT_REG( 0x00, 0x00 );
    uDelay(20);
    DM9000_OUT_REG( 0x00, 0x03 );
    uDelay(20);
    DM9000_OUT_REG( 0x00, 0x00 );
    uDelay(20);

    /* set the internal PHY power - on, GPIOs normal, and wait 20 μs */
```

```
    DM9000_OUT_REG(0x1f,0x01);   /* GPR */
    DM9000_OUT_REG(0x1f,0x00);   /* GPR */
    uDelay(1000);
    uDelay(1000);
    uDelay(1000);
    uDelay(1000);

    DM9000_OUT_REG( 0x02, 0x00 );   /* TX clear */
    DM9000_OUT_REG( 0xff, DM9000_REGFF );   /* Enable TX/RX interrupt mask */
    DM9000_OUT_REG( 0x05, DM9000_REG05 );   /* RX enable */

    /* I/O mode */
    DM9000_IN_REG( 0xfe, tmp );
    dev->io_mode = (tmp & 0xff) >> 6;      /* ISR bit7:6 keeps I/O mode 16bit */

    /* Set PHY Mode */
    set_PHY_mode(dev);
}
```

- 首先开启 DM9000 芯片,设置寄存器 GPR 为 0 来打开芯片;
- 根据 DM9000 设计要求,进行两次软启动,根据芯片的设计要求,要使芯片达到工作状态在上电之后就要对芯片进行两次软启动,软启动是通过设置 DM9000 寄存器 NCR 的 bit[2:0]=0b011(至少 20 μs),设置 NCR(REG_00) bit[2:0]=0b000 来实现的,同样的操作要进行两次;
- 清除 Tx Status 寄存器;
- 设置 IMR(REG_FF)寄存器 PRM bit[0]/PTM bit[1]开启 TX/RX 中断;
- 设置 RCR 寄存器来使能 RX。RX 功能函数的使能是靠设置 RX 控制寄存器(REG_05) RXEN bit[0]=1。

上电后,DM9000 芯片完成上述操作进入到工作状态。如果因为异常而导致芯片重启时,需要再次执行相同的操作过程以确保 DM9000 芯片恢复到正常工作状态。

系统在初始化 DM9000 芯片时,需要设置芯片中的各个控制寄存器。例如需要设置 RX 任务控制寄存器(REG_05H),其中各个标志位的功能描述如表 6.10 所列。RX 控制寄存器是接收任务所要用到的最重要的寄存器,其中包含了设置接收使能的 RX 使能位(bit0 RXEN),系统通过设置这一标志位来实现对接收任务的控制,在禁用芯片的时候要将 RX 使能位清除。bit[4:1]是用来设置对接收包的限制,当有不符合设置的包到达时将被自动丢弃,不产生系统中断。

表 6.10　DM9000 RX 控制寄存器(05H)

位	名称	默认值	功能描述
7	RESERVED	0,RO	保留位
6	WTDIS	0,RW	看门狗禁用位
5	DIS_LONG	0,RW	抛弃大于 1 522 Byte
4	DIS_CRC	0,RW	抛弃
3	ALL	0,RW	忽略所有多播数据包
2	RUNT	0,RW	忽略所有不完整数据包
1	PRMSC	0,RW	混合模式
0	RXEN	0,RW	RX

6.4　网络设备与系统数据交换

网络设备与系统的数据交换实际上就是数据的发送和接收问题,本章内容主要讨论中断方式下的网络数据发送和接收方式。

6.4.1　中断处理原理

系统在网络接口触发中断时,则直接调用注册在系统的中断服务程序。同所有的中断处理程序一样,网络驱动程序的中断服务程序只处理那些占用 CPU 最少的任务,比如状态改变。如果需要处理的任务比较耗时,则将它们排列到网络任务的工作队列中去。

在处理排列任务级的包接收处理工作时,网络中断服务处理程序需要调用 netJobAdd()函数。该函数的原型如下:

```
STATUS  netJobAdd( FUNCPTR routine, int par1, int par2, int par3, int par4, int par5);
```

在调用该函数时,函数参数 routine 表示任务级处理函数的指针;后 5 个参数在将来调用 routine 时传给它的入参。函数 netJobAdd()将把这个函数指针及其参数投递到系统任务 tNetTask 的工作队列中,VxWorks 通过 tNetTask 处理。

tNetTask 调用队列中有如下程序:

- 包接收程序:将接收到的数据包上传到网络缓冲区的堆栈中,然后上传给 MUX。
- 释放所有发送帧程序:程序调用 netClFree()函数释放发送缓冲区中的所有已经发送的数据帧。

netJobAdd()函数把指定的任务处理程序添加到 tNetTask 的工作队列中,自动产生适当的信号量激活 tNetTask 任务；激活后，tNetTask 从它的工作队列中列出函数和分配的参数，然后开始执行队列中的函数；队列中只要还有需要处理的函数排队，tNetTask 就会一直工作，直到函数队列被清空，所有的数据包都上传到 MUX 后，tNetTask 就会进入到睡眠状态。

6.4.2 中断服务程序

中断函数 xxInt()在网络设备中完成数据包的发送和接收，格式如下：

```
void  xxInt( xxEND_DEVICE * pDrvCtrl);
```

xxInt()函数完成几个主要功能：读取网络设备的状态，完成数据接收或者发送的工作（这部分是通过任务来实现的），流程图如图 6.8 所示。

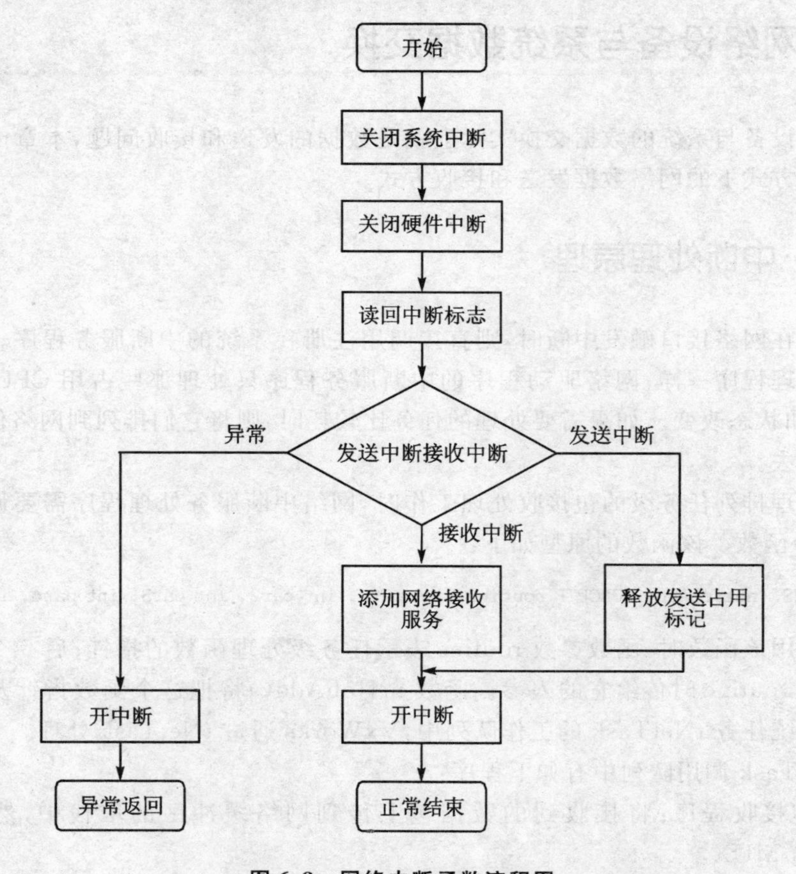

图 6.8　网络中断函数流程图

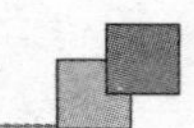

6.4.3　驱动程序与协议层共享缓冲区

网络设备使用环形缓冲描述符，在装载设备驱动程序的时候，为装载函数分配了接收和发送缓冲区。在接收数据包时，首先将数据存放在设备接收缓冲区中，然后将指向接收缓冲区的指针传递给 MUX。在发送数据时，首先将发送缓冲区中的数据包复制到设备发送缓冲区中，然后将缓冲区放入发送队列中，启动发送器，数据包发送后其所占用的缓冲区释放给内存池。

如图 6.9 所示，VxWorks MUX 层的数据包采用的是 mBlk－clBlk－cluster 结构来进行管理。用户主要通过 mBlk 结构来访问和传递由 netPoolnit()函数建立的内存池中存放的数据。mBlk 仅能访问数据，因此网络层不需要复制数据就能实现数据交换；mBlk 结构还具有链接性，用户可以通过 mBlk 的链头传递任意大的数据。

簇中的数据包，如图 6.9 所示。发送时，网卡发送模块处理这样的结构；接收数据时，网卡仍然需要将数据通过这样的结构传递给上层协议。不过，这一切 VxWorks都有相应的规范和函数。另外，mBlk 和 clBlk 可以由 netpool 结构管理，VxWorks 提供一系列接口函数。

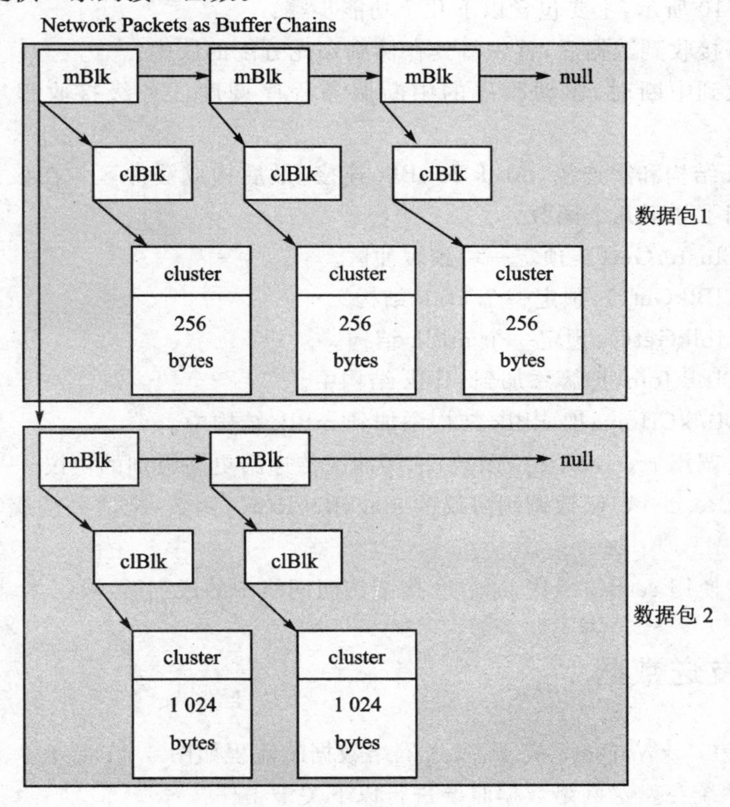

图 6.9　网络数据链

6.4.4 接收数据

实验箱在 VxWork 系统中，接收数据的流程如图 6.10 所示。

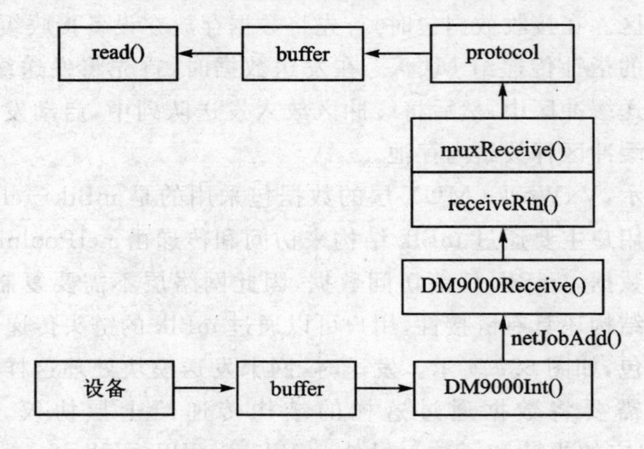

图 6.10 数据接收流程图

如图 6.10 所示，主要包含以下几个功能步骤：

① 设备接收到数据后，首先将其存储到预先分配的簇中。

② 接收到中断时，驱动程序的中断服务程序调度任务级接收程序进行如下操作：

- cBlk 结构和簇连接，mBlk 和 cBlk 连接，最后构成缓冲区。在该过程中包括调用了如下几个函数：

  netClusterGet()：预定一块簇缓冲区。

  netClBlkGet()：预定一个 cBlk 结构。

  netMblkGet()：预定一个 mBlk 结构。

  netClBlkJoin：把簇添加到 clBlk 结构中。

  netMblkClJoin：把 clBlk 结构添加到 mBlk 结构中。

- 通过调用 receiveRtn()函数，把缓冲区传递给更高级别的协议。

③ muxReceive()函数调用协议的 stackRcvRtn()函数，把成列的缓冲区传递给应用。

④ 用户使用 read()函数，通过套接字访问网络中的成列缓冲区。

6.4.5 发送数据

实验箱在 VxWorks 系统下，发送网络数据的流程如图 6.11 所示。

同样，系统在发送网络数据时要进行以下关键工作：

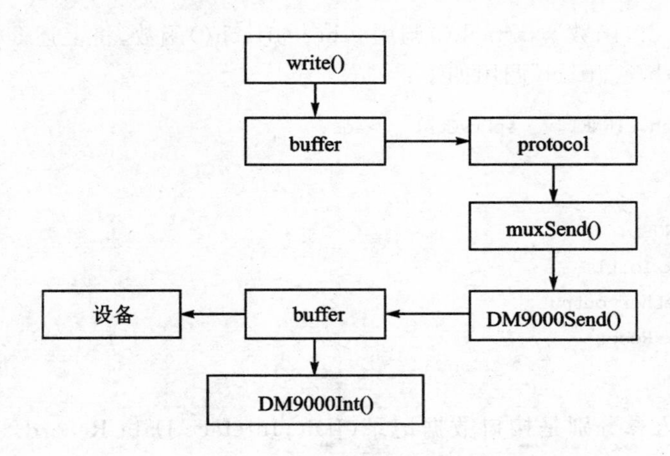

图 6.11 数据发送流程图

① 用户调用 wirte()函数，通过套接字访问网络；

② 网络协议将需要发送的数据复制到网络缓冲区去，然后调用驱动程序的发送程序；

③ 协议驱动程序调用 muxSend()启动发送循环；

④ muxSend()通过调用 send 回调函数，把缓冲区传递给 END；

⑤ 数据发送程序将数据复制到设备缓冲区中，然后将其放置在设备的发送队列中；

⑥ 当产生发送中断时，中断服务程序调度程序丢弃发送完成的数据包，彻底清除发送队列。

6.5 网络程序编写

在 VxWorks 中，网卡驱动程序分为 END(Enhanced Network Driver)和 BSD 两种，有一部分是基于 BSD Unix 4.3 版本的，全部定义在一个全局例程中，即 xxattach 子程序。xxattach 子程序包含 5 个函数指针，如表 6.11 所列，都被映射到 ifnet 结构中，并且在 IP 协议层的任何地方调用。

表 6.11 BSD 网络驱动程序的例程列表

驱动程序指定函数	函数指针	功 能
xxInit()	if_init	初始化接口
xxOutput()	if_output	对要传输的输出分组进行排队
xxIoctl()	if_ioctl	处理 I/O 控制命令
xxReset()	if_reset	复位接口设备
xxWatchdog	if_watchdog(optional)	周期性接口例程

驱动程序入口函数 xxattach()调用 ether_attach()函数，将上述 5 个函数映射到 ifnet 结构中，ether_attach 调用如下：

```
ether_attach((IFNET * ) &pDrvCtrl->idr,
unit,
"xx",
(FUNCPTR) NULL,
(FUNCPTR) xxIoctl,
(FUNCPTR) ether_output,
(FUNCPTR) xxReset
);
```

该函数的入参分别是接口数据记录（IDR，Inteface Data Record）、设备编号、设备名、以及相关驱动程序的函数指针。其中，第一个函数指针是 init()例程，该例程可以设置为 NULL；第二个函数指针是 ioctl()接口，允许上层来控制设备状态；第三个函数指针指向实现将数据包发送到物理层功能的函数；第四个函数指针指向复位函数，用于 TCP/IP 堆栈复位的需求。

```
pDrvCtrl->idr.ac_if.if_start = (FUNCPTR)xxTxStartup;
```

该语句完成的功能是添加数据传输例程到 IDR，在 ether_output()例程调用后，传输一开始例程就被 TCP/IP 协议堆栈调用。

END 驱动程序基于 MUX 模式，是目前在 VxWorks 操作系统上应用最广泛的一种网络驱动程序。该模式下，网络驱动程序分为协议组件和硬件组件。MUX 数据链路层和网络层之间的接口用来管理网络协议接口和底层硬件接口之间的交互，将硬件从网络协议的细节中隔离出来，摒弃了以往采用输入钩子例程来处理收/发数据包的方法。

以下是网络驱动所涉及的相关 END 函数说明。

（1）加载设备函数 endLoad()。

函数 endLoad()是每个网络驱动程序的初始化入口点，该函数的参数由 tUserRoot 任务在调用 muxDevLoad()时传入，muxDevLoad()进而使用该参数调用 endLoad()。该函数原型如下：

```
END-OBJ   * endLoad(char * initString);
```

函数 endLoad()需要处理所有设备专有的初始化，设置 END－OBJ 结构的大部分成员的初始值；该结构由驱动程序和 MUX 共同管理，是二者联系的纽带。函数 endLoad()执行完成后返回一个指向已被初始化的 END－OBJ 结构的指针；如果有错误发生，返回 ERROR。

（2）卸载设备函数 endUnload()。

函数 endUnload()清除所有的局部数据结构，但是不通知系统协议设备已卸载。

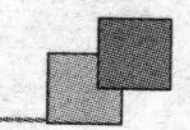

系统在调用 endUnload()前,MUX 会发送一个关闭通知给每个依附于该设备的协议。该函数原型如下:

```
void endUnload(void * pCookie);
```

其中参数 pCookie 传递一个由 endLoad()返回的指向 END-OBJ 结构的指针,应该在 endUnload()中释放与它关联的内存。

(3) 设备控制接口函数 endIoctl()。

函数 endIoctl()处理所有关于设备状态改变的请求,如启动、关闭、打开混杂模式等,该函数原型如下:

```
STATUS endIoctl(void * pCookie,int cmd,caddrt data);
```

函数执行完成后返回 OK,当有错误发生时,则返回 ERROR。参数 pCookie 传递 1 个由 endLoad()返回的指向 END - OBJ 结构的指针,参数 cmd 表示命令码,函数 endIoctl()必须对可能传入的每个命令做出相应的响应;参数 data 传递处理某些 cmd 时需要用到的数据或指针。

(4) 发送数据给设备函数 endSend()。

系统中有数据发送给设备时,MUX 将调用该函数,该函数的原型为:

```
STATUS endSend(void * pCookie,M - BLK - ID pMblk);
```

参数 pCookie 的涵义如前所述;pMblk 指向一个叫做 mBlk 的结构,该结构中包含系统要发送的数据。如果数据包因为资源缺少等原因没有立刻被传送,endSend()必须返回 END - ERROR - BLOCK,此时并不需要从 mBlk 链中释放数据包;但当有错误发生时,数据包则需要从 mBlk 链中释放。

(5) 启动设备函数 endStart()。

函数 endStart()的作用是将设备设置到活动状态,通常它应该调用 sysIntConnect()注册一个中断服务程序(ISR),函数原型如下:

```
STATUS endStart(void pCookie);
```

其中,参数 pCookie 会传递给 ISR,ISR 将它传递给一个负责数据包接收的任务,由数据包接收任务去完成数据的读取和转移工作;中断服务程序只是通知数据包接收任务进行数据包接收,并不进行数据操作,这样可以减少中断处理时间,提高系统的实时性。

(6) 停止设备函数 endStop()。

该函数的作用是使设备处于非活动状态,但并不卸载它。在这个函数里要将系统中注册到设备的中断处理函数的中断向量进行注销,并且关闭硬件的中断响应和任务处理,但是在函数中并不释放所占用的系统资源。

(7) 轮询发送函数 endPollSend()。

函数 endPollSend()首先检查设备是否已被设置成轮询模式(通常由先前的 endIoctl()调用设置)，如果被设置成该模式，则函数 endPollSend()不管输出队列中的其他数据包，而是将当前要发送的包发送到网络上。

(8) 轮询接收函数 endPollReceive()。

函数 endPollReceive()用于任何想要做轮询接收的任务。同样的，函数应该检查设备是否已被设置成轮询模式，如果已处于该模式，endPollReceive()将直接从网络上取得一个数据包。

(9) 组播地址增加函数 endMCastAddrAdd()。

函数 endMCastAddrAdd()功能是增加一个地址到由设备维护的组播表中，并且在特定的硬件中配置这个接口，该配置将使得驱动程序从该组播地址取得帧并将其传递到上层。

(10) 组播地址删除函数 endMCastAddrDel()。

函数 endMCastAddrAdd()从设备维护的组播表中删除一个地址，同时应在特定的硬件中重新配置接口，使得驱动程序不再从该组播地址中接收帧。

(11) 组播地址表获取函数 endMCastAddrGet()。

函数 endMCastAddrGet()获取到组播地址表，并将其存放在一个缓冲区中，同时返回指向该缓冲区的指针。

(12) 复制组织地址函数 endAddressForm()。

函数 endAddressForm()将一个源地址和一个目的地址复制到 mBlk 结构中，同时应调整 mBlk. mBlkHdr. mLen 和 mBlk. mBlkHdr. mData 成员的值。

(13) 获取数据函数 mBlk:endPacketDataGet()。

该函数提供一个去掉了头信息而只包含包数据的 mBlk 的复制。

(14) 返回地址信息函数 endPacketAddrGet()。

该函数返回关联到一个包的地址信息。

上述函数集是整个驱动程序与上层(MUX)的接口。一个完整的驱动程序需要实现上述的全部函数，如图 6.12 所示。在实现过程中，还要用到某些数据结构，这些结构用于驱动程序内部各函数之间或驱动程序与 MUX 接口之间的通信。其中，较为重要的结构有:END - OBJ，DEV - OBJ，NET - FUNCS，M - BLK，这些结构分别在 end. h 和 netBufLib. h 中定义。END - OBJ 的大部分成员在 endLoad()中被设置。END - OBJ 的一个成员 pFuncTable 指向 NET - FUNCS 结构。该结构是一个函数指针表，它的成员均指向前面需要实现的某个驱动程序入口点(如 endStart()函数等)，MUX 通过该表结构来调用驱动程序函数。END - OBJ 另一个成员 devObject 是 DEV - OBJ 结构，驱动程序使用该结构告诉 MUX 设备名和单元号。DEV - OBJ 还包含一个指向驱动程序内部控制结构的指针 pDevice，MUX 并不直接操作该结构。相反的，当 MUX 调用驱动程序入口点时，它会传入一个指向 END - OBJ 的指针 pCookie。通过 pCookie，驱动程序入口点以 pCookie. DevObject. pDevice 的形式访问设备控制结构。还有一个成员 receiveRtn 由 MUX 设置，它指向一个 muxRe-

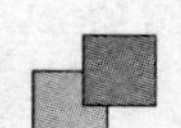

ceive()函数，驱动程序使用该函数指针将数据传递给协议层。END-OBJ 的最后一个成员 pMemPool 指向一个由 netBufLib 管理的内存池，该指针也由 MUX 设置。系统使用 netBufLib 提供的例程从内存池中分配或释放块内存，驱动程序需要使用这样的内存块结构。

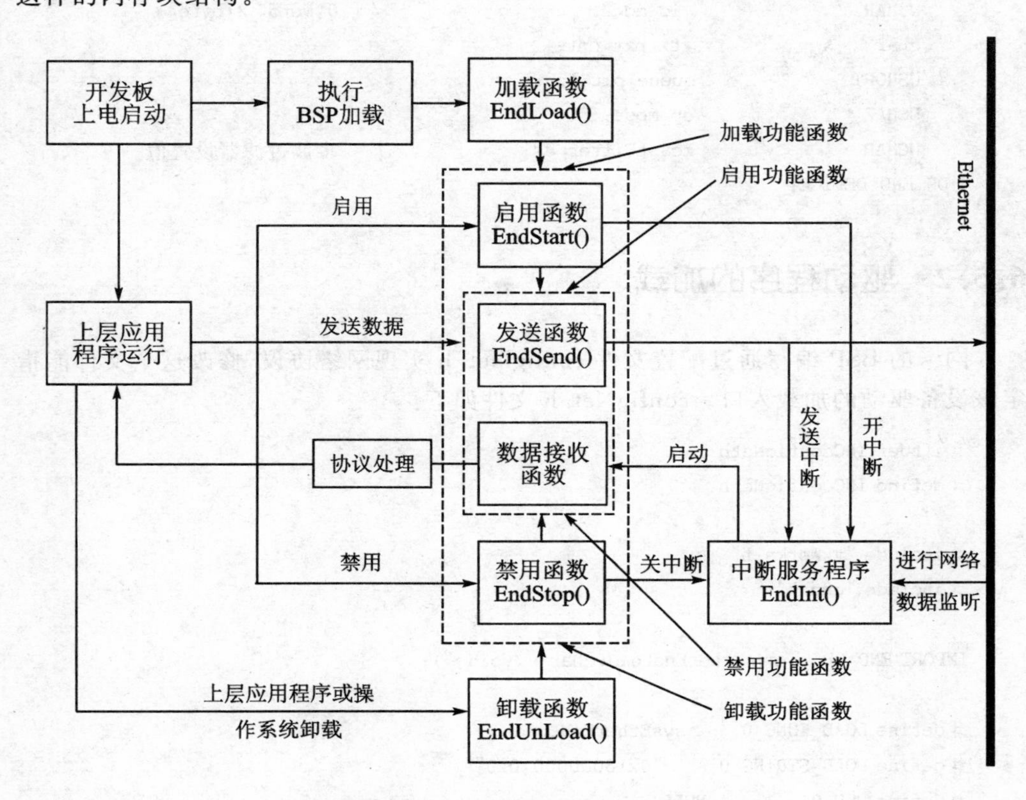

图 6.12 END 函数在系统中关系图

6.5.1 定义设备的描述信息

结构中包含一个 END_OBJ 结构，如下：

```
typedef struct end_device
{
    END_OBJ         end;            /* END_OBJ 对象 */
    int             unit;           /* 网卡单元号 */
    int             ivec;           /* 中断向量号 */
    int             ilevel;         /* 中断号 */
    long            flags;          /* 标志 */
    UCHAR           enetAddr[6];    /* 以太网 IP 地址 */
```

```
    CACHE_FUNCS          cacheFuncs;                  /* cache 功能指针 */
    CL_POOL_ID           pClPoolId;
    BOOL                  rxHandling;
    UINT                        IOBase;
    UCHAR                 io_mode;                    /* 0:word, 2:byte */
    char                  tx_pkt_cnt;
    USHORT               queue_pkt_len;
    UCHAR                op_mode;
    UCHAR                mcastFilter[8];              /* 多波过滤器设置值 */
} DM_END_DEVICE;
```

6.5.2 驱动程序的加载

网卡的 BSP 编写通过配置文件 configNet.h 实现网络协议，修改这个文件能指定该设备驱动的加载入口。configNet.h 文件如下：

```
#ifndef INCconfigNeth
#define INCconfigNeth

#include "vxWorks.h"
#include "end.h"

IMPORT END_OBJ * sysEtherEndLoad(char *,void *);

#define LOAD_FUNC_0      sysEtherEndLoad
#define LOAD_STRING_0       "0x18000000:0x0"
#define BSP_0           NULL

END_TBL_ENTRY endDevTbl [] =
{
    { 0, LOAD_FUNC_0, LOAD_STRING_0, 1, BSP_0, FALSE},
    { 0, END_TBL_END, NULL, 0, NULL, FALSE},
};

#endif /* INCconfigNeth */
```

在该配置文件中，完成对表格 END_TBL_ENTRY 的填写，END_TBL_ENTRY 结构定义如下：

```
typedef struct end_tbl_entry
{
    int unit;                                    /* This device's unit */
```

```
    END_OBJ * (* endLoadFunc) (char *, void *);       /* The Load function. */
    char * endLoadString;                             /* The load string. */
    BOOL endLoan;                                     /* Do we loan buffers? */
    void * pBSP;                                      /* BSP private */
    BOOL processed;                                   /* Has this been processed? */
} END_TBL_ENTRY;
```

END_TBL_ENTRY 声明的数组 endDevTbl 定义了一个网卡的相关信息，以本例定义了"0x18000000:0x0"，包含了网卡基地址，网卡中断号，以及其他有关网卡的参数等。当然，不同的网卡资源字串不同，只要在 endLoad 程序中提供相应的信息即可。

下面是网络驱动程序的加载代码。

```
END_OBJ * sysEtherEndLoad     /* String to be parsed by the driver. */
(
    char * initString,    /* 初始化资源字串 */
    void * pBSP             /* BSP 内部指针 */
)
{
    DM_END_DEVICE   * pDrvCtrl;
    int i;
#ifdef DM_DEBUG_ENABLE
    logMsg("DM9000 loading Start...\n", 0, 0, 0, 0, 0, 0);
#endif
    /* 检查资源字串的合法性 */
if (initString == NULL)
    {
#ifdef DM_DEBUG_ENABLE
        logMsg("sysEtherEndLoad: initString is NULL...\n", 0, 0, 0, 0, 0, 0);
#endif
        return (NULL);
    }

/* 如果输入字串的值为空值，则使用默认值 */
    if (initString[0] == '\0')
    {
        bcopy((char *)DM9000_DEV_NAME, initString, DM9000_DEV_NAME_LEN);
#ifdef DM_DEBUG_ENABLE
        logMsg("sysEtherEndLoad: Returning DM9000 Device Name String...\n",
 0, 0, 0, 0, 0, 0);
#endif
        return ((END_OBJ *)OK);
```

```
    }

    /* 向系统申请空间给设备描述符 */
    pDrvCtrl = (DM_END_DEVICE *)calloc(sizeof (DM_END_DEVICE), 1);
    if (pDrvCtrl == NULL)
    {
#ifdef DM_DEBUG_ENABLE
        logMsg("sysEtherEndLoad: calloc() == NULL\n", 0, 0, 0, 0, 0, 0);
#endif
        goto errorExit;
    }

    /* 配置网卡标志 */
    pDrvCtrl->flags   = 0;

    /* 配置网卡传输模式 */
    pDrvCtrl->op_mode = DM9000_MEDIA_MODE;

    /* 配置网卡字节访问方式 */
    pDrvCtrl->io_mode   = 0;

    /* 解析资源字串,获取硬件资源信息 */
    if (dm9000Parse(pDrvCtrl, initString) == ERROR)
    {
#ifdef DM_DEBUG_ENABLE
        logMsg("sysEtherEndLoad: Parse() == ERROR\n", 0, 0, 0, 0, 0, 0);
#endif
        goto errorExit;
    }

    /* 向系统要求网卡以太网地址 */
    bcopyBytes((char *)dmfEnetAddr,(char *)pDrvCtrl->enetAddr,6);
#ifdef DM_DEBUG_ENABLE
        for(i=0;i<6;i++)
        logMsg(" %x \n",pDrvCtrl->enetAddr[i],0,0,0,0,0);
#endif

    /* 检验网卡芯片是否正常工作 */
    dm9000Chack();

    /* 完成硬件自检之后,进行对象和缓冲的处理 */
    /* 初始化 END_OBJ 结构和 MIB2 接口单元 */
```

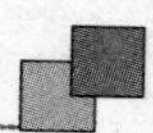

```
    if (END_OBJ_INIT (&pDrvCtrl->end, (DEV_OBJ *)pDrvCtrl, DM9000_DEV_NAME,
                      pDrvCtrl->unit, &dm9000FuncTable,
                      "dm9000 END Driver.") == ERROR
        || END_MIB_INIT (&pDrvCtrl->end, M2_ifType_ethernet_csmacd,
                         (UCHAR *)&(pDrvCtrl->enetAddr[0]),
                         6, 1500,END_SPEED) == ERROR)
    {
#ifdef DM_DEBUG_ENABLE
        logMsg("sysEtherEndLoad: END_OBJ_INIT() or END_MIB_INIT()failed! \n",
               0, 0, 0, 0, 0, 0);
#endif
        goto errorExit;
    }

    /* 缓冲初始化 */
    if (dm9000MemInit (pDrvCtrl) == ERROR)
    {
        goto errorExit;
    }

    /* 复位和配置芯片 */
    dm9000Reset (pDrvCtrl);
    dmfInitChip( pDrvCtrl, 1 );

    /* 设置网卡就绪标志位 */
    END_OBJ_READY (&pDrvCtrl->end,
                   IFF_UP | IFF_RUNNING | IFF_NOTRAILERS | IFF_BROADCAST
                   | IFF_MULTICAST | IFF_SIMPLEX);

#ifdef DM_DEBUG_ENABLE
    logMsg("Dm9000 loading sucessfull! \n", 0, 0, 0, 0, 0, 0);
#endif

    return (&pDrvCtrl->end);

errorExit:
    if (pDrvCtrl != NULL)
    free ((char *)pDrvCtrl);
    return NULL;
}
```

DM9000 网卡加载程序使用 END_OBJ_INIT 来初始化 END_OBJ，主要工作就

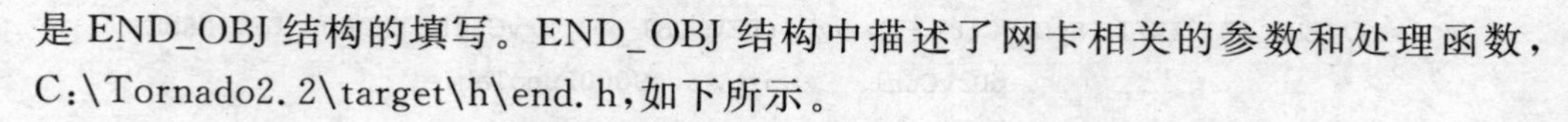

是 END_OBJ 结构的填写。END_OBJ 结构中描述了网卡相关的参数和处理函数，C:\Tornado2.2\target\h\end.h，如下所示。

```
typedef struct end_object
{
    NODE node;
#ifndef _WRS_VXWORKS_5_X
    BOOL nptFlagSpace;                          /* 现在没有使用该标志 */
#endif                                          /* _WRS_VXWORKS_5_X */
    DEV_OBJ devObject;                          /* 继承下来的设备描述符 */
    STATUS (*receiveRtn)();                     /* 用于接收数据的回调函数 */
    struct net_protocol *outputFilter;          /* 可选的输出过滤函数 */
    void* pOutputFilterSpare;                   /* 输出过滤处理时用的指针 */
    BOOL attached;                              /* 网卡是否注册的标志 */
    SEM_ID txSem;                               /* 发送时使用的互斥信号量 */
    long flags;                                 /* 通用的标志 */
    struct net_funcs *pFuncTable;               /* 功能函数表 */
    M2_INTERFACETBL mib2Tbl;                    /* MIBII 计数器 */
    LIST multiList;                             /* 多播地址链表的头 */
    int    nMulti;                              /* 多播地址链表节点的数量 */
    LIST protocols;                             /* 协议链表 */
    int snarfCount;                             /* 是否阻止数据传递给低优先级的协议 */
    NET_POOL_ID pNetPool;                       /* MUX 缓冲信息 */
#ifndef _WRS_VXWORKS_5_X
    void * pNptCookie;
#endif /* _WRS_VXWORKS_5_X */
    M2_ID * pMib2Tbl;                           /* 兼容 2233 MIB 接口的对象指针 */
} END_OBJ;
```

一般来说，用户通常选择系统提供的默认配置；net_funcs 结构体变量 pFuncTable 里面规定了具体网卡的操作功能，需要用户进行填写，net_funcs 结构体如下。

```
typedef struct net_funcs
{
    STATUS (*start)(END_OBJ*);                                  /* 驱动 start 函数 */
    STATUS (*stop)(END_OBJ*);                                   /*驱动 stop 函数*/
    STATUS (*unload)(END_OBJ*);                                 /* 驱动 unload 函数 */
    int (*ioctl)(END_OBJ*, int, caddr_t);                       /* 驱动 ioctl 函数 */
    STATUS (*send)(END_OBJ* , M_BLK_ID);                        /* 驱动 send 函数 */
    STATUS (*mCastAddrAdd)(END_OBJ*, char*);                    /* 多播地址添加函数 */
    STATUS (*mCastAddrDel)(END_OBJ*, char*);                    /* 多播地址删除函数 */
    STATUS (*mCastAddrGet)(END_OBJ*, MULTI_TABLE*);             /* 多播地址获取函数 */
```

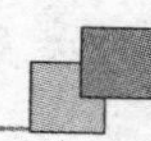

```
    STATUS ( * pollSend) (END_OBJ * , M_BLK_ID);           /* 轮询方式发送函数 */
    STATUS ( * pollRcv) (END_OBJ * , M_BLK_ID);            /* 轮询方式接收函数 */
    M_BLK_ID ( * formAddress) (M_BLK_ID, M_BLK_ID, M_BLK_ID, BOOL);
                                                           /* 地址信息检查函数 */
    STATUS ( * packetDataGet) (M_BLK_ID, LL_HDR_INFO * );
                                                           /* 数据获取函数 */
    STATUS ( * addrGet) (M_BLK_ID, M_BLK_ID, M_BLK_ID, M_BLK_ID, M_BLK_ID);
                                                           /* 数据包地址获取函数 */
    int ( * endBind) (void * , void * , void * , long type);

} NET_FUNCS;
```

DM9000 网卡驱动定义一个结构,用来实现驱动程序的接口需要,定义如下。

```
static NET_FUNCS dm9000FuncTable =
{
    (FUNCPTR) dm9000Start,
    (FUNCPTR) dm9000Stop,
    (FUNCPTR) dm9000Unload,
    (FUNCPTR) dm9000Ioctl,
    (FUNCPTR) dm9000Send,
    (FUNCPTR) dm9000MCastAdd,

    (FUNCPTR) dm9000MCastDel,
    (FUNCPTR) dm9000MCastGet,
    (FUNCPTR) dm9000PollSend,
    (FUNCPTR) dm9000PollRcv,
    endEtherAddressForm,
    endEtherPacketDataGet,
    endEtherPacketAddrGet
};
```

在完成上述定义之后,DM9000 驱动将用 endObjInit 函数来进行 END_OBJ 对象的初始化,函数在 C:\Tornado2.2\target\src\drv\end\endLib.c 文件中,如下所示。

```
STATUS endObjInit
(
    END_OBJ *     pEndObj,              /* END_OBJ 对象指针 */
    DEV_OBJ *     pDevice,              /* END_OBJ 对象指针 */
    char *        pBaseName,            /* 网卡名称 */
    int           unit,                 /* 网卡编号 */
    NET_FUNCS *   pFuncTable,           /* 网卡操作函数列表 */
```

```
    char *          pDescription        /* 网卡描述信息 */
)
{
    pEndObj->devObject.pDevice = pDevice;

    /* 创建发送处理时使用的互斥信号量 */
    pEndObj->txSem = semMCreate ( SEM_Q_PRIORITY |
                                  SEM_DELETE_SAFE |
                                  SEM_INVERSION_SAFE);


    if (pEndObj->txSem == NULL)
    {
        return (ERROR);
    }

    /* 初始化相关协议列表 */
    pEndObj->flags = 0;
    lstInit (&pEndObj->protocols);

    /* 检查和控制网卡名称的长度 */
    if (strlen(pBaseName) > sizeof(pEndObj->devObject.name))
        pBaseName[sizeof(pEndObj->devObject.name - 1)] = EOS;

    /* 保存网卡名称 */
    strcpy (pEndObj->devObject.name, pBaseName);

    /* 检查和控制网卡描述字符串的长度 */
    if (strlen(pDescription) > sizeof(pEndObj->devObject.description))
        pDescription[sizeof(pEndObj->devObject.description - 1)] = EOS;
    strcpy (pEndObj->devObject.description, pDescription);

    /* 设置网卡编号 */
    pEndObj->devObject.unit = unit;
    /* 设置处理函数 */
    pEndObj->pFuncTable = pFuncTable;

    /* 清除多播地址信息 */
    lstInit (&pEndObj->multiList);
    pEndObj->nMulti = 0;

    pEndObj->snarfCount = 0;
    return OK;
```

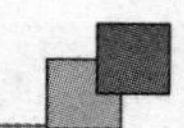

```
}
```

完成了 END_OBJ 的初始化后,DM9000 的大部分操作通过在 pFuncTable 里面定义的函数来实现。程序在加载网卡驱动函数 endLoad()时,通过 dm9000MemInit 来完成内存的初始化。网卡驱动中的内存管理是基于内存池的,每个 END 单元都需要自身的内存池。通常内存池由 mBlk 结构,clBlk 结构和内存块组成,下面就是 mBlk 和 clBlk 的结构定义。

```
typedef struct clBlk
{
    CL_BLK_LIST      clNode;              /* 指向下一个 clBlk 缓冲区指针 */
    UINT          clSize;                 /* 簇的大小 */
    int              clRefCnt;            /* 引用簇计数 */
    FUNCPTR          pClFreeRtn;          /* 释放例程 */
    int              clFreeArg1;          /* 释放例程参数 1 */
    int              clFreeArg2;          /*释放例程参数 2 */
    int              clFreeArg3;          /*释放例程参数 3 */
    struct netPool *     pNetPool;        /* netPool 指针 */
} CL_BLK;

typedef struct mBlk
{
    M_BLK_HDR      mBlkHdr;
    M_PKT_HDR     mBlkPktHdr;
    CL_BLK *      pClBlk;
} M_BLK;

typedef union clBlkList
{
    struct clBlk *       pClBlkNext;
    char *          pClBuf;
} CL_BLK_LIST;
```

在系统中要访问缓冲 BUFF,则首先要通过 MBLK,再根据 CLBLK 链表读取对应的缓冲区,排列图如图 6.13 和图 6.14 所示。

函数 dm9000MemInit()首先设置内存池的参数,然后申请内存空间,最后使用 netPoolInit()来形成具体的内存池。DM9000 网卡驱动程序在任何需要内存分配的地方,都会在这个内存池进行申请。

申请的方法是:

- 使用 netClusterGet 申请缓冲区;
- 使用 netClBlkGet 申请 CLBLK;

MBLK	CLBLK	BUFFER
MBLK	CLBLK	BUFFER
MBLK	CLBLK	BUFFER
MBLK	CLBLK	BUFFER
MBLK	CLBLK	BUFFER
MBLK	CLBLK	BUFFER
MBLK	CLBLK	BUFFER
MBLK	CLBLK	BUFFER
…	…	…

图 6.13　内存池中的缓冲组织

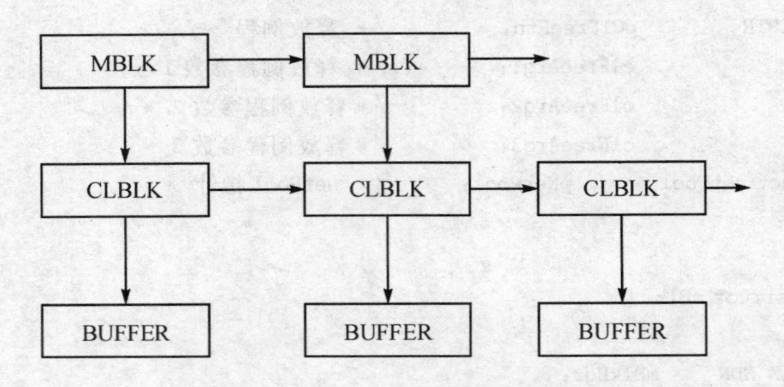

图 6.14　MBLK 和 CLBLK 链表示意图

● 填写数据后，通过 netClBlkJoin 和 netMblkClJoin 将缓冲区、CLBLK、MBLK；

● 传递 MBLK 给处理函数。

由于一个 CLBLK 与多个 MBLK 可以关联，在不同协议之间传输数据只需要传递指针，并不需要复制其中的数据，提高了处理效率。dm9000MemInit()程序如下。

```
static STATUS dm9000MemInit
(
    DM_END_DEVICE * pDrvCtrl /* 设备指针 */
)
{


    if ((pDrvCtrl->end.pNetPool = malloc (sizeof(NET_POOL))) == NULL)
        return (ERROR);

    /* 计算所有的 M-Blks 和 CL-Blks 的内存 */
    dm9000MclBlkConfig.memSize = (dm9000MclBlkConfig.mBlkNum *
```

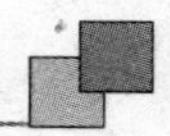

```
                                    (M_BLK_SZ + sizeof (long))) +
                                    (dm9000MclBlkConfig.clBlkNum *
                                    (CL_BLK_SZ + sizeof(long)));

    if ((dm9000MclBlkConfig.memArea = (char *) memalign (sizeof(long),
                        dm9000MclBlkConfig.memSize)) == NULL)
        return (ERROR);

    /* 计算所有簇的内存 */
    dm9000ClDescTbl[0].memSize = (dm9000ClDescTbl[0].clNum * (2048 + 8))
                                    + sizeof(int);

    /* 分配内存 */
    dm9000ClDescTbl[0].memArea =
                    (char *) cacheDmaMalloc (dm9000ClDescTbl[0].memSize) ;

    if (dm9000ClDescTbl[0].memArea == NULL)
    {
#ifdef DM_DEBUG_ENABLE
        logMsg("dm9000MemInit: Unit = %d system memory unavailable! \n",
            pDrvCtrl->unit, 0, 0, 0, 0, 0);
#endif
        return (ERROR);
    }

    /* 初始化内存池 */
    if (netPoolInit(pDrvCtrl->end.pNetPool, &dm9000MclBlkConfig,
                &dm9000ClDescTbl[0], dm9000ClDescTblNumEnt,NULL) == ERROR)
    {
#ifdef DM_DEBUG_ENABLE
        logMsg("dm9000MemInit: Unit = %d Could not init buffering! \n",
            pDrvCtrl->unit, 0, 0, 0, 0, 0);
#endif
        return (ERROR);
    }

    /* 保存内存池 */
    if ((pDrvCtrl->pClPoolId = netClPoolIdGet (pDrvCtrl->end.pNetPool,2048,
(BOOL)FALSE)) == NULL)
        /* (int)END_BUFSIZ, (BOOL)FALSE)) == NULL) */
    {
#ifdef DM_DEBUG_ENABLE
```

```
            logMsg("dm9000MemInit: Unit = %d Could not memory cluster ID! \n",
                pDrvCtrl->unit,0,0,0,0,0);
#endif
            return (ERROR);
        }
#ifdef DM_DEBUG_ENABLE
    logMsg("dm9000MemInit: Unit = %d  Memory setup complete! \n",
        pDrvCtrl->unit, 0, 0, 0, 0, 0);
#endif
    return OK;
}
```

6.5.3 驱动程序清单

1. 函数 dm9000Ioctl()

Ioctl 系列函数主要完成网卡的启动、停止，MAC 地址设置等功能，由函数 dm9000Ioctl()实现，具体如下。

```
static int dm9000Ioctl
(
    DM_END_DEVICE *pV,              /* 设备指针 */
    int cmd,                        /* ioctl 命令码 */
    caddr_t data
)
{
    int status = 0;
    long value;

    DM_END_DEVICE *pDrvCtrl;
    pDrvCtrl = (DM_END_DEVICE *)pV;

    switch (cmd)
    {
        case EIOCSADDR:
            if (data == NULL)
                return (EINVAL);
            bcopy ((char *)data, (char *)END_HADDR(&pDrvCtrl->end),
                END_HADDR_LEN(&pDrvCtrl->end));
            break;
```

```
        case EIOCGADDR:
            if  (data == NULL)
                return (EINVAL);
            bcopy ((char * )END_HADDR(&pDrvCtrl->end), (char * )data,
                END_HADDR_LEN(&pDrvCtrl->end));
            break;

        case EIOCSFLAGS:
            value = (long)data;
            if (value < 0)
            {
                value = -(--value);
                END_FLAGS_CLR (&pDrvCtrl->end, value);
            }
            else
            {
                END_FLAGS_SET (&pDrvCtrl->end, value);
            }
            dm9000Config (pDrvCtrl);
            break;
        case EIOCGFLAGS:
            *(int * )data = END_FLAGS_GET(&pDrvCtrl->end);
            break;
        case EIOCMULTIADD:
            status = dm9000MCastAdd ((void * )pDrvCtrl, (char * ) data);
            break;
        case EIOCMULTIDEL:
            status = dm9000MCastDel((void * )pDrvCtrl, (char * ) data);
            break;
        case EIOCMULTIGET:
            status = dm9000MCastGet ((void * )pDrvCtrl, (MULTI_TABLE * ) data);
            break;
        case EIOCPOLLSTART:    /* 开始查询操作 */
#ifdef DM_DEBUG_ENABLE
        logMsg("dm9000Ioctl: begin polled operation\n", 0, 0, 0, 0, 0, 0);
#endif
            dm9000PollStart ((void * )pDrvCtrl);
            break;

        case EIOCPOLLSTOP:   /* 结束查询操作 */
#ifdef DM_DEBUG_ENABLE
        logMsg ("dm9000Ioctl: end polled operation\n", 0, 0, 0, 0, 0, 0);
```

```
#endif
            dm9000PollStop ((void *)pDrvCtrl);
            break;

        case EIOCGMIB2:        /* 返回 MIB 信息 */
            if (data == NULL)
                return (EINVAL);
            bcopy((char *)&pDrvCtrl->end.mib2Tbl,(char *)data,
                sizeof(pDrvCtrl->end.mib2Tbl));
            break;
        case EIOCGFBUF:
            if (data == NULL)
                return (EINVAL);
            *(int *)data = DM9000_MIN_FBUF;
            break;
        case EIOCGHDRLEN:
                if(data == NULL) return EINVAL;
                *(int *)data = EH_SIZE;
                break;

        default:
            status = EINVAL;
    }
    return (status);
}
```

2. 函数 dm9000Send()

实现数据发送时，先要获取发送互斥信号量，然后将数据复制到硬件发送缓冲区中，最后启动发送即可，程序如下。

```
static STATUS dm9000Send
(
    DM_END_DEVICE *pV,   /* 设备指针 */
    M_BLK_ID      pMblk     /* 待发送的数据 */
)
{
    static UCHAR txBuf[DM_CHIP_FRAME_BUF_SIZE];
    int           oldLevel = 0;
    int           length = 0;
    int           TxStatus = 0;

    DM_END_DEVICE *pDrvCtrl;
```

```
    pDrvCtrl = (DM_END_DEVICE *)pV;


# ifdef DM_DEBUG_ENABLE
    logMsg("dm9000Send()\n", 0, 0, 0, 0, 0, 0);
# endif

   /* 判断发送方式是否为轮询方式,如果不是则需要获取信号量 */
   if (! (pDrvCtrl->flags & DM9000_POLLING))
   {
       END_TX_SEM_TAKE (&pDrvCtrl->end, WAIT_FOREVER);
   }

    /* 从 Mblk 复制数据到发送缓冲区 */
    length = netMblkToBufCopy (pMblk, (char *)txBuf, NULL);

    /* 查询是否是轮询方式,如果不是则锁住中断 */
    if (! (pDrvCtrl->flags & DM9000_POLLING))
    {
        oldLevel = intLock ();  /* protect dm9000Int */
    }

    /* 调用函数发送数据帧 */
    TxStatus = dmfCopyTxFrame(pDrvCtrl, txBuf, length);

    /* 解锁中断 */
    if (! ((pDrvCtrl->flags) & DM9000_POLLING))
        intUnlock (oldLevel);

    /* 发送数据之后,判断是否为非轮询状态,如果为非,则释放信号量 */
    if (! ((pDrvCtrl->flags) & DM9000_POLLING))
        END_TX_SEM_GIVE (&pDrvCtrl->end);

    /* 判断发送是否失败 */
    if(TxStatus == ERROR)
    {
        logMsg("dm9000Send: Transmit packet is failed! \n", 0, 0, 0, 0, 0, 0);
        netMblkClChainFree (pMblk);
        END_ERR_ADD (&pDrvCtrl->end, MIB2_OUT_ERRS, +1);
        return(END_ERR_BLOCK);
    }
```

```
    /* 更新发送失败统计数据 */
    END_ERR_ADD (&pDrvCtrl->end, MIB2_OUT_UCAST, +1);

    /*
     * Cleanup.  The driver must either free the packet now or
     * set up a structure so it can be freed later after a transmit
     * interrupt occurs.
     */
    netMblkClChainFree (pMblk);
    return (OK);
}
```

3. 函数 dm9000Start()

启动函数 dm9000Start()所要完成的功能有注册中断处理程序和打开中断，程序如下。

```
static STATUS dm9000Start
(
    END_DEVICE *pV    /* 设备结构的指针 */
)
{
     END_DEVICE *pDrvCtrl;
     pDrvCtrl = (END_DEVICE *)pV;

     s3cExtIntPinEnable ();

    /*注册 DM9000 硬件中断 dm9000Int 函数到 END 驱动*/
    if(intConnect(INUM_TO_IVEC(pDrvCtrl->ivec), dm9000Int, (int)pDrvCtrl)
        == ERROR)
    {
#ifdef DM_DEBUG_PRINT
    logMsg("dm9000Start:Can not connect interrupt! \n",
                    0, 0, 0, 0, 0, 0);
#endif
            return ERROR;
    }

     /* 使能 DM9000 中断 */
    if(intEnable(pDrvCtrl->ilevel) == ERROR)
    {
#ifdef DM_DEBUG_PRINT
    logMsg("dm9000Start: Can not enable interrupt! \n ",
```

```
            0, 0, 0, 0, 0, 0);
    #endif
        return ERROR;
        }


  /* 设置接口为 up 和 running 状态 */
  END_FLAGS_SET (&pDrvCtrl->end, (IFF_UP | IFF_RUNNING));

  /* 激活 DM9000 芯片 */
  DM9000_OUT_REG( 0x05, DM9000_REG05 );                /* RX 接收使能 */
  DM9000_OUT_REG( 0xff, DM9000_REGFF );                /* 使能接收和发送中断 */

  return (OK);
}
```

4. 函数 dm9000Int()

中断处理函数 dm9000Int()在启用函数 dm9000Start()被注册到 DM9000 相关的中断向量上去,所以在 DM9000 有中断产生时,系统将会自动调度 dm9000Int()函数来处理 DM9000 的实时中断任务。

```
static void dm9000Int
(
    END_DEVICE    * pDrvCtrl   /* 设备结构的指针 */
)
{

    UCHAR stat, reg_save,isr_status,TX_comple_status,tx_status;

    /* clear s3c2410 external intpand reg */
    intDisable (pDrvCtrl->ilevel);
    rpINTMSK | = 1;
    rpINTPND | = 1;
    rpSRCPND | = 1;
    /* save previous register address */
    DM9000_IN_ADDR( reg_save );
    DM9000_OUT_REG( 0xff, DM9000_REGFF_OFF );

    /* Read and Clear ISR status */
    DM9000_IN_REG( 0xfe, isr_status );
    DM9000_OUT_REG( 0xfe, isr_status );

#ifdef DM_DEBUG_PRINT
```

```
        logMsg("dm9000Int: got interrupt! \n", 0, 0, 0, 0, 0, 0);
    #endif

        /*
         * enable interrupts, clear receive and/or transmit interrupts, and clear
         * any errors that may be set.
         */
        if( isr_status & 0x02 )
        {

            /* Clear Tx Packet Complete Status */
            DM9000_IN_REG(0x01, TX_comple_status);

            if (TX_comple_status & 0xc)
            {
                if (TX_comple_status & 0x4) pDrvCtrl->tx_pkt_cnt--;
                  if (TX_comple_status & 0x8) pDrvCtrl->tx_pkt_cnt--;
            }
            else
            {
                DM9000_IN_REG( 0x02, tx_status );
                    while(tx_status & 0x1){DM9000_IN_REG( 0x02, tx_status );}
            }

            /* if not complete tranmission */
            if(pDrvCtrl->tx_pkt_cnt>0)
            {
                /* Set TX length to DM9000 */
                DM9000_OUT_REG( 0xfc, pDrvCtrl->queue_pkt_len & 0xff );
                DM9000_OUT_REG( 0xfd, (pDrvCtrl->queue_pkt_len >> 8) & 0xff );

                /* Issue TX polling command */
                DM9000_OUT_REG( 0x2, 0x01 );    /* Cleared after TX complete */
            }
        }

        /* Have netTask handle any input packets */
        if (isr_status & 0x01 )
        {
            if ( pDrvCtrl->rxHandling != TRUE )
            {
                pDrvCtrl->rxHandling = TRUE;
```

```
            netJobAdd ((FUNCPTR)dm9000Recv, (int)pDrvCtrl, 0, 0, 0, 0);
        }
    }

    /* Re-enable interrupt mask */
    DM9000_OUT_REG( 0xff, DM9000_REGFF );
    DM9000_OUT_REG( 0x05, 0x11 );

    /* Restore previous register address */
    DM9000_OUT_ADDR( reg_save );

    rpINTMSK &= ~1;
    intEnable (pDrvCtrl->ilevel);
}
```

5. 函数 dm9000Stop()

函数 dm9000Stop()的参数是所要停止设备控制结构的指针，由它来确定停止哪一个设备。停止函数所必须执行的操作有：

- 停用硬件设备；
- 关闭硬件上的所有中断请求；
- 在系统中注销与中断服务程序关联的中断向量。

当系统要停用网卡设备时会通过 MUX 接口层来调用此函数完成关闭设备的操作。系统在执行完 dm9000Stop()函数之后，驱动程序并没有释放掉在加载函数中所申请的系统资源，只是将 DM9000 在系统中的工作功能进行了禁用。

```
static STATUS dm9000Stop
(
    END_DEVICE * pV /* 设备结构的指针 */
)
{
    END_DEVICE * pDrvCtrl;
    pDrvCtrl = (END_DEVICE *)pV;

    intDisable(pDrvCtrl->ilevel);

    /* TODO - stop/disable the device. */
    dmfe_Stop_Chip( pDrvCtrl );

    return (OK);
}
```

6. 函数 dm9000Recv()

```
static STATUS dm9000Recv
(
    END_DEVICE * pDrvCtrl   /* 设备结构的指针 */
)
{
    char *          pNewCluster = NULL;
    M_BLK_ID     pMblk;
    CL_BLK_ID    pClBlk;
    int DataLen;
    UCHAR  status;

    while((status = dmfe_Get_NextPacket()) != 0)
    {
        /* 获取一个接收帧缓存 */
        pNewCluster = netClusterGet (pDrvCtrl->end.pNetPool,
                                     pDrvCtrl->pClPoolId);

        /* 从内存池中获取 clBlk 结构 */
        pClBlk = netClBlkGet (pDrvCtrl->end.pNetPool, M_DONTWAIT);

        pMblk = mBlkGet (pDrvCtrl->end.pNetPool, M_DONTWAIT, MT_DATA);

        if(pNewCluster == NULL || pClBlk == NULL || pMblk == NULL)
        {
            if(pNewCluster)
            {
                netClFree(pDrvCtrl->end.pNetPool,pNewCluster);
            }

            if(pClBlk)
            {
                netClBlkFree (pDrvCtrl->end.pNetPool, pClBlk);
            }

            if(pMblk)
            {
                netMblkFree(pDrvCtrl->end.pNetPool,pMblk);
            }

            /* 将输入错误计数器加 1 */
```

```
        END_ERR_ADD (&pDrvCtrl->end, MIB2_IN_ERRS, +1);

        return;
      }

      END_ERR_ADD (&pDrvCtrl->end, MIB2_IN_UCAST, +1);

      DataLen = dmfe_Copy_RxFrame(pDrvCtrl, pNewCluster + 2, status);

      /* 将 clBlk 与存有数据的 cluster 联接起来 */
      netClBlkJoin (pClBlk, pNewCluster, DataLen, NULL, 0, 0, 0);
      /* 将 mBlk 与 clBlk - ckuster 结构联接起来 */
      netMblkClJoin (pMblk, pClBlk);

      pMblk->mBlkHdr.mData += 2;
      pMblk->mBlkHdr.mLen = DataLen;
      pMblk->mBlkHdr.mFlags |= M_PKTHDR;
      pMblk->mBlkPktHdr.len = DataLen;

      /* 调用顶层的接收程序 */
      END_RCV_RTN_CALL(&pDrvCtrl->end, pMblk);

      pNewCluster = NULL;
      if(pNewCluster) netClFree(pDrvCtrl->end.pNetPool,pNewCluster);
    }
    pDrvCtrl->rxHandling = FALSE;
    return (OK);

cleanRXD:
    pDrvCtrl->rxHandling = FALSE;
    return (ERROR);
}
```

7. 函数 dm9000PollRcv()

```
static STATUS dm9000PollRcv
(
    END_DEVICE *pV,    /*设备结构的指针*/
    M_BLK_ID    pMblk
)
{
```

```
u_short stat;
char * pPacket;
int len = 64;

END_DEVICE * pDrvCtrl;
pDrvCtrl = (END_DEVICE * )pV;

# ifdef DM_DEBUG_PRIMT
        logMsg("dm9000PollRcv()\n", 0, 0, 0, 0, 0, 0);
# endif

stat = dm9000StatusRead (pDrvCtrl);
/* 如果没有可接收包,可迅速返回 */
if( ! (stat&DM9000_RXRDY) )
{
    logMsg("dm9000PollRcv: no data.\n", 0, 0, 0, 0, 0, 0);
    return (EAGAIN);
}

if ((pMblk->mBlkHdr.mLen < len) || (! (pMblk->mBlkHdr.mFlags & M_EXT)))
{
    logMsg("dm9000PollRcv: PRX bad mblk.\n", 0, 0, 0, 0, 0, 0);
    return (EAGAIN);
}

END_ERR_ADD (&pDrvCtrl->end, MIB2_IN_UCAST, +1);

while(dmfe_Get_NextPacket() != 0)
{
    len = dmfe_Copy_RxFrame( pDrvCtrl, pMblk->mBlkHdr.mData + 2, 0 );
}

if(len == ERROR)
{
    logMsg("dm9000PollRcv: packet receive FAIL.\n", 0, 0, 0, 0, 0, 0);
    return (EAGAIN);
}

pMblk->mBlkHdr.mData += 2;
/* 设置包头 */
pMblk->mBlkHdr.mFlags |= M_PKTHDR;
```

```
    /* 设置数据长度 */
    pMblk->mBlkHdr.mLen = len;
    pMblk->mBlkPktHdr.len = len;

    return (OK);
}
```

8. 函数 dm9000PollSend()

```
static STATUS dm9000PollSend
(
    END_DEVICE *pV,              /* device to be polled */
    M_BLK_ID    pMblk            * packet to send */
)
{
    static UCHAR txBuf[DM9000_FRAME_BUFSIZE];
    int          len,lastone;
    u_short      stat;

    END_DEVICE *pDrvCtrl;
    pDrvCtrl = (END_DEVICE *)pV;

    #ifdef DM_DEBUG_PRIMT
    logMsg("dm9000PollSend: dm9000PollSend()\n", 0, 0, 0, 0, 0, 0);
    #endif

    /* TODO - test to see if tx is busy */
    stat = dm9000StatusRead (pDrvCtrl);               /* dummy code */
    if ((stat & (DM9000_TINT|DM9000_TFULL)) == 0)
        return ((STATUS) EAGAIN);

    /* Get data from Mblk to tx buffer. */
    len = netMblkToBufCopy (pMblk, (char *)txBuf, NULL);
    len = max (len, ETHERSMALL);

    /* transmit packet */
    lastone = intLock();
    dmfe_Copy_TxFrame(pDrvCtrl, txBuf ,len);
    /* check a Completion Flag */
    intUnlock( lastone );
    /* Bump the statistic counter. */
    END_ERR_ADD (&pDrvCtrl->end, MIB2_OUT_UCAST, +1);
```

```
    /* Free the data if it was accepted by device */
    netMblkClFree (pMblk);

    #ifdef  DM_DEBUG_PRIMT
        logMsg("dm9000PollSend: leaving dm9000PollSend.\n", 0, 0, 0, 0, 0, 0);
    #endif
        return (OK);
    }
```

9. 函数 dm9000PollStart( )

```
static STATUS dm9000PollStart
(
    END_DEVICE * pV /* device to be polled */
)
{
    int          lastone;
    END_DEVICE * pDrvCtrl;

    pDrvCtrl = (END_DEVICE *)pV;

    lastone = intLock ();

    /* TODO - turn off interrupts */
    (pDrvCtrl->flags) |= DM9000_POLLING;

    dm9000Reset( pDrvCtrl );
    dm9000Config( pDrvCtrl );          /* reconfigure device */

    intUnlock (lastone);               /* now dm9000Int won't get confused */

    logMsg("dm9000PollStart Poll Mode Start.\n", 0, 0, 0, 0, 0, 0);

    return (OK);
}
```

10. 函数 dm9000PollStop( )

```
static STATUS dm9000PollStop
(
    END_DEVICE * pV   /* device to be polled */
```

```
)
{
    int           lastone;
    END_DEVICE * pDrvCtrl;

    pDrvCtrl = (END_DEVICE * )pV;

    lastone = intLock ();  /* disable ints during register updates */
    (pDrvCtrl->flags) &= ~DM9000_POLLING;

    dm9000Reset( pDrvCtrl );
    dm9000Config( pDrvCtrl );          /* reconfigure device */

    intUnlock (lastone);

    logMsg ("dm9000PollStart Poll Mode Stop.\n", 0, 0, 0, 0, 0, 0);

    return (OK);
}
```

11. 函数 dm9000Stop()

```
static STATUS dm9000Stop
(
    END_DEVICE * pV   /* device to be stopped */
)
{
    END_DEVICE * pDrvCtrl;
    pDrvCtrl = (END_DEVICE * )pV;

    intDisable(pDrvCtrl->ilevel);

    /* TODO - stop/disable the device. */
    dmfe_Stop_Chip( pDrvCtrl );

    return (OK);
}
```

12. 函数 dm9000Unload()

卸载函数 dm9000Unload()的作用就是从系统中将加载的驱动程序卸载掉，在函数执行时要调用 END_OBJECT_UNLOAD(&pDrvCtrl－>end)，将驱动使用的

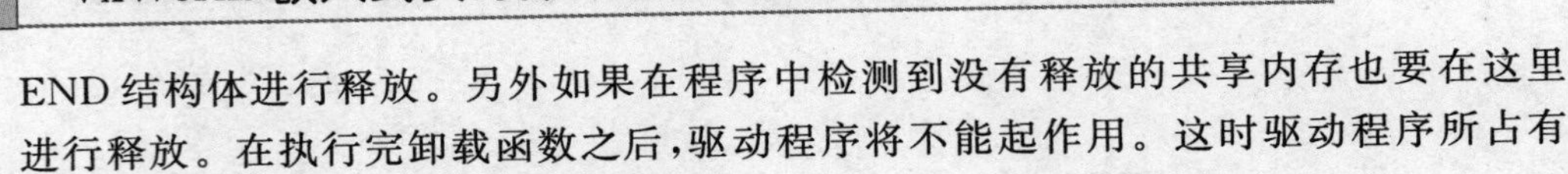

END 结构体进行释放。另外如果在程序中检测到没有释放的共享内存也要在这里进行释放。在执行完卸载函数之后，驱动程序将不能起作用。这时驱动程序所占有的资源得到释放，驱动程序真正终止。

```
static STATUS dm9000Unload
(
    END_DEVICE * pV    /* device to be unloaded */
)
{
    END_DEVICE * pDrvCtrl;
    pDrvCtrl = (END_DEVICE *)pV;

#ifdef DM_DEBUG_PRIMT
    logMsg("dm9000Unload()\n", 0, 0, 0, 0, 0, 0);
#endif

    END_OBJECT_UNLOAD (&pDrvCtrl->end);

    /* TODO - Free any shared DMA memory */
    return (OK);
}
```

第7章 LCD液晶设备驱动程序设计

7.1 WindML简介

WindML(Wind Media Library,媒体库),是WRS公司提供的VxWorks库的一部分,支持多媒体程序运行于嵌入式操作系统。WindML提供基本的图形、图像和音频的支持。WindML API函数库提供对多种CPU结构和操作系统适用的图形硬件接口。同时WindML还提供操作输入设备和处理输入设备事件的功能。

WindML具备以下特点:

- 便捷性,WindML为用户提供一个灵活的图形源语集和基本的视频和音频功能;
- 硬件适用性,能够支持在多种CPU体系结构上应用;
- 操作系统适用性,支持多种RTOS系统上应用;
- 驱动程序定制,提供给开发者一个定制设备的驱动程序。

7.1.1 WindML结构

WindML由两部分组成:软件开发工具包SDK和驱动程序开发工具包DDK。SDK为用户提供应用程序代码和底层硬件驱动程序的接口,包括图形、输入句柄、多媒体、内存管理等API函数,使开发人员能够在不同硬件平台上开发与底层硬件无关的应用代码;DDK则是对通用硬件设备提供完整的驱动程序,整体层次结构如图7.1所示。

1. 软件开发工具包SDK

软件开发工具包SDK组件被用来开发应用程序。它提供了一个全面的API集图形,输入处理,多媒体,字体和内存管理,能够为硬件平台编写独立的便携式硬件

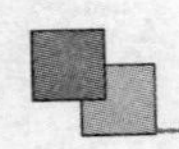

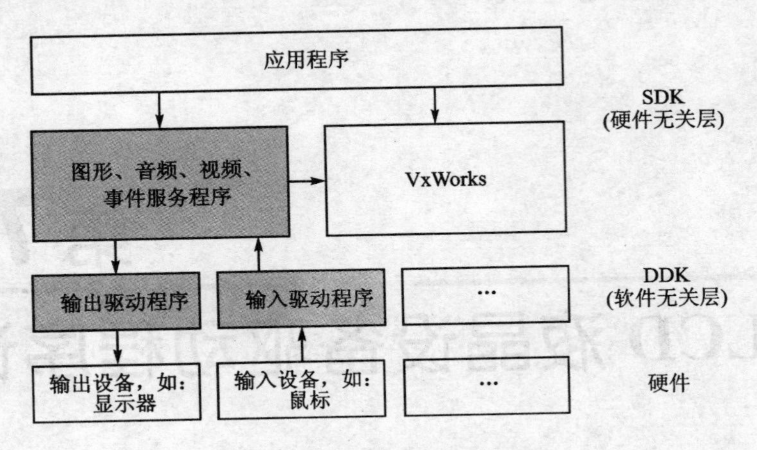

图 7.1　WindML 层次结构图

代码。

SDK 提供了下列 API 集：

(1) 图形芯片的初始化程序；

(2) 多媒体 API,包括：

- 2D 图形；
- 区域管理；
- 窗口；
- 颜色管理；
- 视频支持；
- JPEG 支持；
- 音频；
- 事件服务；
- 内存管理；
- 扩展 API;
- 设备管理。

2. 驱动程序开发工具包 DDK

DDK 组件为通用的硬件配置提供了一套完整的参考驱动程序，以及一个 API 集。因此,用户可以快速地从所提供通用驱动程序中引导出新的驱动程序。DDK 是可扩展和定制的。WindML 提供完整的源代码库,用户可以进一步定制。

如图 7.1 所示,WindML 组件在图中标示为阴影部分。WindML DDK 中有一个多层次的架构,是夹在高级别 SDK 和硬件之间的中间层。WindML DDK 可直接与如显示器,视频,音频的硬件设备,键盘和鼠标连接。WindML 定义并规定了下列驱动程序类的支持：

（1）图形驱动

此驱动程序为图形设备分配颜色，执行原始图纸操作，内存分配，管理和覆盖页面，比如 VGA，VESABIOS 等。

（2）视频驱动

WindML DDK 作为一种扩展驱动程序来实现图形驱动程序。该驱动器提供的视频功能包括开始、停止和流操作，例如 IGS 的视频扩展驱动程序的操作。

（3）字体驱动

该驱动器提供文本渲染操作。它使用图形驱动程序绘制文本。

（4）输入驱动

此驱动程序从输入设备读取数据，对其进行适当的格式化，并为它们安排队列以便应用程序从中读取。键盘、鼠标和触摸屏是典型的输入驱动。

（5）音频驱动

此驱动程序支持播放和记录音频，兼容声音系统（OSS）技术。

（6）窗口管理

WindML 具备窗口管理器的设计功能，可被用于设计用户定义的扩展窗口管理器，在基础配置中类似设备驱动程序。

7.1.2　WindML 源码架构

1. 驱动 WindML 目录内容

WindML 源代码树内容和目录的建立与特定架构或者特定操作系统的分支无关，为用户提供一个统一的目录结构进行移植工作。

src/ugl 目录是 WindML 顶层源目录。该 WindML 源代码被组织成多个组件目录。如图 7.2 所示，src 目录包括以下子目录：

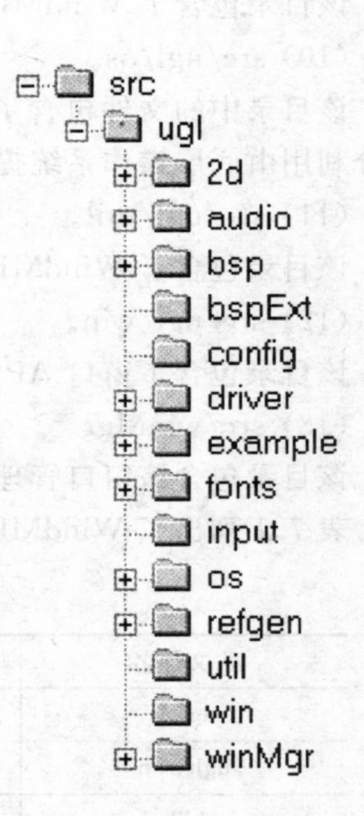

图 7.2　WindML 源代码目录

（1）src/ugl/2d。

该目录中的文件组成 WindML 的 2－D API 层。这些库和子程序支持在 WindML 上运行的应用程序。

（2）src/ugl/audio。

该目录包含音频组件的 WindML 代码。

（3）src/ugl/bsp。

该目录包含 BSP 的信息，一个 BSP 的自述文件，以及使用 WindML 的 BSP 所需的所有补丁。

(4) src/ugl/bspExt。

该目录包含了由 WindML 要求的 BSP 需要的扩展代码。此目录仅包含 tornad 2.2 版本之前的 tornado 版本。

(5) src/ugl/config。

该目录包含用于 WindML 的配置文件。

(6) src/ugl/driver。

该目录包含 WindML 上运行的硬件平台的驱动程序代码。这包括硬件无关的通用源代码 WindML 驱动程序和硬件相关的驱动程序。图形目录包含源代码，输出显示数据的驱动程序。键盘和鼠标的目录包含源代码的驱动，从输入设备上读取数据给 WindML。字体目录包含了字体驱动程序源代码以及实现 2－D 层字体的相关要求。在每个子目录下有一个设备供应商目录。例如，由英特尔公司开发的图形处理装置，存在 src/ugl/driver/graphics/intel 目录。此外，在图形目录有一个通用的子目录，在帧缓冲器按照线性内存组织时提供基本型画面缓冲器的操作例程。

(7) src/ugl/example。

该目录包含使用 WindML 的示例程序。

(8) src/ugl/fonts。

该目录包含了字体的代码。

(9) src/ugl/input。

该目录包含了 WindML 的事件处理和输入服务代码 API。

(10) src/ugl/os。

该目录中的文件包含了操作系统的 WindML API 实现的各种平台。WindML 充分利用由底层操作系统提供的最优特性。

(11) src/ugl/util。

该目录包含了 WindML 库和驱动程序使用的实用功能。

(12) src/ugl/win。

该目录包含了窗口 API 的 WindML 层代码。

(13) src/winMgr。

该目录包含该窗口管理器的代码。

表 7.1 列出了 WindML 头文件在 h/ugl 目录中的定义。

表 7.1 h/ugl 目录头文件定义

头文件名	定 义
ugl. h	所有的 WindML 通用定义
uglclr. h	与 WindML 颜色管理 API 有关的定义
ugldib. h	设备无关位图定义
uglfont. h	WindML 字体 API 相关定义

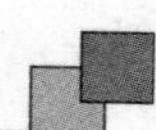

续表 7.1

头文件名	定义
uglinfo. h	WindML 驱动信息查询 API 相关定义
uglinput. h	WindML 输入设备 API 相关定义
uglkbdmap. h	设定键盘映射支持不同语言的相关定义
ugllog. h	WindML 错误日志相关定义
uglmem. h	内存管理相关定义
uglmode. h	图形模式设定的相关定义
uglMsgTypes. h	WindML 消息类型的相关定义
uglos. h	OS 相关定义
uglpage. h	双缓冲定义
uglRegion. h	WindML 区域处理相关定义
ugltypes. h	WindML 所有数据类型的相关定义
uglucode. h	WindML 统一编码的相关定义
uglugi. h	WindML 输出通用显卡接口 API 相关定义
uglwin. h	WindML 窗口支持相关定义

如图 7.3 所示，以下是 h/ugl 下的部分子目录。

(14) h/ugl/audio。

该目录包含定义了 API 和声音处理流音频的头文件。

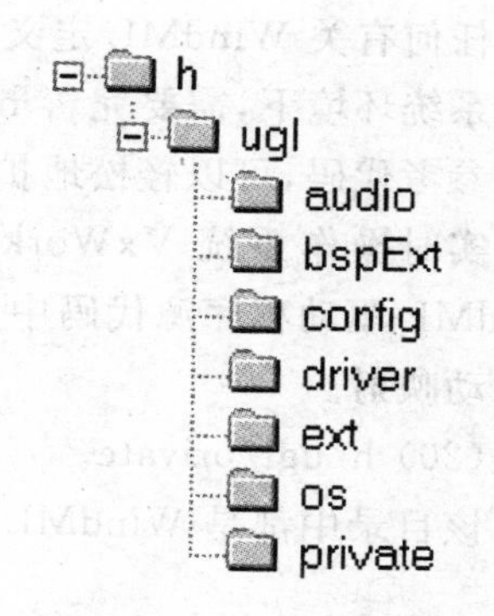

图 7.3 WindML 源代码目录

(15) h/ugl/bspExt。

该目录包含用于板级功能或 WindML BSP 所要求的扩展头文件。

(16) h/ugl/config。

该目录包含配置 WindML 库特定的硬件和操作系统的头文件。

(17) h/ugl/driver。

WindML 驱动程序定义包括以下子目录：

① h/ugl/driver/keyboard。

② h/ugl/driver/pointer。

输入设备驱动程序的定义由在这些目录下的文件组成。如果要开发新的 WindML 输入驱动程序，需要将所有必要的定义在头文件中完成，并将其放置在相应的目录下。例如在这些目录中的驱动文件有：

- uglms. h，Microsoft 鼠标驱动程序的定义；
- uglpckbd. h，PC8042 键盘驱动程序的定义；

● uglps2. h,PS/2 鼠标驱动程序的定义；

● uglasbt. h,StrongARM Assabet 触摸屏的定义。

③ h/ugl/driver/graphics。

该目录包含了图形驱动的头文件。子目录以不同的硬件平台来组织。例如：芯片(芯片和技术图形驱动程序的定义),vga(VGA 模式图形驱动程序的定义)。如果开发一个新的 WindML 制造商图形驱动程序,需要创建一个新的目录放置驱动头文件。

④ h/ugl/driver/ext。

该目录放置所有 WindML 扩展的头文件。提供 JPEG 和视频扩展的相关定义。

⑤ h/ugl/driver/font。

该目录是特定的字体或字体引擎头文件被放置的位置。提供 BMF 字体格式的头文件。

⑥ h/ugl/driver/audio。

该目录是特定的音频驱动程序头文件目录。提供英特尔音频编解码器 97 和 IGS 的音频硬件头文件。

(18) h/ugl/ext。

该目录包含了 WindML 扩展能力的头文件,如 JPEG 和视频扩展。

(19) h/ugl/os。

任何有关 WindML 定义的操作系统都在该目录下。要移植 WindML 到另一个操作系统环境下,需要完善 h/ugl/uglos. h 文件。目前,提供 VxWorks 操作系统移植层参考代码,可以轻松地扩展到其他环境。有关详细信息,在 udvxw. h 文件中定义的实时操作系统 VxWorks 平台。不需要在驱动程序中包括所有的头文件,在 WindML 驱动程序源代码中,通过包含 h/ugl/ugltypes. h 文件,正确的头文件将会被自动映射。

(20) h/ugl/private。

该目录中都是 WindML 内部文件。

7. 1. 3 WindML 图形设备驱动介绍

WindML 图形驱动程序架构提供了一个框架,能够实现让驱动快速启动和运行,以及创建一个具有高效定制的驱动程序。WindML 图形驱动提供以下特性：

● 通过优化的底层驱动程序实现一套较为完整驱动程序的综合功能,利用极少量附加代码,就可以用于大多数类型的硬件平台上；

● 一个功能丰富的 API,实现高加速和优化的图形驱动。

WindML 包括多种类型的可参考硬件驱动程序。用户在使用时,可以根据需要直接使用 WindML 优化后的驱动程序,或者按照自己的硬件平台来重新编写。

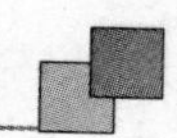

如图 7.4 所示，WindML 提供了一个简化的图形控制器框图，标明处理器和图形显示之间的关系。

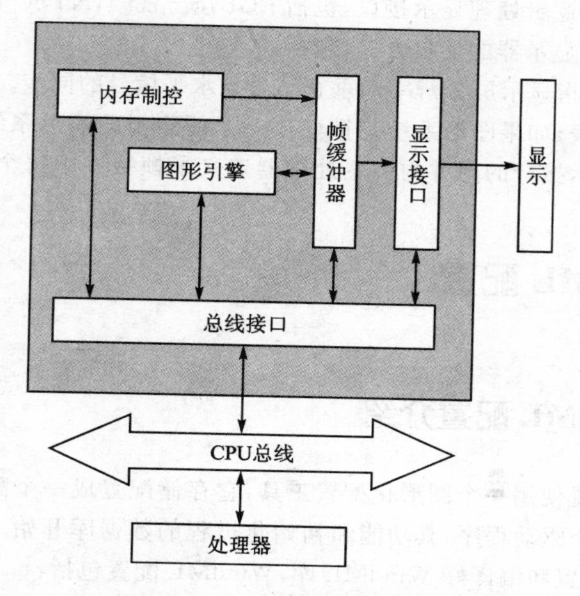

图 7.4　图形控制器框图

在设计开发图形设备驱动时，一般必须配置以下内容：

1. 帧缓冲

提供存储，用来存储显示的图像。以系列像素为单位，每个像素在屏幕上对应一个颜色和位置。帧缓冲区中的每个像素是由一组比特位组成，该比特位数取决于每个像素的颜色位数。例如，一个有 8 位色彩深度像素的设备能够显示 256 种颜色。如果像素有 16 位色彩深度的话，则能显示 65 536 种颜色。如果像素只有一位深度，则只能显示两种颜色。像素色彩深度的选择取决于成本、性能和应用类型等因素。帧缓冲存储器通常属于线性存储器，像素(0,0)储存在第一个地址中，像素(0,1)则是位于接下来的第二个地址，以此类推。像素(0,0)位于屏幕的左上角，帧缓冲区的最后一个像素则位于屏幕右下角。

2. 内存控制器

定义帧缓冲存储器访问时序。

3. 图形引擎

图形引擎是一个特殊用途的处理器，用于加速图形渲染。它承担从主处理器绘画到帧缓冲区的任务。如果一个图形设备没有图形引擎，则主处理器承担所有图形绘制的工作。

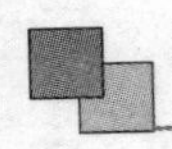

4. 显示接口

显示接口提供逻辑处理显示接口，包括 RGB，S－视频，NTSC 和 LCD。同时，还定义了图形设备与显示器的接口方式，包括：

- 显示时间，指显示屏分辨率和垂直信号与水平信号的同步。
- 颜色查找表，如果图形系统是被建成一个能索引的颜色系统，那么查找表就保留在显示接口的附件内。主处理器定义了帧缓冲中每个像素值所代表的颜色组件。

7.2 WindML 配置

7.2.1 WindML 配置介绍

WindML 配置使用一个图形化配置工具，它存储配置成一个配置文件。这个配置工具从描述每个驱动程序，其功能和初始化过程的数据库开始。在第一次使用之前，用户必须先配置和编译好 WindML 库，WindML 配置包括：

- 输出驱动程序的选择和硬件设置，比如显示器等；
- 输入驱动程序的选择和硬件设置，比如键盘或者触摸屏；
- 音频驱动程序的选择和硬件设置；
- 应用程序使用的字体。

除了这些配置需要之外，WindML 可以被定制成一个指定应用环境的应用。定制部分包括内存管理定制和硬件定制。配置好后，WindML 库必须被编译和连接到应用程序或者连接到 VxWorks Image 中。

配置 WindML 有两种方法：

- 使用 WindML 配置工具（Tornado→Tools→WindML 菜单选项），这是配置 WindML 和相关驱动程序的主要方法；
- 通过直接编辑配置头文件和源文件，实现配置工具中无法实现的定制。

在实际应用中，根据目标程序需求来决定采用何种配置方法。WindML 的标准配置包含一个图形设备、一个键盘设备和一个指示设备。配置工具允许用户配置这个标准的设备集，如果要使用多个设备，则需要修改配置文件来实现。完成了 WindML 配置工作后，可以使用 Tornado 工程来添加 WindML 到 VxWorks Image。

表 7.2 列出了两种配置方法的相应特性。用户在第一次安装和配置时，推荐使用配置工具进行配置工作。

表7.2 配置方法比较

配置方法	选择标准
配置工具:(基于GUI)	选择设备驱动程序可选项
	强制有效配置
	选择特殊字体
	存储多个配置
编辑配置文件	只允许标准WindML配置
	编辑配置文件允许非标准配置
	修改设备驱动程序可选项
	选择特殊字体
	配置定制内存管理器

7.2.2 WindML标准配置

标准配置使用的是WindML提供的驱动程序,包括:

① 1个单一的图形设备;

② 1个指示设备;

③ 1个键盘设备;

④ 1个音频设备;

⑤ 1个支持的字体引擎;

⑥ 默认的内存管理器。

如果使用非标准的硬件设备或者超出标准配置的设备,就必须做一个非标准配置。如果使用配置工具来配置WindML,那么配置工具在build过程中产生的文件是用来编译和连接WindML的;如果不使用配置工具,就必须编辑文件target/src/ugl/config/uglInit. h。

1. 配置图形设备

配置图形设备,需要设置下列基本配置选项:

① 图形设备类型,如MediaGx;

② 显示分辨率;

③ 帧缓冲颜色深度,如4,8或16位;

④ 显示器的刷新率;

⑤ 输出设备类型,即CRT或LCD显示器。

除以上配置选项外,WindML还支持可裁剪的图形设备功能(需要硬件设备支持):

① software cursors;

② overlay surfaces；

③ video；

④ JPEG；

⑤ alpha blending；

⑥ double buffers。

2. 配置键盘设备

配置键盘设备，需要设置下列配置选项：

- 键盘设备类型；
- 设备名，默认的设备名是/keyboard/0。

3. 配置指示设备

配置指示设备，需要设置下列选项：

① 指示设备类型；

② 设备名，缺省的指示设备名取决于设备类型：

- PS2 是/pointer/0；
- 串口设备指示器/tyCo/0；
- 触摸屏设备是/touchscreen/0。

4. 配置字体

WindML 需要配置字体引擎来显示文本，系统提供的字体引擎是 bitmap（位图）字体引擎，其他可用的字体引擎都是来自第三方。通常来说，必须进行下面的操作：

- 选择字体；
- 选择字体引擎类型配置选项。

如果采用命令行来配置字体的话，则必须修改字体引擎文件 uglFontengineCfg.c（在 target/src/ugl/config 目录）来定义要包含的字体和字体引擎特性，文件名中"Fontengine"是代表字体引擎名。比如，对于位图字体引擎，这个文件名就是 uglBmfCfgf.c。

5. 配置音频

配置音频，需要定义下列选项：

- 音频设备类型；
- 音频通道。

6. 其他配置项

WindML 还要配置以下项目：

(1) Event Queue Size（事件队列大小）。

设置 WindML 应用程序能够支持的事件数，缺省的事件队列大小是 100 个

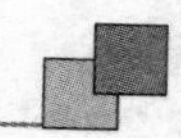

事件。

(2) Memory Manager(内存管理器)。

WindML可以使用专有内存池,也可以使用Vxworks系统内存池。当指定使用专有内存池时,WindML执行的所有内存分配都来自于专有内存池;同样,当指定使用VxWorks系统内存池时,WindML执行的所有内存分配都来自于VxWorks系统内存池。

(3) Special Processor Requirements(特殊处理器需求)。

部分处理器类型需要进行特殊设置,比如PowerPC有两个内存模型,PowerPC Reference Platform(PreP)和Common Hardware Reference Platform(CHRP)。因此,在配置如同PowerPC这样的处理器时,就必须定义正确的内存模型。

7.2.3 采用配置工具配置

1. 定义一个新的配置

以Windows操作系统为例,从Tornado菜单中选择Tornado→Tools→WindML,弹出配置窗口,如图7.5所示。

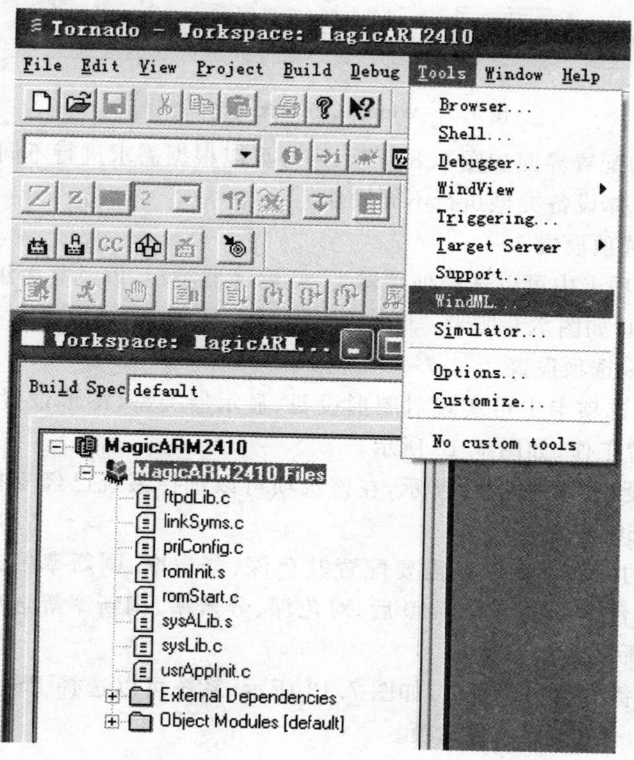

图7.5 WindML操作图

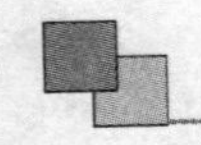

用户可以从 WindML Configuration 的下拉列表中选择已有的驱动配置，如图 7.6 所示。

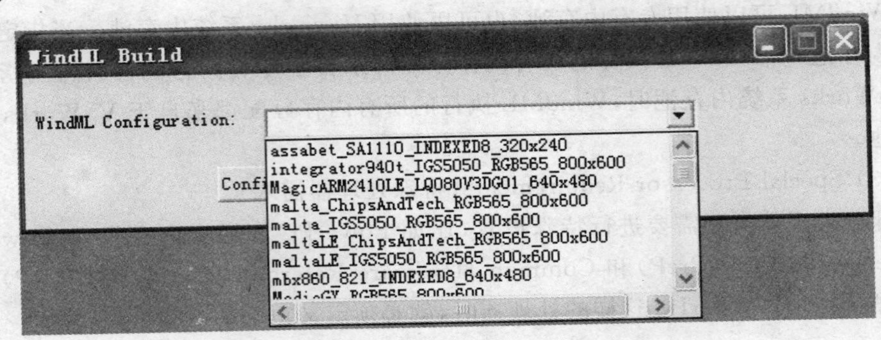

图 7.6　WindML 已有驱动配置下拉框图

用户亦可根据需求重新配置一个设备，如图 7.7 所示。

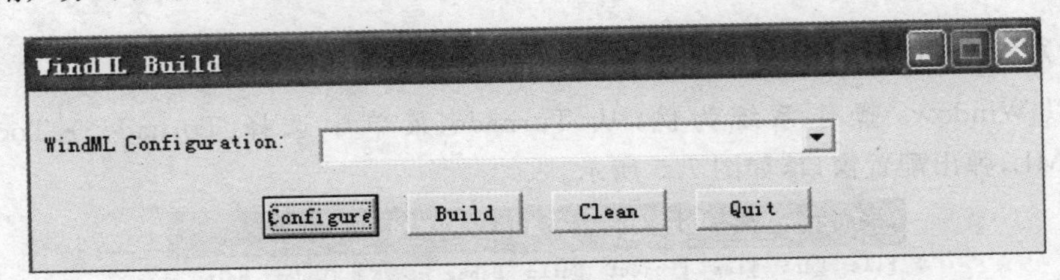

图 7.7　WindML 新建配置对话框图

WindML 的配置界面如图 7.8 所示，用户可以根据需求进行不同的选择。

用户根据具体设备类型执行不同操作。

(1) Build 选项设置。

在 Build 选项卡中可以选择处理器类型、编译器类型，以及是否编译 Debug 版本和附加编译选项，如图 7.9 所示。

(2) Devices 选项设置。

在 Devices 选项卡中可实现对图形设备、显示器类型、图形设置、鼠标、键盘、音频等选项的设置工作，如图 7.10 所示。

选择图形设备，如图 7.11 所示，在该选项可以选择系统已经支持的各种图形设备，如 IGS－5050 等。

选择特定的图形设备后还需要配置其色深、分辨率、刷新率等选项。以本书为例，选择开发平台 MagicARM2410 后，对色深、分辨率、刷新率等选项分别进行了设定，如图 7.12 所示。

Pointer 设备类型和设备名，如图 7.13 所示，系统可以选择 PS/2 Style Pointer，USB Mouse，No Pointer 3 种选项。

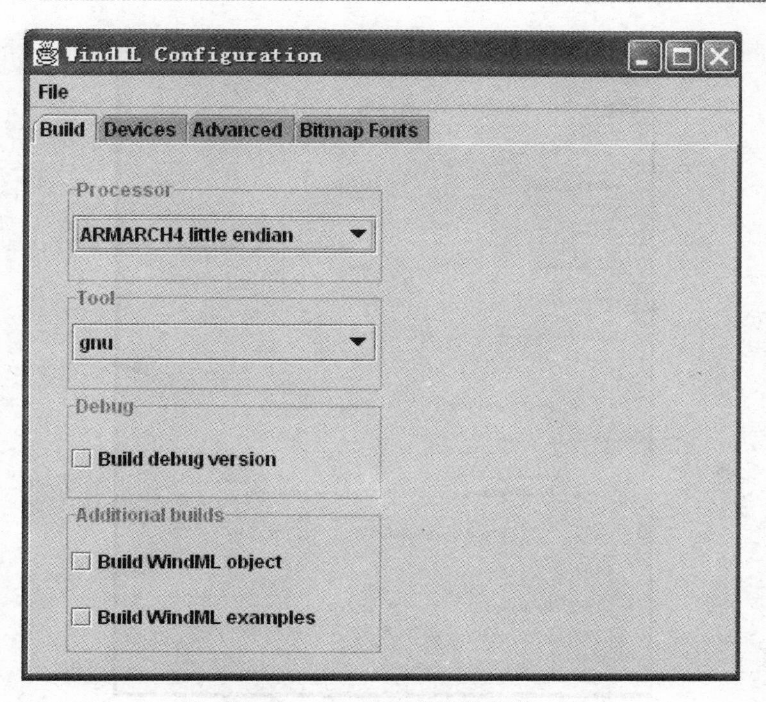

图 7.8 WindML 配置界面图

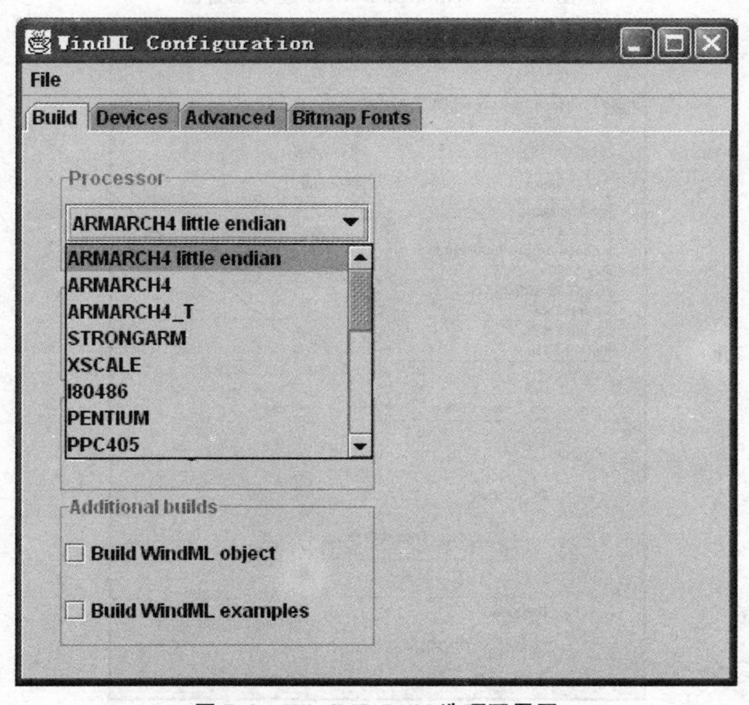

图 7.9 WindML Build 选项配置图

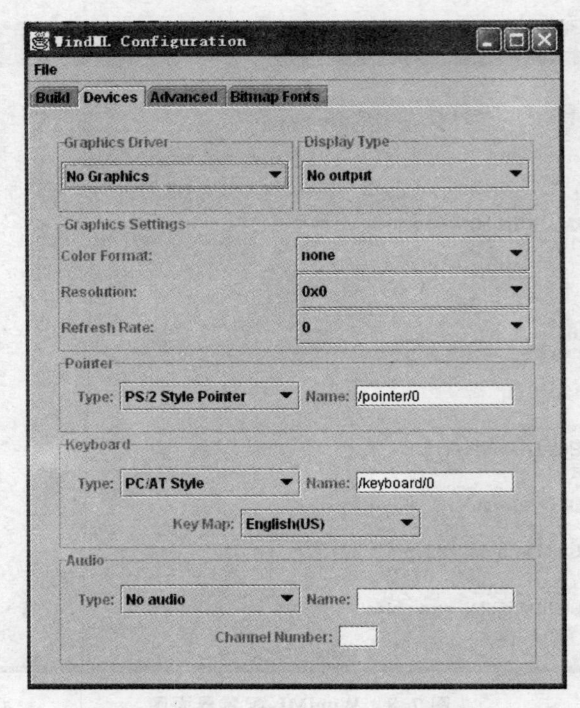

图 7.10　WindML Devices 选项配置图

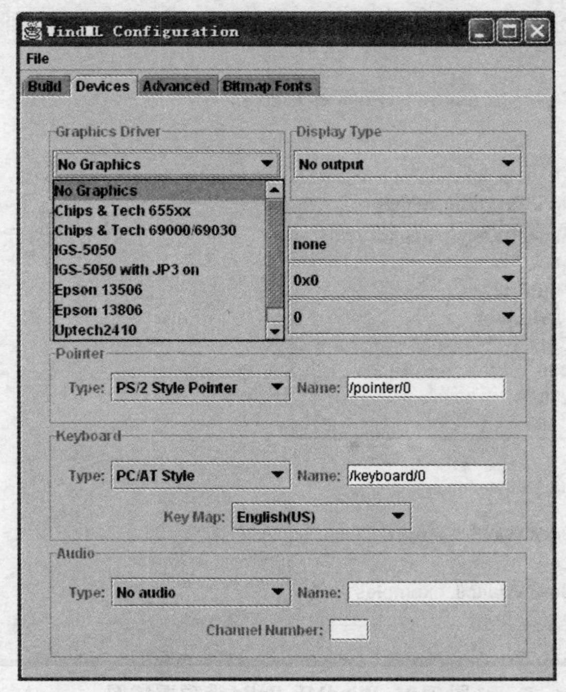

图 7.11　WindML Devices 图形设备选项配置图

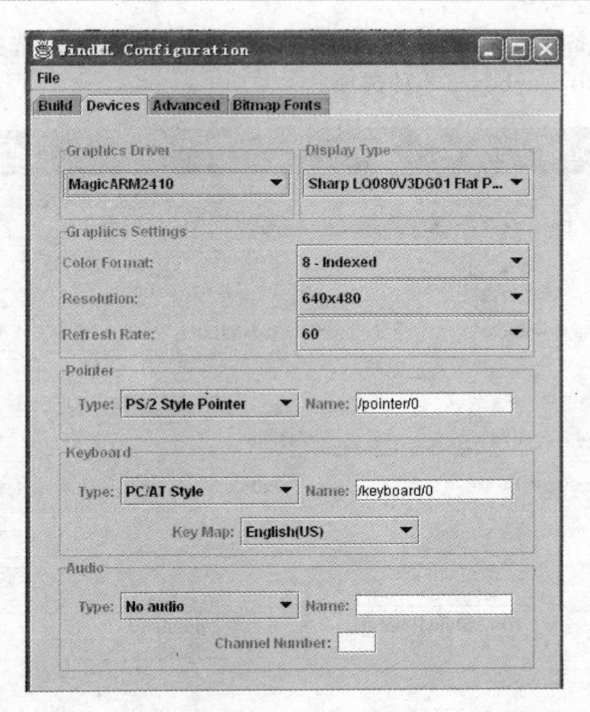

图 7.12　WindML Devices MagicARM2410 配置图

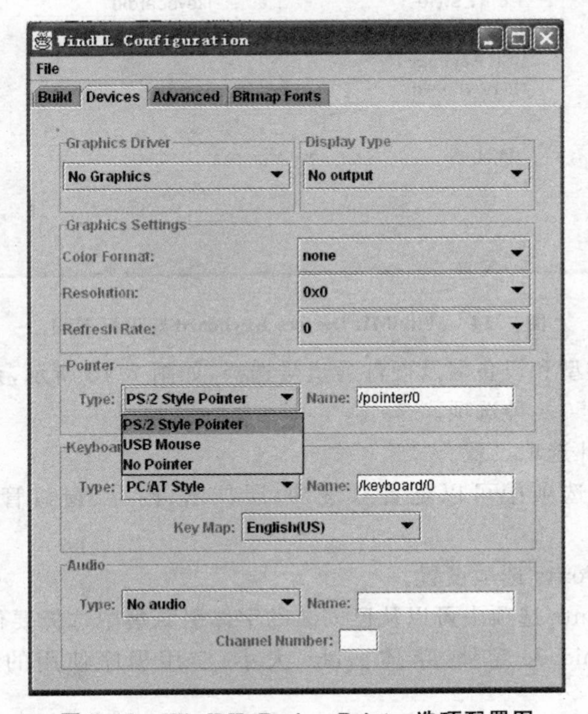

图 7.13　WindML Devices Pointer 选项配置图

Keyboard 设备类型和设备名，如图 7.14 所示，系统可以选择 PC/AT Style，USB Keyboard，No Keyboard 3 种选项。

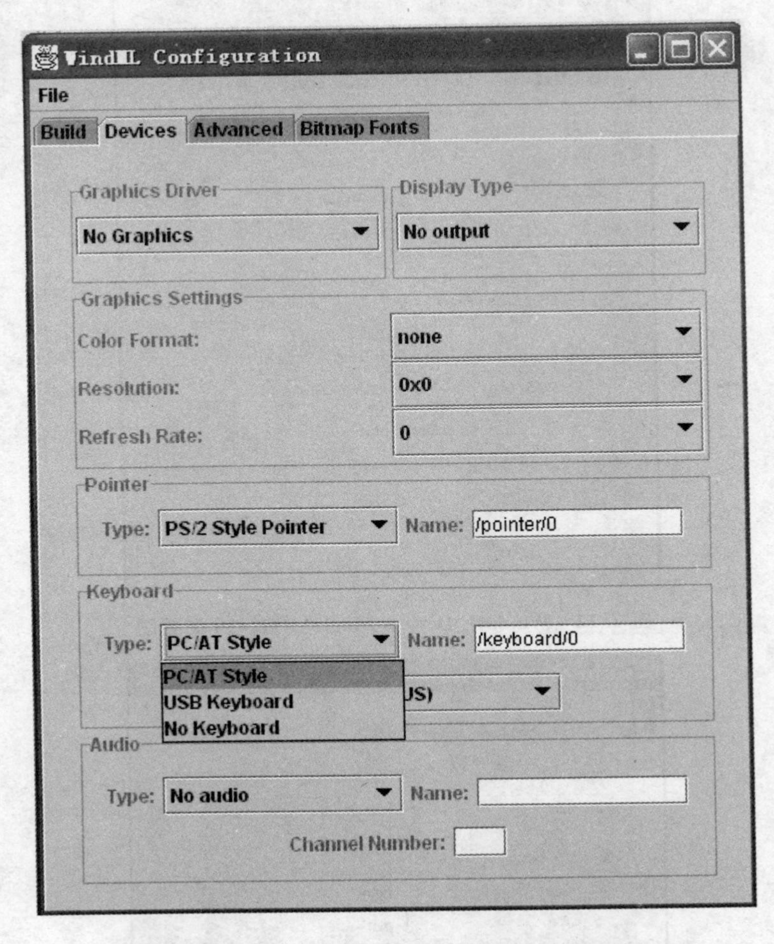

图 7.14　WindML Devices Keyboard 选项配置图

Audio 设备类型和设备名及硬件通道号选择，如图 7.15 所示，系统可以选择 No Audio 和 IGS－5050 2 种选项。

(3) Advanced 选项设置。

在 Advanced 选项中可以配置双缓冲、层技术、内存、窗口管理等，如图 7.16 所示。

(4) Bitmap Fonts 选项设置。

在 Bitmap Fonts 选项中可以从已安装的字体引擎集中选择要使用的字体引擎，以及是否使用 Unicode 字体、字体 cache 大小、应用程序使用的字体，如图 7.17 所示。

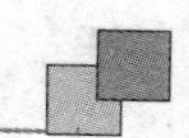

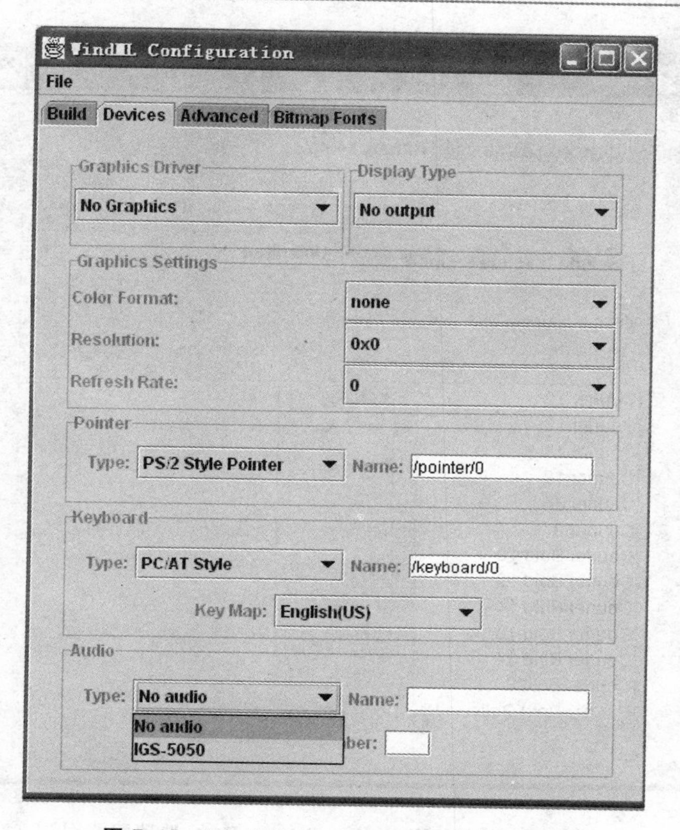

图 7.15　WindML Devices Audio 选项配置图

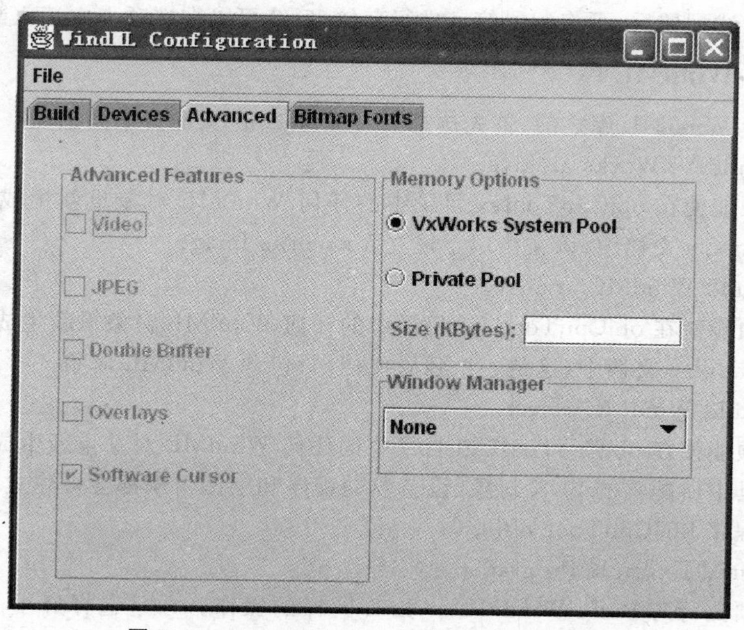

图 7.16　WindML Devices Advanst 选项配置图

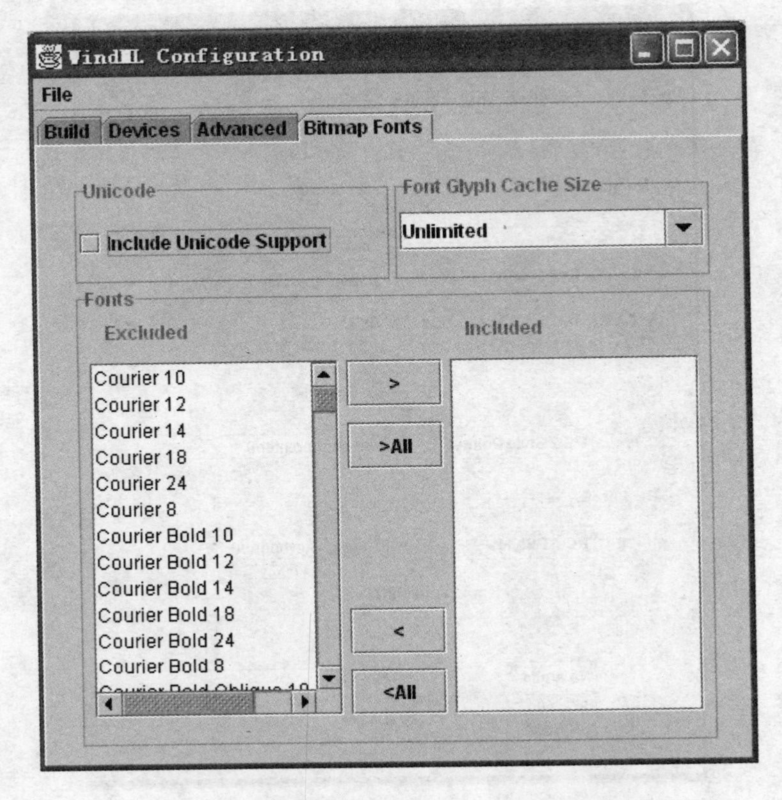

图 7.17 WindML Devices Bitmap Fonts 选项配置图

完成上述配置后，可将配置信息进行存储，后续可以在选项卡中读入默认配置。

2. 建立 WindML 库

在建立 WindML 库之前，首先选择 Miscellaneous 页配置下列选项：

(1) Build VxWorks archive。

该选项能使在 objCpuToolvx 目录中的任何 WindML 对象重新生成，并加入到 libCpuToolvx. a 文档中，该文档用于建立 VxWorks Image。

(2) Build WindML archive。

该选项能使在 objCpuToolUgl 目录中的任何 WindML 对象重新生成，并加入到 libCpuToolUgl. a 文档中，这是一个独立文档，只包含 WindML 对象。

(3) Build WindML object。

该选项能使在 objCpuToolUgl 目录中的任何 WindML 对象重新生成之后，所有在配置工具中所配置的 SDK 对象、设备驱动程序和字体引擎都会被建立在一个可下载的目标程序 lib/CpuTool. o 中。

(4) Build Example Programs。

该选项执行与 Build WindML archive 同样的操作。一旦选择好了 build 选项，

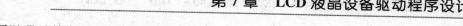

就可以通过单击“Build”按钮来建立 WindML 库。

3. 导出 WindML 目标文件

单击“Clean”按钮，将清除掉所选择的目标。比如，如果选择 Build Examples Programs 选项，则在 objCpuToolUgl 中目标就会随着 libCpuToolUgl. a 一起被清除掉。如果选择 Build VxWorks archive 选项，所有 objCpuToolvx 目录下的 WindML 目标都会被清除掉。

7.2.4 命令行配置方法

通常情况下，不推荐使用命令行配置。如果要使用多个图形设备、指示设备或者键盘，则需要修改源文件，利用命令行进行 WindML 配置。通常至少需要修改 uglInit. h 文件和所选择的字体引擎相关字体配置文件，有时也需要修改 uglInit. c 文件，这两个文件都在 target/src/ugl/config 目录下。

uglInit. h 文件指定了 WindML 的基本配置。

uglInit. c 文件控制了 WindML 库的初始化，包括函数 uglInitialize()和 uglDeinitialize()。该文件提供了处理标准 WindML 系统的函数，包括 1 个图形设备、1 个键盘、1 个指示器、1 个字体引擎和 1 个音频设备。通常，在需要处理多个设备(如多个图形设备等)时要修改这个文件。

uglInit. h 文件实现配置包括图形、键盘、指示器和音频、字体引擎、内存管理器和 miscellaneous 目标/处理器项目，文件被分成以下几个部分：

- 设备驱动程序选择；
- 字体引擎选择；
- 输入设备配置；
- 图形设备配置；
- 音频设备配置；
- 字体引擎配置；
- 内存管理器配置；
- Miscellaneous target/processor 配置。

uglInit. h 文件第一部分是选择要使用的设备驱动程序，每个所支持的设备驱动程序通过 INCLUDE_ * 来标记。要选择指定的设备驱动程序，定义(# define)相关 INCLUDE_ *，而在相关子部分其他的 INCLUDE_都必须 undefined。按照以下主要步骤进行命令行的配置。

1. 设备驱动程序

```
/*定义图形设备*/
#undef INCLUDE_BIOS_GRAPHICS
```

```
#undef INCLUDE_CHIPS_GRAPHICS
#undef INCLUDE_CUSTOM_GRAPHICS                    /* 用户定义的图形设备 */
#undef INCLUDE_IGS_GRAPHICS
#undef INCLUDE_MEDIAGX_GRAPHICS
#undef INCLUDE_SA11XX_GRAPHICS
#undef INCLUDE_SIMULATOR_GRAPHICS
#undef INCLUDE_Q2SD_GRAPHICS
#undef INCLUDE_M821_GRAPHICS
#define INCLUDE_VGA_GRAPHICS                      /* 选择 VGA 图形设备 */

/* 定义键盘类型 */
#define INCLUDE_PC_AT_KEYBOARD                    /* 选择标准的 PC AT 键盘 */
#undef INCLUDE_CUSTOM_KEYBOARD
#undef INCLUDE_SIMULATOR_KEYBOARD

/* 定义指示类设备 */
#undef INCLUDE_ASSABET_POINTER
#undef INCLUDE_CUSTOM_POINTER                     /* 用户定义的指示类设备 */
#define INCLUDE_MS_POINTER                        /* 选择了微软串口鼠标 */
#undef INCLUDE_PS2_POINTER                        /* PS-2 鼠标 */
#undef INCLUDE_SIMULATOR_POINTER                  /* 仿真指示类设备 */

/* 定义音频硬件设备 */
#define INCLUDE_IGS_AUDIO                         /* 选择了 IGS 音频设备 */
#undef INCLUDE_CUSTOM_AUDIO
```

以上这个例子，配置包括 VGA 图形设备、一个标准 PC AT 型键盘、一个微软串口鼠标和 IGS 音频设备。

2. 选择字体引擎

如同上述例子，下面建立要使用的字体引擎。

```
/* 定义字体引擎 */
#define INCLUDE_BMF_FONTS
```

本例则定义了位图字体引擎。

3. 配置输入设备

WindML 支持两种输入设备类型，键盘和指示器设备。指示器设备包括鼠标、跟踪球、触摸屏、光笔等。

默认的设备名可以通过添加相关宏来改变，比如，要改变串口鼠标名为/tyCo/1，在文件 uglInit.h 中添加下列行，这行应该在 #include <ugl/config/uglDepend.h> 前面。

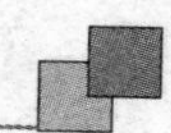

```
#define SYS_POINTER_NAME "/tyCo/1"
#include <ugl/config/uglDepend.h>
```

可以定义键盘映射类型,如下:

```
/*定义键盘映射*/
#define INCLUDE_KMAP_ENGLISH_US         /*选择美制英语键盘映射*/
#undef INCLUDE_KMAP_ENGLISH_UK
#undef INCLUDE_KMAP_GERMAN
#undef INCLUDE_KMAP_ITALIAN
#undef INCLUDE_KMAP_FRENCH
#undef INCLUDE_KMAP_SWEDISH
#undef INCLUDE_KMAP_NONE
```

4. 配置图形设备

图形配置的各个部分包括图像的分辨率、刷新率、帧缓冲格式以及可选图形设备组件内容。

图形设备配置在文件uglInit.h中包含下列内容:

```
/*定义图形显示特性*/
#define UGL_DISPLAY_WIDTH 800           /*定义显示分辨率水平显示800像素*/
#define UGL_DISPLAY_HEIGHT 600          /*定义显示分辨率垂直显示600像素*/
#define UGL_REFRESH_RATE 60             /*定义显示刷新率为60*/

/*定义帧缓冲格式*/
#undef INCLUDE_UGL_MONO
#undef INCLUDE_UGL_GREYSCALE2
#undef INCLUDE_UGL_GRAYSCALE4
#undef INCLUDE_UGL_GRAYSCALE8
#undef INCLUDE_UGL_INDEXED4
#define INCLUDE_UGL_INDEXED8
#undef INCLUDE_UGL_ARGB4444
#undef INCLUDE_UGL_RGB565
#undef INCLUDE_UGL_ARGB8888

/*定义图形驱动可选组建*/
#undef INCLUDE_UGL_ALPHA                /*透明混合处理*/
#undef INCLUDE_UGL_DOUBLE_BUFFERING     /* 双重缓冲区处理 */
#define INCLUDE_UGL_JPEG                /* JPEG扩展*/
#undef INCLUDE_UGL_OVERLAY              /* 视频渲染支持 */
#define INCLUDE_UGL_SW_CURSOR           /* 软件光标 */
#undef INCLUDE_UGL_VIDEO                /* Video扩展 */
```

上面相关代码配置如下：

- 显示分辨率 800×600；
- 刷新率为 60 Hz；
- 帧缓冲格式为每像素 8 位；
- 支持 JPEG 扩展；
- 支持软件光标。

5. 配置字体引擎

字体配置机制功能在 uglFontengineCfg. c 文件中实现，配置的主要方法就是指定字体引擎的类型。比如，要配置位图字体引擎，那么就需要在 uglInit. h 文件中添加相关定义。设置位图字体引擎的代码如下：

```
/ * 定义 cache 尺寸 * /
#define BMF_FONT_GLYPH_CACHE_SIZE? UGL_BMF_GLYPH_CACHE_SIZE_MAX

/ * 字形 cache 所用内存池 * /
#define BMF_FONT_GLYPH_CACHE_MEM_POOL UGL_DEFAULT_MEM_POOL_ID

/ * 包含 unicode 字体 * /
#define INCLUDE_UGL_BMF_UNICODE
```

要选择指定的字体，需要编辑 uglFontengineCfg. c 文件。位图配置文件 uglBm-fCfg. c 包含了一个数据结构用来定义应用程序使用的字体。

```
extern const UGL_BMF_FONT_DESC uglBMFFont_Lucida_Sans_12;
extern const UGL_BMF_FONT_DESC uglBMFFont_Helvetica_Bold_12;
extern const UGL_BMF_FONT_DESC uglBMFFont_Lucida_Sans_8;
extern const UGL_BMF_FONT_DESC uglBMFFont_Courier_12;
const UGL_BMF_FONT_DESC * uglBMFFontData[ ] =
{
    &uglBMFFont_Lucida_Sans_12,
    &uglBMFFont_Helvetica_Bold_12,
    &uglBMFFont_Lucida_Sans_8,
    &uglBMFFont_Courier_12,
    NULL
};
```

用户要使用不同的位图字体，则需要修改上述的 uglBMFFontData 数据结构，以包含 WindML 应用程序所需要的位图字体，如果是外部引用字体则必须添加到外部引用列表中。例如，要添加一个间距 18 的 Courier Bold Oblique 字体，首先添加下列外部引用：

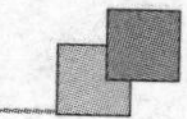

```
extern const UGL_BMF_FONT_DESC uglBMFFont_Courier_Bold_Oblique_18;
```

然后在数据结构 uglBMFFontData 中添加如下代码：

```
&uglBMFFont_Courier_Bold_Oblique_18,
```

目录 target/src/ugl/fonts/bmf 包含 WindML 发布可用的字体，如果需要其他的字体则需要添加到这个目录。

7.2.5 修改 VxWorks BSP

通常情况下，用户必须修改单板的 VxWorks BSP 来满足图形设备和输入设备的使用需要。由于目标板和处理器种类繁多，本书以通常情况下需要修改的步骤为例进行分析。

1. 图形设备内存映射

1 个图形设备有 2 个基本组件：

- 帧缓冲器；
- 1 个或多个控制器。

帧缓冲器是一个内存块，用来存储要显示的图形数据。控制器包括图形处理器、RAMDACs 和时钟芯片。

帧缓冲器和每个控制器对于处理器必须是可见的，根据不同的处理器结构，能够实现内存访问、IO 访问或者二者结合来访问的模式。要使得处理器能够访问图形设备，必须修改 BSP 的两个部分：

- 使能设备物理映射；
- 使能逻辑或虚映射。

需要根据处理器、图形设备和总线结构类型来做出准确的修改。

(1) 物理映射。

物理映射通过地址映射器来解码图形设备所占用的物理地址。图形控制器的映射取决于图形控制器的类型、处理器结构和 VxWorks BSP。

WindML 没有限制图形设备所用的总线结构，总线可以是 CPU 内部总线、ISA、PCI、AGP、VME 总线等。

(2) 虚拟映射。

虚拟映射负责处理器内存管理单元(MMU)：

执行需要的虚拟到物理的映射；

指定合适的缓存方法。

2. Tornado 工程中 WindML 组件的使用

Tornado 工程工具具有 VxWorks 中添加或者删除 WindML 的功能。在

WindML 安装完成后，Tornado 工程工具中将有一个 WindML components 的文件夹被创建，如图 7.18 所示。

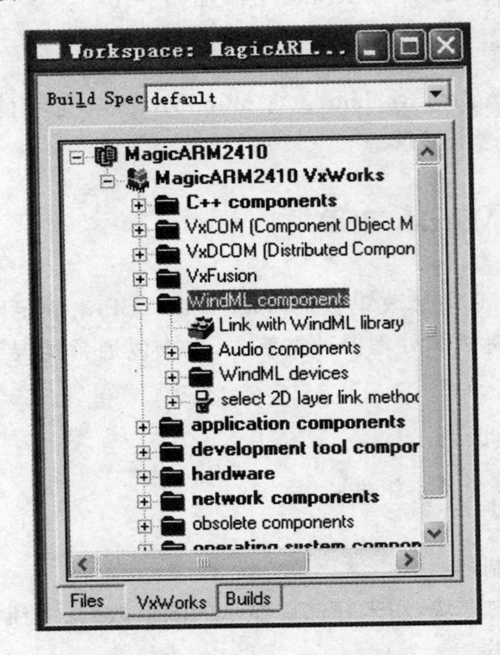

图 7.18　Tornado 工程工具 WindML components

在 WindML components 目录下有 4 个组件：Link with WindML library，Audio components，WindML devices 和 select 2D layer link method。

用户可采用 INCLUDE 方式，根据需求来配置目标工程。用户可按照以下步骤来增加 WindML 组件，如图 7.19 所示。

各组件的相关说明如下：

(1) Link with WindML library。

包含该组件后，系统才能将 WindML 编译进 VxWork 映像。

(2) Audio components。

包含该组件后，系统支持 AU、WAVE 一类的音频文件。

(3) WindML devices。

● WindML input device。

包含该组件后，系统支持 PS2 或 USB 的鼠标和键盘设备。

● graphics devices support。

包含该组件后，系统支持 PCI 类型的图形设备。

(4) selet 2D layer link method。

WindML 2D 库可以通过两种方式和 VxWorks image 相连。

● complete 2D library。

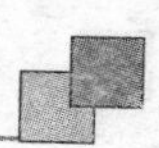

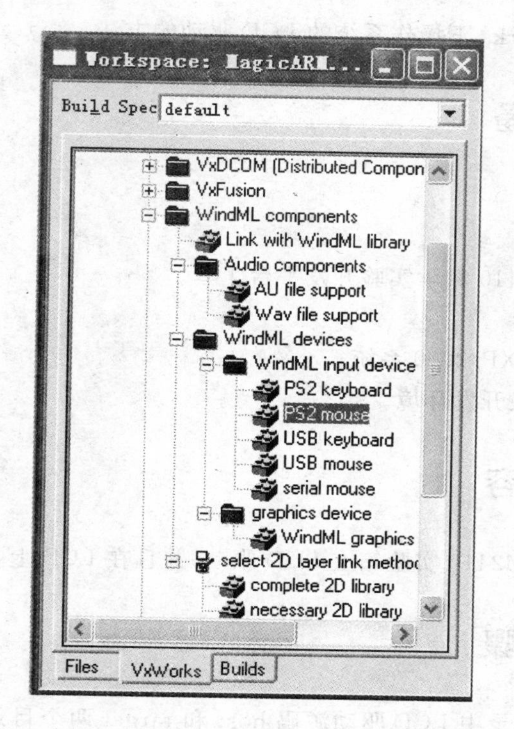

图 7.19 WindML components 组件示意图

实现将全部 2D 库连接到 VxWorks image 中，它允许 WindML 应用程序在目标机启动后下载。这种模式在 Vxworks/WindML 包含了全部的 WindML 2D 功能后，允许任何的 WindML 应用程序动态的下载。

● necessary 2D library。

系统只将必需的 2D 库连接到 VxWorks image 中，该模式需要将 WindML 编译进 VxWorks，除 WindML 应用程序之外。

注意，如果已经连接了一个更高一层的图形产品如 Zinc 或者 Personal JWorks，系统应该选择 necessary 2D library 而不是 complete 2D library。

此外如果组件已经包含进来了，可以通过相同的方式移除他们。在需要移除的组件上单击鼠标右键，弹出菜单，选择 Exclude 选项即可。

7.3 LCD 液晶驱动程序设计实验

7.3.1 实验目的

熟悉 S3C2410A 处理器的 LCD 模块，熟悉 WindML 组件的使用方法。

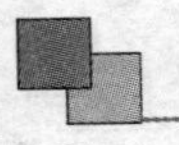

熟悉建立 VxWorks 下操作系统的 LCD 驱动的方法。

7.3.2 实验设备

(1) 硬件。

- PC 1 台;
- MagicARM2410 教学实验开发平台 1 套。

(2) 软件。

- Windows98/XP/2000 系统;
- Tornado 交叉开发环境。

7.3.3 实验内容

编写 MagicARM2410 实验箱上 LCD 驱动,并且在 LCD 上显示字符和图像。

7.3.4 实验步骤

(1) 复制本书附录中 LCD 驱动源码 host 和 target 两个目录到 Tornado 安装目录,提示是否覆盖时,选择"Yes"。

(2) 运行 Tools→WindML,选择 uptech2410LE_LQ080V3DG01_640×480,如图 7.20 所示进行配置,逐步操作,直至按 Build 进行编译。

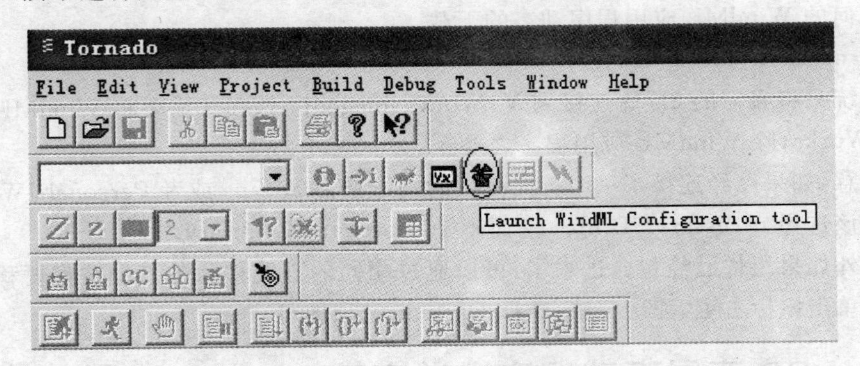

图 7.20 启动 WindML 配置工具

如图 7.21 所示,在下拉列表中选择 MagicARM2410LE_LQ080V3DG01_640×480 选项。

此时,点击 Configure 按键后,界面如图 7.22 所示。

如图 7.22 所示,处理器选择 ARMARCH4 little endian,Tool 选择 diab,Debug 选择 Build debug version,在 Additional builds 中选择 Build WindML object。然后

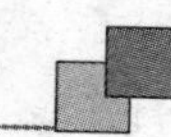

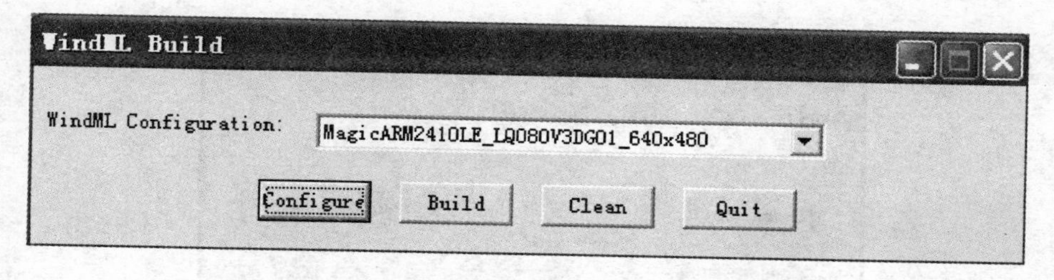

图 7.21　WindML 配置对象图

File
Build　Devices　Advanced　Bitmap Fonts
Processor
ARMARCH4 little endian
Tool
diab
Debug
Build debug version
Additional builds
Build WindML object
Build WindML examples

图 7.22　WindML Build 配置图

选择 Devices 页面，如图 7.23 所示。

在 Devices 配置页中，如图完成相应配置。Graphics Driver 选项选择 MagicARM2410，Graphics Settings 中的 Color-Format 选择 16 - RGB565，Resolution 选择 640×480，Refresh Rate 选择 60。Display Type 选择 Sharp LQ080V3DG01 Flat Panel。其余配置选项都选择 None 即可。

接下来选择 Advanced 配置页，如图 7.24 所示。

如图 7.24 所示，在该配置页中共有 3 大项。在 Advanced Features 选择 JPEG 和 Software Cursor 两项；在 Memory Options 中选择 VxWorks System Pool 项；在 Window Manager 中选择 WWM 项。完成该页配置后，选择 Bitmap Fonts 配置页，如图 7.25 所示。

如图 7.25 所示，共有 3 大项：Unicode 项不做选择；Font Glyph Cache Size 选择 Unlimited；在 Fonts 配置项中，选择 Times Roman 18，完成该页配置。

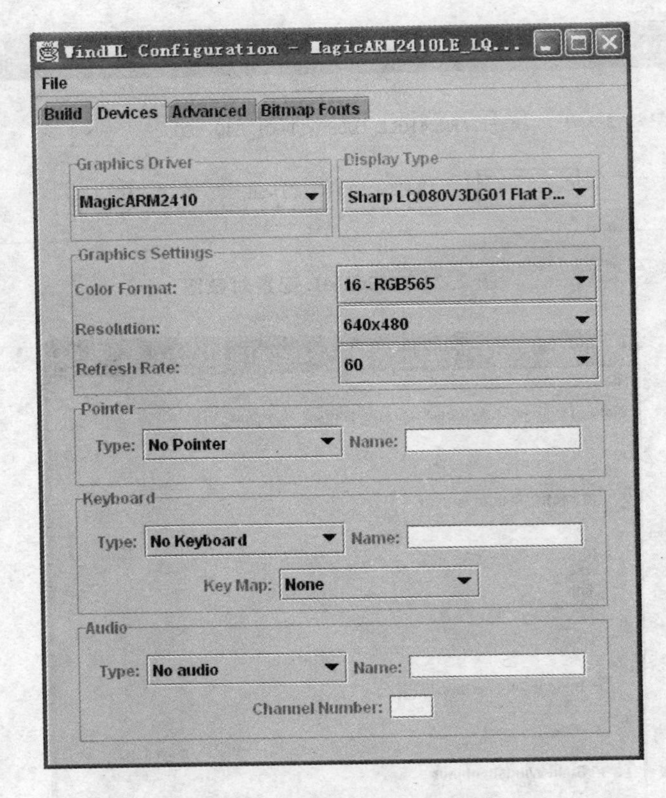

图 7.23　WindML Devices 配置图

图 7.24　WindML Advanced 配置图

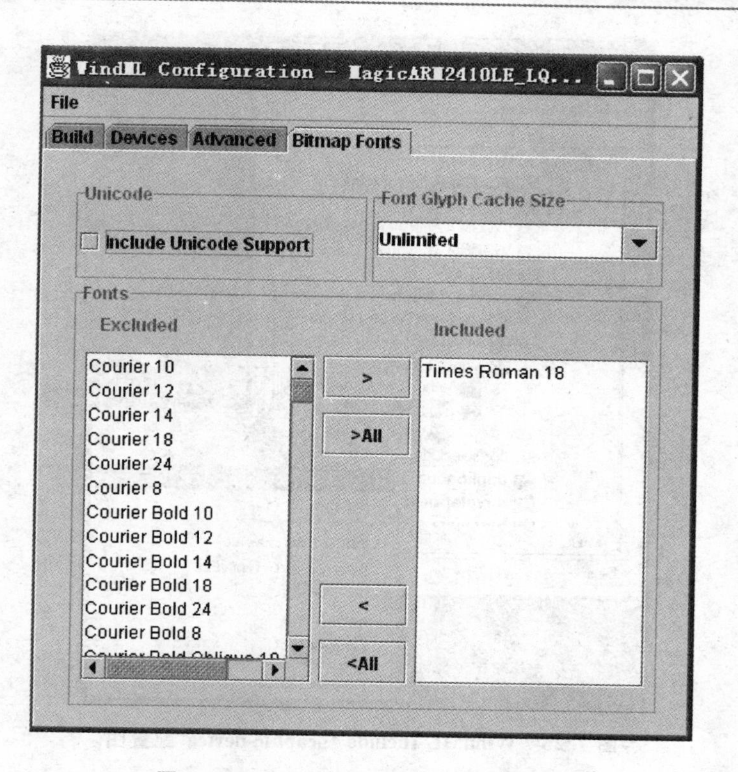

图 7.25 WindML Bitmap Fonts 配置图

配置完毕后，关闭配置界面并保存自动产生的配置文件。单击 Build 按钮，编译 LCD 驱动，编译后会在 C:\Tornado2.2\target\lib\arm\ARMARCH4\diab 目录下产生 libwndml.a 库。编译的过程大概在 10 到 20 分钟之间，请耐心等待。

(3) 在 MagicARM2410 的 Project 中添加 WindML 组件，过程如图 7.26 和图7.27 所示：

(4) 重新编译 VxWorks 镜像，包含 LCD 驱动的 VxWorks 镜像大小在 4 MB 左右。

(5) 按照前文所述方法，下载包含 LCD 驱动的 VxWorks 镜像。

(6) 在 MagicARM2410 的 Project 中添加 ugldemo.c 文件并编译，通过 Target Server 把 ugldemo.o 下载到系统中去。

(7) 在 WShell 中执行“ugldemo(-1)”。

(8) 观察 LCD 屏上的输出，显示如图 7.28 所示，则完成实验。

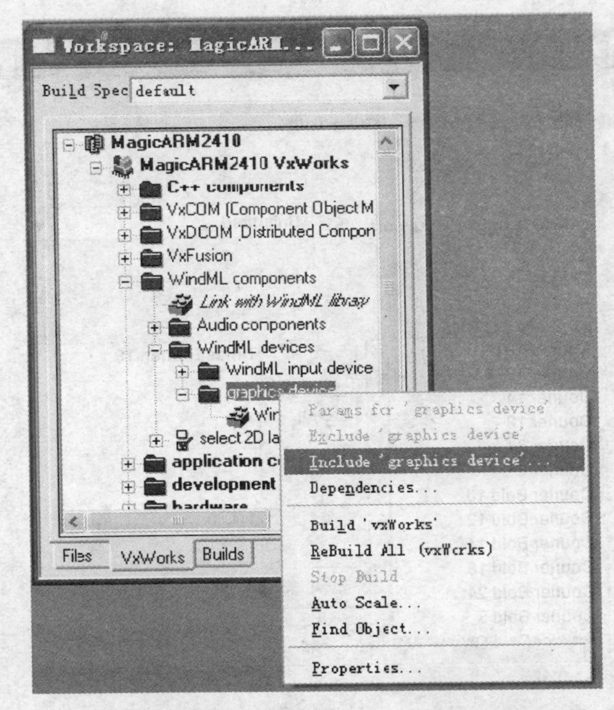

图 7.26　WindML Include 'graphic device'配置图

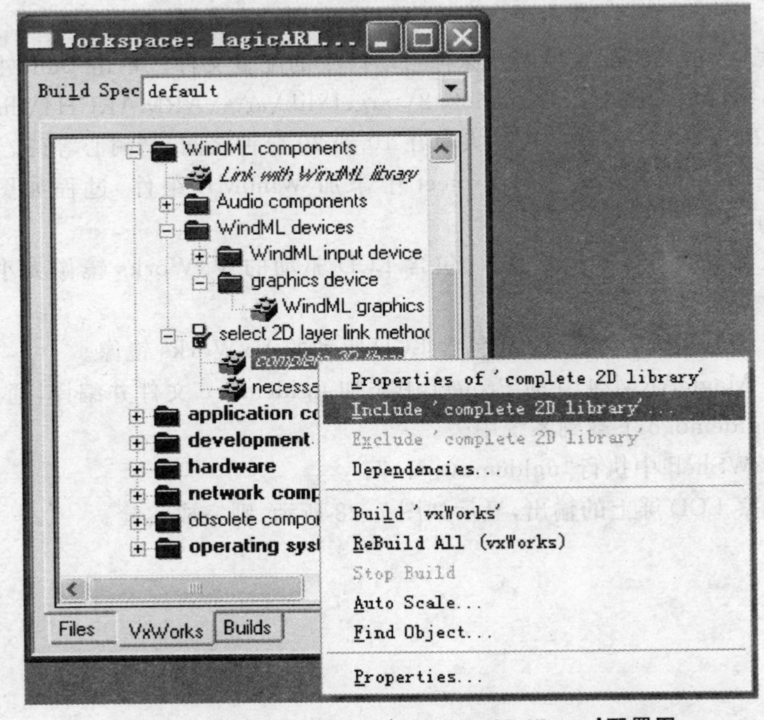

图 7.27　WindML Include 'complete 2D library'配置图

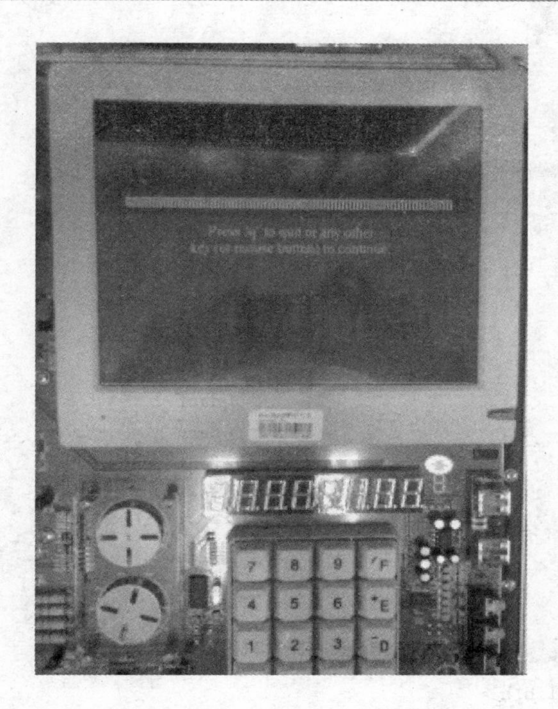

图 7.28　MagicARM2410 运行效果图

7.3.5　程序清单

具体程序如下所示。实验所给出的 LCD 目录程序如图 7.29 所示。

```
/* ugldemo.c - Graphics primitives demonstration program */

#if ! defined(WINDML_NATIVE)
#include <vxWorks.h>
#elif defined(__unix__)
#include <unistd.h>
#endif

#include <ugl/ugl.h>
#include <ugl/uglos.h>
#include <ugl/uglMsg.h>
#include <ugl/uglfont.h>
#include <ugl/uglinput.h>
#include <ugl/ugldib.h>
#include <stdio.h>
```

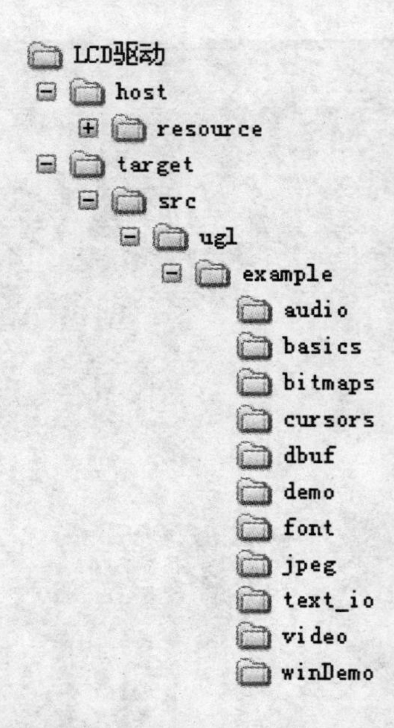

图 7.29　MagicARM2410 LCD 程序目录

```
#include <stdlib.h>
#include <ioLib.h>

UGL_DEVICE_ID devId;
UGL_GC_ID gc;
static UGL_INPUT_SERVICE_ID inputServiceId;
static UGL_REGION_ID regionId;
static UGL_FONT_ID fontDialog;
static UGL_FONT_ID fontSystem;
static UGL_FONT_ID fontFixed;
static UGL_FONT_DRIVER_ID fontDrvId;
static UGL_DDB_ID stdDdb;
static UGL_MDDB_ID patternDdb;
static UGL_CDDB_ID cursorDdb;
static UGL_TDDB_ID transDdb;
static int * randomData;
static UGL_COLOR * colorData;
static int displayHeight, displayWidth;

int fd;
void windMLDemo (int mode);
```

```
/*
 * The color table is where we define the colors we want
 * to have available.  The format is an array of
 * ARGB values paired with their allocated uglColor.  As
 * of this writing, we don't need to worry about Alpha
 * ("A") values unless we are using video.
 */
struct _colorStruct
{
    UGL_RGB rgbColor;
    UGL_COLOR uglColor;
}
colorTable[] =
{
    { UGL_MAKE_RGB(0, 0, 0), 0},
    { UGL_MAKE_RGB(0, 0, 168), 0},
    { UGL_MAKE_RGB(0, 168, 0), 0},
    { UGL_MAKE_RGB(0, 168, 168), 0},
    { UGL_MAKE_RGB(168, 0, 0), 0},
    { UGL_MAKE_RGB(168, 0, 168), 0},
    { UGL_MAKE_RGB(168, 84, 0), 0},
    { UGL_MAKE_RGB(168, 168, 168), 0},
    { UGL_MAKE_RGB(84, 84, 84), 0},
    { UGL_MAKE_RGB(84, 84, 255), 0},
    { UGL_MAKE_RGB(84, 255, 84), 0},
    { UGL_MAKE_RGB(84, 255, 255), 0},
    { UGL_MAKE_RGB(255, 84, 84), 0},
    { UGL_MAKE_RGB(255, 84, 255), 0},
    { UGL_MAKE_RGB(255, 255, 84), 0},
    { UGL_MAKE_RGB(255, 255, 255), 0}
};

/*
 * This is the data for a user defined fill pattern that
 * can be used with various drawing primitives.  This data
 * will be used to create a monochrome DIB (MDIB), which
 * will be used in turn to create a monochrome "bitmap".
 * It is the bitmap that is used by the drawing primitives.
 *
 * Solid fills do not need to have a pattern defined, just
 * pass an UGL_NULL value in the pattern parameter to
 * revert (it is the default) to solid fills.
 */
```

```
struct
{
    int width;
    int height;
    unsigned char data[32];
} patternData =
 {
     16, 16,
     {
       0xFF, 0xFF,
       0x00, 0x01,
       0x00, 0x01,
       0x00, 0x01,
       0x00, 0x01,
       0x00, 0x01,
       0x00, 0x01,
       0x00, 0x01,
       0xFF, 0xFF,
       0x01, 0x00,
       0x01, 0x00,
       0x01, 0x00,
       0x01, 0x00,
       0x01, 0x00,
       0x01, 0x00,
       0x01, 0x00
     }
 };

/* Label the colors we defined */
#define BLACK              (0)
#define BLUE               (1)
#define GREEN              (2)
#define CYAN               (3)
#define RED                (4)
#define MAGENTA            (5)
#define BROWN              (6)
#define LIGHTGRAY          (7)
#define DARKGRAY           (8)
#define LIGHTBLUE          (9)
#define LIGHTGREEN         (10)
```

```
#define LIGHTCYAN          (11)
#define LIGHTRED           (12)
#define LIGHTMAGENTA       (13)
#define YELLOW             (14)
#define WHITE              (15)
#define TRANS              (255)
#define INVERT             (254)

/* Definition of a smiley face cursor */
UGL_ARGB cursorClut[] =
{
    UGL_MAKE_RGB(0, 0, 0),
    UGL_MAKE_RGB(255, 255, 84),
};

UGL_UINT8 cursorData[] =
{
    #define B 0,
    #define Y 1,
    #define T 255,
    #define I 254,
    T T T T T T T T T T T T Y Y Y Y Y Y Y T T T T T T T T T T T T T
    T T T T T T T T T Y Y Y Y Y Y Y Y Y Y Y Y Y Y T T T T T T T T T
    T T T T T T T Y Y Y Y Y Y Y Y Y Y Y Y Y Y Y Y Y Y T T T T T T T
    T T T T T T Y Y Y Y Y Y Y Y Y Y Y Y Y Y Y Y Y Y Y Y T T T T T T
    T T T T T Y Y Y Y Y Y Y Y Y Y Y Y Y Y Y Y Y Y Y Y Y Y T T T T T
    T T T T Y Y Y Y Y Y Y Y Y Y Y Y Y Y Y Y Y Y Y Y Y Y Y Y T T T T
    T T T Y Y Y Y Y Y Y Y Y Y Y Y Y Y Y Y Y Y Y Y Y Y Y Y Y Y T T T
    T T Y Y Y Y Y Y Y Y Y Y Y Y Y Y Y Y Y Y Y Y Y Y Y Y Y Y Y Y T T
    T T Y Y Y Y Y Y Y T T T Y Y Y Y Y Y Y Y I I I Y Y Y Y Y Y Y T T
    T Y Y Y Y Y Y Y T T T T T Y Y Y Y Y Y I I I I I Y Y Y Y Y Y Y T
    T Y Y Y Y Y Y Y T T T T T Y Y Y Y Y Y I I I I I Y Y Y Y Y Y Y T
    T Y Y Y Y Y Y Y T T T T T Y Y Y Y Y Y I I I I I Y Y Y Y Y Y Y T
    Y Y Y Y Y Y Y Y Y T T T Y Y Y Y Y Y Y Y I I I Y Y Y Y Y Y Y Y Y
    Y Y Y Y Y Y Y Y Y Y Y Y Y Y Y Y Y Y Y Y Y Y Y Y Y Y Y Y Y Y Y Y
    Y Y Y Y Y Y Y Y Y Y Y Y Y Y Y Y Y Y Y Y Y Y Y Y Y Y Y Y Y Y Y Y
    Y Y Y Y Y Y Y Y Y Y Y Y Y Y Y Y Y Y Y Y Y Y Y Y Y Y Y Y Y Y Y Y
    Y Y Y Y Y Y Y Y Y Y Y Y Y Y Y Y Y Y Y Y Y Y Y Y Y Y Y Y Y Y Y Y
    Y Y Y Y Y Y Y Y Y Y Y Y Y Y Y Y Y Y Y Y Y Y Y Y Y Y Y Y Y Y Y Y
    Y Y Y Y Y Y Y Y Y Y Y Y Y Y Y Y Y Y Y Y Y Y Y Y Y Y Y Y Y Y Y Y
    Y Y Y Y Y B B Y Y Y Y Y Y Y Y Y Y Y Y Y Y Y Y Y Y B B Y Y Y Y Y
    T Y Y Y Y B B Y Y Y Y Y Y Y Y Y Y Y Y Y Y Y Y Y Y B B Y Y Y Y T
```

```
    T Y Y Y Y B B B Y Y Y Y Y Y Y Y Y Y Y Y Y Y Y Y B B B Y Y Y Y T
    T Y Y Y Y Y B B Y Y Y Y Y Y Y Y Y Y Y Y Y Y Y Y B B Y Y Y Y Y T
    T T Y Y Y Y B B B Y Y Y Y Y Y Y Y Y Y Y Y Y Y B B B Y Y Y Y T T
    T T Y Y Y Y Y B B B B Y Y Y Y Y Y Y Y Y Y B B B B Y Y Y Y Y T T
    T T T Y Y Y Y Y B B B B B B Y Y Y Y B B B B B B Y Y Y Y Y T T T
    T T T T Y Y Y Y Y Y B B B B B B B B B B B B Y Y Y Y Y Y T T T T
    T T T T T Y Y Y Y Y Y Y Y B B B B B B Y Y Y Y Y Y Y Y T T T T T
    T T T T T T Y Y Y Y Y Y Y Y Y Y Y Y Y Y Y Y Y Y Y Y T T T T T T
    T T T T T T T Y Y Y Y Y Y Y Y Y Y Y Y Y Y Y Y Y Y T T T T T T T
    T T T T T T T T T Y Y Y Y Y Y Y Y Y Y Y Y Y Y T T T T T T T T T
    T T T T T T T T T T T T Y Y Y Y Y Y Y Y T T T T T T T T T T T T
    #undef B
    #undef Y
    #undef T
    #undef I
};

UGL_UINT8 transparentData[] =
{
    #define O BLACK,
    #define _ BLACK,
    #define I YELLOW,
    O O O O O O O O O O O O I I I I I I I I O O O O O O O O O O O O
    O O O O O O O O O I I I I I I I I I I I I I I O O O O O O O O O
    O O O O O O O I I I I I I I I I I I I I I I I I I O O O O O O O
    O O O O O O I I I I I I I I I I I I I I I I I I I I O O O O O O
    O O O O O I I I I I I I I I I I I I I I I I I I I I I O O O O O
    O O O O I I I I I I I I I I I I I I I I I I I I I I I I O O O O
    O O O I I I I I I I I I I I I I I I I I I I I I I I I I I O O O
    O O I I I I I I I I I I I I I I I I I I I I I I I I I I I I O O
    O O I I I I I I I _ _ _ I I I I I I I I _ _ _ I I I I I I I O O
    O I I I I I I I _ _ _ _ _ I I I I I I _ _ _ _ _ I I I I I I I O
    O I I I I I I I _ _ _ _ _ I I I I I I _ _ _ _ _ I I I I I I I O
    O I I I I I I I _ _ _ _ _ I I I I I I _ _ _ _ _ I I I I I I I O
    I I I I I I I I I _ _ _ I I I I I I I I _ _ _ I I I I I I I I I
    I I I I I I I I I I I I I I I I I I I I I I I I I I I I I I I I
    I I I I I I I I I I I I I I I I I I I I I I I I I I I I I I I I
    I I I I I I I I I I I I I I I I I I I I I I I I I I I I I I I I
    I I I I I I I I I I I I I I I I I I I I I I I I I I I I I I I I
    I I I I I I I I I I I I I I I I I I I I I I I I I I I I I I I I
    I I I I I I I I I I I I I I I I I I I I I I I I I I I I I I I I
    I I I I I _ _ I I I I I I I I I I I I I I I I I I _ _ I I I I I
```

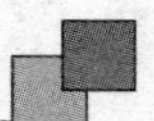

```
    OIIII__IIIIIIIIIIIIIIIIII__IIIIO
    OIIII___IIIIIIIIIIIIIIII___IIIIO
    OIIIII__IIIIIIIIIIIIIIII__IIIIIO
    OOIIII___IIIIIIIIIIIIII___IIIIOO
    OOIIIII____IIIIIIIIII____IIIIIOO
    OOOIIIII______IIIII_____IIIIIOOO
    OOOOIIIIII____________IIIIIIOOOO
    OOOOOIIIIIIII______IIIIIIIIOOOOO
    OOOOOOIIIIIIIIIIIIIIIIIIIIOOOOOO
    OOOOOOOIIIIIIIIIIIIIIIIIIOOOOOOO
    OOOOOOOOOIIIIIIIIIIIIIIOOOOOOOOO
    OOOOOOOOOOOOIIIIIIIIOOOOOOOOOOOO
    #undef _
    #undef O
    #undef I
};

UGL_UINT8 transparentMask[] =
{
    0x00, 0x0F, 0xF0, 0x00,
    0x00, 0x7F, 0xFE, 0x00,
    0x01, 0xFF, 0xFF, 0x80,
    0x03, 0xFF, 0xFF, 0xC0,
    0x07, 0xFF, 0xFF, 0xE0,
    0x0F, 0xFF, 0xFF, 0xF0,
    0x1F, 0xFF, 0xFF, 0xF8,
    0x3F, 0xFF, 0xFF, 0xFC,
    0x3F, 0xFF, 0xFF, 0xFC,
    0x7F, 0xFF, 0xFF, 0xFE,
    0x7F, 0xFF, 0xFF, 0xFE,
    0x7F, 0xFF, 0xFF, 0xFE,
    0xFF, 0xFF, 0xFF, 0xFF,
    0xFF, 0xFF, 0xFF, 0xFF,
    0xFF, 0xFF, 0xFF, 0xFF,
    0xFF, 0xFF, 0xFF, 0xFF,
    0xFF, 0xFF, 0xFF, 0xFF,
    0xFF, 0xFF, 0xFF, 0xFF,
    0xFF, 0xFF, 0xFF, 0xFF,
    0xFF, 0xFF, 0xFF, 0xFF,
    0x7F, 0xFF, 0xFF, 0xFE,
    0x7F, 0xFF, 0xFF, 0xFE,
    0x7F, 0xFF, 0xFF, 0xFE,
```

```
    0x3F, 0xFF, 0xFF, 0xFC,
    0x3F, 0xFF, 0xFF, 0xFC,
    0x1F, 0xFF, 0xFF, 0xF8,
    0x0F, 0xFF, 0xFF, 0xF0,
    0x07, 0xFF, 0xFF, 0xE0,
    0x03, 0xFF, 0xFF, 0xC0,
    0x01, 0xFF, 0xFF, 0x80,
    0x00, 0x7F, 0xFE, 0x00,
    0x00, 0x0F, 0xF0, 0x00
};

/************************************************************************
** Function name: flushQ
** Descriptions: 读取消息队列直至队列清空
** Input: 无
**
** Output: 无
************************************************************************/
static void flushQ (void)
{
    UGL_MSG msg;
    UGL_STATUS status;

    do
    {
        status = uglInputMsgGet (inputServiceId, &msg, UGL_NO_WAIT);
        if (msg.type == MSG_POINTER)
        {
            uglCursorMove(devId,
                          msg.data.pointer.position.x,
                          msg.data.pointer.position.y);
        }
    } while (status != UGL_STATUS_Q_EMPTY);
}

/************************************************************************
** Function name: pauseDemo
** Descriptions: 读取消息队列直至队列清空
** Input: mode 0  配置输入设备等待输入;
**
**
** Output: mode 为"q" 或者"Q" 时,返回值为 -1,其他输入返回值为 0
```

```
***********************************************************************/
static int pauseDemo
(
    int mode                          /* operating mode */
)
{
    static UGL_CHAR * message = "Press 'q' to quit or any other";
    static UGL_CHAR * message2 = "key (or mouse button) to continue.";
    int textWidth, textHeight;
    UGL_MSG msg;
    UGL_STATUS status;
    int retVal = 0;
    char ch;

    if (mode == 0 && inputServiceId != UGL_NULL)
    {
        uglBackgroundColorSet(gc, colorTable[BLACK].uglColor);
        uglForegroundColorSet(gc, colorTable[LIGHTRED].uglColor);
        uglFontSet(gc, fontSystem);

        uglTextSizeGet(fontSystem, &textWidth, &textHeight, -1, message);

        uglTextDraw(gc, (displayWidth - textWidth) / 2,
                    (displayHeight - textHeight) / 2 - textHeight, -1, mes-
                    sage);

        uglTextSizeGet(fontSystem, &textWidth, &textHeight, -1, message2);

        uglTextDraw (gc, (displayWidth - textWidth) / 2,
                    (displayHeight - textHeight) / 2, -1, message2);

        flushQ();

        UGL_FOREVER
        {
            status = uglInputMsgGet (inputServiceId, &msg, UGL_WAIT_FOREVER);

          if (msg.type == MSG_KEYBOARD)
          {
              if (msg.data.keyboard.modifiers & UGL_KBD_KEYDOWN)
              {
                  if (msg.data.keyboard.key == 'q' ||
```

```
                    msg.data.keyboard.key == 'Q')
                    retVal = -1;

                break;
            }
        }
        else if (msg.type == MSG_POINTER)
        {
            uglCursorMove (devId, msg.data.pointer.position.x,
                          msg.data.pointer.position.y);
            if ((msg.data.pointer.buttonChange &
                 msg.data.pointer.buttonState) != 0)
                break;
        }
    }
}
else if (mode > 0)
{
    uglOSTaskDelay (mode * 1000);
}
else if(mode == -1)
{
    ioctl (fd, FIOFLUSH, 0);
    uglBackgroundColorSet(gc, colorTable[BLACK].uglColor);
    uglForegroundColorSet(gc, colorTable[LIGHTRED].uglColor);
    uglFontSet(gc, fontSystem);

    uglTextSizeGet(fontSystem, &textWidth, &textHeight,
                   -1, message);

    uglTextDraw(gc, (displayWidth - textWidth) / 2,
                (displayHeight - textHeight) / 2  - textHeight, -1, mes-
                sage);

    uglTextSizeGet(fontSystem, &textWidth, &textHeight,
                   -1, message2);

    uglTextDraw(gc, (displayWidth - textWidth) / 2,
                (displayHeight - textHeight) / 2, -1, message2);


    /* no input device and check the console port input, if input q or Q,
```

```
exit */
        while(read(fd, &ch, 1) > 0)
         {
            if((ch == 'q') || (ch == 'Q'))
            {
                retVal = -1;
                break;
            }
             break;
         }
      }
    return(retVal);
  }

  /***********************************************************
  ** Function name: ClearScreen
  ** Descriptions: 清屏
  ** Input:
  **
  **
  ** Output: 无
  ***********************************************************/
  static void ClearScreen
  (
      UGL_GC_ID gc                    /* Graphics context */
  )
  {
      uglBackgroundColorSet(gc, colorTable[BLACK].uglColor);
      uglForegroundColorSet(gc, colorTable[BLACK].uglColor);
      uglLineStyleSet(gc, UGL_LINE_STYLE_SOLID);
      uglLineWidthSet(gc, 1);
      uglRectangle(gc, 0, 0, displayWidth - 1, displayHeight - 1);
  }

  /***********************************************************
  ** Function name: cleanUp
  ** Descriptions: 在停止前清除所有资源
  ** Input: mode  操作模式
  **
  **
  ** Output: 无
  ***********************************************************/
```

```
static void cleanUp
(
    int mode                          /* Operating mode */
)
{
    if (mode >= 0 && inputServiceId != UGL_NULL)
    {
        uglCursorBitmapDestroy (devId, cursorDdb);
        uglCursorDeinit (devId);
    }

    uglTransBitmapDestroy (devId, transDdb);
    uglBitmapDestroy(devId, stdDdb);
    UGL_FREE(colorData);
    uglMonoBitmapDestroy (devId, patternDdb);
    uglRegionDestroy(regionId);
    UGL_FREE (randomData);
    uglFontDestroy (fontFixed);
    uglFontDestroy (fontDialog);
    uglFontDestroy (fontSystem);
    uglGcDestroy (gc);

    uglDeinitialize();
}


/*************************************************************************
** Function name: main
** Descriptions:
** Input: 无
**
**
** Output: 无
*************************************************************************/
int main (int argc, char *argv [])
{
    int mode = 0;
    if (argc > 1)
        mode = atoi (argv [1]);

    windMLDemo (mode);
    return 0;
```

```
}
#elif defined(WINDML_NATIVE) && defined(_WIN32)


/*************************************************************
** Function name: WinMain
** Descriptions:
** Input: 无
**
**
** Output: 无
*************************************************************/
int WINAPI WinMain(HINSTANCE hInstance, HINSTANCE hPrevInstance,
                LPSTR lpCmdLine, int nShowCmd)
{
    uglWin32Parameters(hInstance, hPrevInstance, lpCmdLine, nShowCmd);
    windMLDemo(0);
    return (0);
}
#else

/*************************************************************
** Function name: ugldemo
** Descriptions:
** Input: 无
**
**
** Output: 无
*************************************************************/
void ugldemo
(
    int mode                    /* Operating mode */
)
{
    fd = open("/tyCo/0", 2, 0644);
    ioctl (fd, FIOSETOPTIONS, OPT_ECHO | OPT_CRMOD | OPT_TANDEM | \
          OPT_MON_TRAP | OPT_7_BIT | OPT_ABORT );

    uglOSTaskCreate("tWindMLDemo", (UGL_FPTR)windMLDemo, 110, 0, 10000,
                  mode,0,0,0,0);
}
#endif
```

```
/*************************************************************
** Function name: windMLDemo
** Descriptions:
** Input: 无
**
**
** Output: 无
*************************************************************/
void windMLDemo
(
    int mode                        /* Operating mode */
)
{
    UGL_REG_DATA *pRegistryData;
    UGL_DIB transDib;
    UGL_MDIB transMdib;
    UGL_MDIB patternDib;
    UGL_CDIB  cursorDib;
    UGL_FONT_DEF systemFontDef;
    UGL_FONT_DEF dialogFontDef;
    UGL_FONT_DEF fixedFontDef;
    UGL_ORD textOrigin = UGL_FONT_TEXT_UPPER_LEFT;
    int numRandomPoints;
    int i, index, y, textpage, tmp;
    char *fontTestText =
    "ABCDEFGHIJKLMNOPQRSTUVWXYZabcdefghijklmnopqrstuvwxyz0123456789";
    char regionMessage[23];
    int textWidth, textHeight;
    UGL_FB_INFO fbInfo;
    UGL_RECT rect;

    /* Initialize UGL */
    if (uglInitialize() == UGL_STATUS_ERROR)
        return;

    /* Obtain display device identifier */
    pRegistryData = uglRegistryFind (UGL_DISPLAY_TYPE, 0, 0, 0);
    if (pRegistryData == UGL_NULL)
    {
        printf("Display not found. Exiting.\n");
        uglDeinitialize();
```

```
        return;
    }

    devId = (UGL_DEVICE_ID)pRegistryData->id;

    if (mode >= 0)
    {

        /* obtain the input service identifier. */
        pRegistryData = uglRegistryFind (UGL_INPUT_SERVICE_TYPE, 0, 0, 0);
        if (pRegistryData == UGL_NULL)
        {
            printf("Input service not found. Exiting.\n");
            uglDeinitialize();

            return;
        }

        inputServiceId = (UGL_INPUT_SERVICE_ID)pRegistryData->id;
    }

    /* Create a graphics context */
    gc = uglGcCreate(devId);

    /* Create Fonts */
    pRegistryData = uglRegistryFind (UGL_FONT_ENGINE_TYPE, 0, 0, 0);
    if (pRegistryData == UGL_NULL)
    {
        printf("Font engine not found. Exiting.\n");
        uglDeinitialize();

        return;
    }

    fontDrvId = (UGL_FONT_DRIVER_ID)pRegistryData->id;

    uglFontDriverInfo(fontDrvId, UGL_FONT_TEXT_ORIGIN, &textOrigin);

    uglFontFindString ( fontDrvId, " familyName = Lucida; pixelSize = 12 ",
                &systemFontDef);

    if ((fontSystem = uglFontCreate(fontDrvId, &systemFontDef)) == UGL_NULL)
```

```
{
    printf("Font not found. Exiting.\n");
    return;
}

uglFontFindString ( fontDrvId, " familyName = Helvetica; pixelSize = 18 ",
                    &dialogFontDef);

if ((fontDialog = uglFontCreate(fontDrvId, &dialogFontDef)) == UGL_NULL)
{
    printf("Font not found. Exiting.\n");
    return;
}

uglFontFindString ( fontDrvId, " familyName = Courier; pixelSize = 12 ",
                    &fixedFontDef);

if ((fontFixed = uglFontCreate(fontDrvId, &fixedFontDef)) == UGL_NULL)
{
    printf("Font not found. Exiting.\n");
    return;
}

/* Obtain the dimensions of the display */
uglInfo(devId, UGL_FB_INFO_REQ, &fbInfo);
displayWidth = fbInfo.width;
displayHeight = fbInfo.height;

/* Setup random points */
srand(6);
numRandomPoints = 2000;
randomData = (int *)UGL_MALLOC(2 * numRandomPoints * sizeof(int));
for (i = 0; i < numRandomPoints * 2; i += 2)
{
    randomData[i] = (rand() % displayWidth);
    randomData[i + 1] = (rand() % displayHeight);
}

/* Initialize colors. */
uglColorAlloc(devId, &colorTable[BLACK].rgbColor, UGL_NULL,
              &colorTable[BLACK].uglColor, 1);
uglColorAlloc(devId, &colorTable[BLUE].rgbColor, UGL_NULL,
```

```
            &colorTable[BLUE].uglColor, 1);
uglColorAlloc(devId, &colorTable[GREEN].rgbColor, UGL_NULL,
            &colorTable[GREEN].uglColor, 1);
uglColorAlloc(devId, &colorTable[CYAN].rgbColor, UGL_NULL,
            &colorTable[CYAN].uglColor, 1);
uglColorAlloc(devId, &colorTable[RED].rgbColor, UGL_NULL,
            &colorTable[RED].uglColor, 1);
uglColorAlloc(devId, &colorTable[MAGENTA].rgbColor, UGL_NULL,
            &colorTable[MAGENTA].uglColor, 1);
uglColorAlloc(devId, &colorTable[BROWN].rgbColor, UGL_NULL,
            &colorTable[BROWN].uglColor, 1);
uglColorAlloc(devId, &colorTable[LIGHTGRAY].rgbColor, UGL_NULL,
            &colorTable[LIGHTGRAY].uglColor, 1);
uglColorAlloc(devId, &colorTable[DARKGRAY].rgbColor, UGL_NULL,
            &colorTable[DARKGRAY].uglColor, 1);
uglColorAlloc(devId, &colorTable[LIGHTBLUE].rgbColor, UGL_NULL,
            &colorTable[LIGHTBLUE].uglColor, 1);
uglColorAlloc(devId, &colorTable[LIGHTGREEN].rgbColor, UGL_NULL,
            &colorTable[LIGHTGREEN].uglColor, 1);
uglColorAlloc(devId, &colorTable[LIGHTCYAN].rgbColor, UGL_NULL,
            &colorTable[LIGHTCYAN].uglColor, 1);
uglColorAlloc(devId, &colorTable[LIGHTRED].rgbColor, UGL_NULL,
            &colorTable[LIGHTRED].uglColor, 1);
uglColorAlloc(devId, &colorTable[LIGHTMAGENTA].rgbColor, UGL_NULL,
            &colorTable[LIGHTMAGENTA].uglColor, 1);
uglColorAlloc(devId, &colorTable[YELLOW].rgbColor,  UGL_NULL,
            &colorTable[YELLOW].uglColor, 1);
uglColorAlloc(devId, &colorTable[WHITE].rgbColor,  UGL_NULL,
            &colorTable[WHITE].uglColor, 1);

/* Create Region */
regionId = uglRegionCreate ();
uglBackgroundColorSet(gc, colorTable[BLACK].uglColor);
uglForegroundColorSet(gc, colorTable[LIGHTGREEN].uglColor);

sprintf(regionMessage,"Welcome to WindML %d.%d.%d", uglVersionMajor,
        uglVersionMinor,
        uglVersionPatch);

uglTextSizeGet(fontDialog, &textWidth, &textHeight,
            -1, regionMessage);
uglFontSet(gc, fontDialog);
```

```
uglTextDraw(gc, (displayWidth - textWidth) / 2,
            (displayHeight - textHeight) / 3, -1, regionMessage);

rect.left = rect.top = 0;
rect.right = displayWidth - 1;
rect.bottom = displayHeight - 1;
uglRegionRectInclude (regionId, &rect);
rect.right = ((displayWidth - textWidth) / 2) + textWidth + 5;
rect.bottom = ((displayHeight - textHeight) / 3) + textHeight + 5;
rect.left = ((displayWidth - textWidth) / 2) - 6;
rect.top = ((displayHeight - textHeight) / 3) - 6;
uglRegionRectExclude (regionId, &rect);
uglClipRegionSet (gc, regionId);

/* Create the brick pattern */
patternDib.width = patternDib.stride = patternData.width;
patternDib.height = patternData.height;
patternDib.pImage = patternData.data;
patternDdb = uglMonoBitmapCreate(devId, &patternDib,
                                 UGL_DIB_INIT_DATA, 0, UGL_NULL);

/* Create standard and transparent DDBs */
colorData = (UGL_COLOR *)UGL_MALLOC(32 * 32 * sizeof(UGL_COLOR));

for (i = 0; i < 32 * 32; i++)
    colorData[i] = colorTable[transparentData[i]].uglColor;

transDib.pImage = (void *)colorData;
transDib.colorFormat = UGL_DEVICE_COLOR_32;
transDib.clutSize = 0;
transDib.pClut = UGL_NULL;
transDib.imageFormat = UGL_DIRECT;
transDib.width = transDib.height = transDib.stride = 32;

stdDdb = uglBitmapCreate(devId, &transDib, UGL_DIB_INIT_DATA, 0,
                         UGL_NULL);

transMdib.width = transMdib.stride = transMdib.height = 32;
transMdib.pImage = transparentMask;

transDdb = uglTransBitmapCreate(devId, &transDib, &transMdib,
                                UGL_DIB_INIT_DATA, 0, UGL_NULL);
```

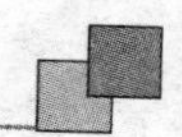

```
    /* Create the cursor */
    if (mode >= 0 && inputServiceId != UGL_NULL)
    {
        uglCursorInit (devId, 32, 32, displayWidth / 2, displayHeight / 2);
        cursorDib.width = cursorDib.height = cursorDib.stride = 32;
        cursorDib.hotSpot.x = cursorDib.hotSpot.y = 16;
        cursorDib.pImage = cursorData;
        cursorDib.clutSize = 2;
        cursorDib.pClut = cursorClut;
        cursorDdb = uglCursorBitmapCreate(devId, &cursorDib);
        uglCursorImageSet (devId, cursorDdb);

        uglCursorOn(devId);
    }

    /* Initialization finished, drawing begins */
    ClearScreen(gc);

    if(pauseDemo(mode) < 0)
    {
        ClearScreen(gc);
        cleanUp(mode);
        return;
    }

    /* BitmapWrite test */
    uglBitmapWrite(devId, &transDib, 0,0,31,31,UGL_DISPLAY_ID,100,100);

    if(pauseDemo(mode) < 0)
    {
        ClearScreen(gc);
        cleanUp(mode);
        return;
    }

    /* DDB blt */
    ClearScreen(gc);

    uglBatchStart(gc);

    for (i = 0; i < 1000; i++)
```

```
        uglBitmapBlt(gc, stdDdb,0,0,31,31,UGL_DEFAULT_ID, randomData[i],random-
                     Data[i+1]);

    uglBatchEnd(gc);

    if(pauseDemo(mode) < 0)
    {
        ClearScreen(gc);
        cleanUp(mode);

        return;
    }

    /* TDDB blt */
    ClearScreen(gc);

    uglBatchStart(gc);

    for (i = 0; i < 1000; i++)
        uglBitmapBlt(gc, transDdb,0,0,31,31,UGL_DEFAULT_ID, randomData[i],ran-
                     domData[i+1]);

    uglBatchEnd(gc);

    if(pauseDemo(mode) < 0)
    {
        ClearScreen(gc);
        cleanUp(mode);

        return;
    }

    /* Simple lines */
    ClearScreen(gc);

    uglBatchStart(gc);

    index = 0;
    for (i = 0; i < numRandomPoints / 2; i++)
    {
        uglForegroundColorSet(gc, colorTable[ i % 16].uglColor);
        uglLine(gc, randomData[index], randomData[index + 1],
```

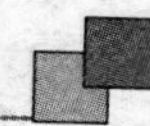

```
            randomData[index + 2], randomData[index + 3]);

        index += 4;
    }

    uglBatchEnd(gc);

    if(pauseDemo(mode) < 0)
    {
        ClearScreen(gc);
        cleanUp(mode);

        return;
    }

    /* Dashed lines */
    ClearScreen(gc);

    uglBatchStart(gc);

    uglLineStyleSet(gc, UGL_LINE_STYLE_DASHED);
    index = 0;
    for (i = 0; i < numRandomPoints / 2; i++)
    {
        uglForegroundColorSet(gc, colorTable[ i % 16].uglColor);
        uglLine(gc, randomData[index], randomData[index + 1],
                randomData[index + 2], randomData[index + 3]);
        index += 4;
    }

    uglBatchEnd(gc);

    if(pauseDemo(mode) < 0)
    {
        ClearScreen(gc);
        cleanUp(mode);

        return;
    }

    /* Wide solid lines */
    ClearScreen(gc);
```

```
uglBatchStart(gc);

index = 0;
for (i = 0; i < numRandomPoints / 2; i++)
{
    uglForegroundColorSet(gc, colorTable[ i % 15 + 1].uglColor);
    uglLineWidthSet(gc, (i % 6) + 1);
    uglLine(gc, randomData[index], randomData[index + 1],
            randomData[index + 2], randomData[index + 3]);
    index += 4;
}

uglBatchEnd(gc);

if(pauseDemo(mode) < 0)
{
    ClearScreen(gc);
    cleanUp(mode);

    return;
}

/* Wide dashed lines */
ClearScreen(gc);

uglBatchStart(gc);

uglLineStyleSet(gc, UGL_LINE_STYLE_DASHED);
index = 0;
for (i = 0; i < numRandomPoints / 2; i++)
{
    uglForegroundColorSet(gc, colorTable[ i % 15 + 1].uglColor);
    uglLineWidthSet(gc, (i % 6) + 1);
    uglLine(gc, randomData[index], randomData[index + 1],
            randomData[index + 2], randomData[index + 3]);

    index += 4;
}

uglBatchEnd(gc);
```

```
    if(pauseDemo(mode) < 0)
    {
        ClearScreen(gc);
        cleanUp(mode);

        return;
    }

    /* Filled rectangles */
    ClearScreen(gc);

    uglBatchStart(gc);

    index = 0;
    for (i = 0; i < numRandomPoints / 2; i++)
    {
        int left = min(randomData[index], randomData[index + 2]);
        int right = max(randomData[index], randomData[index + 2]);
        int top = min(randomData[index + 1], randomData[index + 3]);
        int bottom = max(randomData[index + 1], randomData[index + 3]);
        uglForegroundColorSet(gc, colorTable[ i % 15 + 1].uglColor);
        uglBackgroundColorSet(gc, colorTable[ 15 - (i % 15)].uglColor);
        uglLineWidthSet(gc, (i % 6) + 1);
        uglRectangle(gc, left, top , right, bottom);
        index += 4;
    }

    uglBatchEnd(gc);

    if(pauseDemo(mode) < 0)
    {
        ClearScreen(gc);
        cleanUp(mode);

        return;
    }

    /* Rectangles filled with a pattern */
    ClearScreen(gc);

    uglBatchStart(gc);
```

```
uglFillPatternSet(gc, patternDdb);

index = 0;
for (i = 0; i < numRandomPoints / 15; i++)
{
      int left = min(randomData[index], randomData[index + 2]);
      int right = max(randomData[index], randomData[index + 2]);
      int top = min(randomData[index + 1], randomData[index + 3]);
      int bottom = max(randomData[index + 1], randomData[index + 3]);
      uglForegroundColorSet(gc, colorTable[ i % 15 + 1].uglColor);
      uglBackgroundColorSet(gc, colorTable[ 15 - (i % 15)].uglColor);
      uglLineWidthSet(gc, i % 6 + 1);
      uglRectangle(gc, left, top , right, bottom);
      index += 4;
}

uglFillPatternSet(gc, 0);
uglBatchEnd(gc);

if(pauseDemo(mode) < 0)
{
      ClearScreen(gc);
      cleanUp(mode);

      return;
}

/* Filled polygons */
ClearScreen(gc);

uglBatchStart(gc);

index = 0;
for (i = 0; i < numRandomPoints / 10; i++)
{
      uglForegroundColorSet(gc, colorTable[ i % 15 + 1].uglColor);
      uglBackgroundColorSet(gc, colorTable[ 15 - (i % 15)].uglColor);
      randomData[index + 18] = randomData[index];
      randomData[index + 19] = randomData[index + 1];
      uglPolygon(gc, 10, &randomData[index]);
      index += 20;
}
```

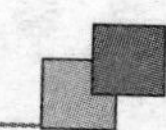

```
    uglBatchEnd(gc);

    if(pauseDemo(mode) < 0)
    {
        ClearScreen(gc);
        cleanUp(mode);

        return;
    }

    /* Polygons filled with a pattern */
    ClearScreen(gc);

    uglBatchStart(gc);

    uglFillPatternSet(gc, patternDdb);

    index = 0;
    for (i = 0; i < numRandomPoints / 22; i++)
    {
        uglForegroundColorSet(gc, colorTable[ i % 15 + 1].uglColor);
        uglBackgroundColorSet(gc, colorTable[ 15 - (i % 15)].uglColor);
        uglLineWidthSet(gc, i % 4 + 1);
        randomData[index + 18] = randomData[index];
        randomData[index + 19] = randomData[index + 1];
        uglPolygon(gc, 10, &randomData[index]);
        index += 20;
    }

    uglFillPatternSet(gc, 0);
    uglBatchEnd(gc);

    if(pauseDemo(mode) < 0)
    {
        ClearScreen(gc);
        cleanUp(mode);

        return;
    }

    /* Text */
```

```
ClearScreen(gc);

uglBatchStart(gc);

y = 0;
textpage = 0;
tmp = 0;

for (i = 0; i < 1000; i++)
{
    uglForegroundColorSet(gc, colorTable[ i % 15 + 1].uglColor);
    uglBackgroundColorSet(gc, colorTable[ 15 - (i % 15)].uglColor);

    switch (i % 3)
    {
        case 0:
            uglFontSet(gc, fontSystem);
            uglTextSizeGet(fontSystem, UGL_NULL, &tmp, -1, fontTestText);
            break;

        case 1:
            uglFontSet(gc, fontDialog);
            uglTextSizeGet(fontDialog, UGL_NULL, &tmp, -1, fontTestText);
            break;

        case 2:
            uglFontSet(gc, fontFixed);
            uglTextSizeGet(fontFixed, UGL_NULL, &tmp, -1, fontTestText);
            break;

    }

    uglTextDraw(gc, 0, y, -1, fontTestText);
    y += tmp;

    if (y >= displayHeight)
    {
        y = 0;
        textpage++;
    }
}
```

```
uglBatchEnd(gc);

if(pauseDemo(mode) < 0)
{
      ClearScreen(gc);
      cleanUp(mode);

      return;
}

/* Filled Ellipses */
ClearScreen(gc);

uglBatchStart(gc);

index = 0;
for (i = 0; i < numRandomPoints / 2; i++)
{
      int left = min(randomData[index], randomData[index + 2]);
      int right = max(randomData[index], randomData[index + 2]);
      int top = min(randomData[index + 1], randomData[index + 3]);
      int bottom = max(randomData[index + 1], randomData[index + 3]);

      uglForegroundColorSet(gc, colorTable[ i % 16].uglColor);
      uglBackgroundColorSet(gc, colorTable[ 15 - (i % 15)].uglColor);
      uglEllipse(gc, left, top, right, bottom, 0, 0, 0, 0);
      index += 4;
}
uglBatchEnd(gc);

if(pauseDemo(mode) < 0)
{
      ClearScreen(gc);
      cleanUp(mode);
      return;
}

/* Pie Shapes */
ClearScreen(gc);

uglBatchStart(gc);
```

```
index = 0;
for (i = 0; i < numRandomPoints / 4; i++)
{
    int left = min(randomData[index], randomData[index + 2]);
    int right = max(randomData[index], randomData[index + 2]);
    int top = min(randomData[index + 1], randomData[index + 3]);
    int bottom = max(randomData[index + 1], randomData[index + 3]);

    uglForegroundColorSet(gc, colorTable[ i % 16].uglColor);
    uglBackgroundColorSet(gc, colorTable[ 15 - (i % 15)].uglColor);
    uglEllipse(gc, left, top, right, bottom,
               randomData[index + 4], randomData[index + 5],
                 randomData[index + 6], randomData[index + 7]);
    index += 8;
}
uglBatchEnd(gc);

if(pauseDemo(mode) < 0)
{
    ClearScreen(gc);
    cleanUp(mode);
    return;
}

/* Stretch Blits */
ClearScreen(gc);

uglBitmapStretchBlt(gc,stdDdb,0,0,31,31,UGL_NULL,100,100,200,150);

if(pauseDemo(mode) < 0)
{
    ClearScreen(gc);
    cleanUp(mode);

    return;
}

ClearScreen(gc);

uglBitmapStretchBlt(gc,stdDdb,0,0,31,31,UGL_NULL,100,100,150,200);

if(pauseDemo(mode) < 0)
```

```
    {
        ClearScreen(gc);
        cleanUp(mode);

        return;
    }

    ClearScreen(gc);

    uglBitmapStretchBlt(gc,stdDdb,0,0,31,31,UGL_NULL,0,0,
                        displayWidth - 1, displayHeight - 1);

    if(pauseDemo(mode) < 0)
    {
        ClearScreen(gc);
        cleanUp(mode);

        return;
    }

    ClearScreen(gc);

    /* Clean Up */
    cleanUp(mode);

    return;
}
```

第 8 章

I^{2}C 设备驱动程序设计

8.1　I^{2}C 总线概述

随着大规模集成电路技术的发展，将 CPU 和工作系统所必需的 ROM、RAM、I/O端口、A/D、D/A 等外围电路集成在一个 IC 上的工艺越来越成熟。通常称这种集成了多种外围电路的 IC 为微控制器或者单片机。但是，在实际应用中需求多种多样，微控制器无法满足用户的各种需求。此时，就需要在微控制器或者单片机的外围进行扩展。扩展方法有两种：并行总线和串行总线。并行总线传输效率高，但是所用线缆数目繁多，耗费资源。相对于并行总线，串行总线连线少，结构简单，可大大简化系统的硬件设计和节省硬件资源。

I^{2}C(Inter-Integrated Circuit)总线是 PHILIPS 公司在 20 世纪 80 年代开发的一种两线式串行总线。I^{2}C 是一种多向控制型总线，即在同一总线结构下能够连接多个芯片，每个芯片都可以作为数据传输的控制源。因此，利用该总线可实现多主机系统所需要的裁决和高低速设备同步等功能。由于 I^{2}C 总线具有特点突出、应用简便等特性，所以 I^{2}C 快速崛起成为业界标准之一。I^{2}C 总线已经成为嵌入式应用的一种标准解决方案，被广泛地应用在各式各样的专业类产品、消费类产品中，实现高性能的串行数据传输。

8.2　I^{2}C 总线原理

I^{2}C 总线支持多主系统：系统在同一总线结构下允许多个 I^{2}C 节点设备接入，同时满足多主系统要求，任何一个 I^{2}C 接入设备都能够作为 I^{2}C 主设备，任一时刻只能有一个 I^{2}C 主设备生效，I^{2}C 具有总线仲裁功能，保证系统正确运行。I^{2}C 总线是两线制：双向传输的数据线 SDA；双向传输的时钟线 SCL。

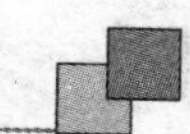

I²C 总线系统组成部分：

- 发送器（Transmitter）：发送数据到总线的器件。
- 接收器（Receiver）：从总线接收数据的器件。
- 主机（Master）：初始化发送、产生时钟信号和终止发送的器件。
- 从机（Slave）：被主机寻址的器件。

I²C 是双向传输的总线，I²C 主设备和 I²C 从设备都有可能担任发送器或者接收器的角色。如果是 I²C 主设备向 I²C 从设备发送数据，则 I²C 主设备是发送器，而 I²C 从设备是接收器；如果是 I²C 主设备从 I²C 从设备读取数据，则 I²C 主设备是接收器，而 I²C 从设备是发送器。I²C 总线的通信速率由 I²C 主设备设置，I²C 总线上数据的传输速率在标准模式（Standard - mode）下最高可达 100 kbps。

如图 8.1 所示，构建 I²C 系统的基本要求：

- 每个节点设备必须具有 I²C 接口功能。
- 每个节点设备必须共地。
- SDA 和 SCL 信号线必须连接上拉电阻。

在 I²C 系统中，I²C 主设备在 SCL 线上产生时钟信号、在 SDA 线上产生地址信号和控制信号，所选择的 I²C 从设备则通过 SDA 线与 I²C 主设备进行通信。

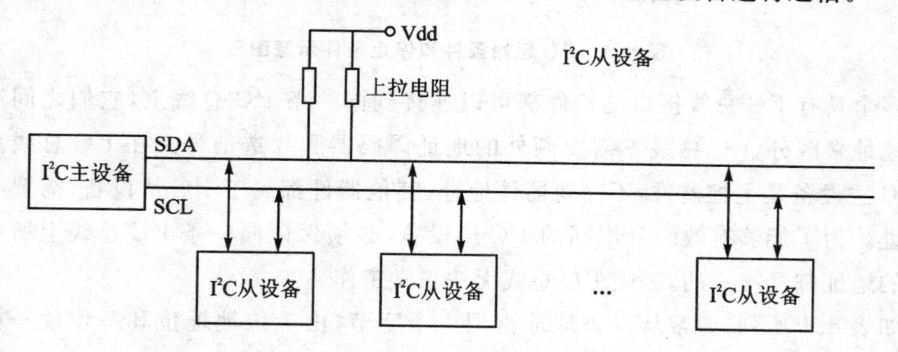

图 8.1 I²C 总线连接示意图

I²C 总线传输数据需要数据线 SDA 和时钟线 SCL 在时序上配合传输，如图 8.2 所示。数据线 SDA 的电平状态必须在时钟线 SCL 处于高电平期间保持稳定不变；SDA 的电平状态只有在 SCL 处于低电平期间才允许改变。但是，在 I²C 总线处于起始和结束时例外。与某些其他的串行总线协议规定数据在时钟信号的边沿（上升沿或下降沿）有效不同，I²C 总线是电平有效。

I²C 起始条件：如图 8.3 所示，当 SCL 处于高电平时，SDA 从高电平向低电平跳变时产生起始条件。总线在起始条件产生后便处于忙碌状态。起始条件通常简记为 S。

I²C 停止条件：当 SCL 处于高电平期间时，SDA 从低电平向高电平跳变时产生停止条件。总线在停止条件产生后处于空闲状态。停止条件通常简记为 P。

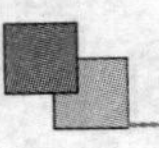

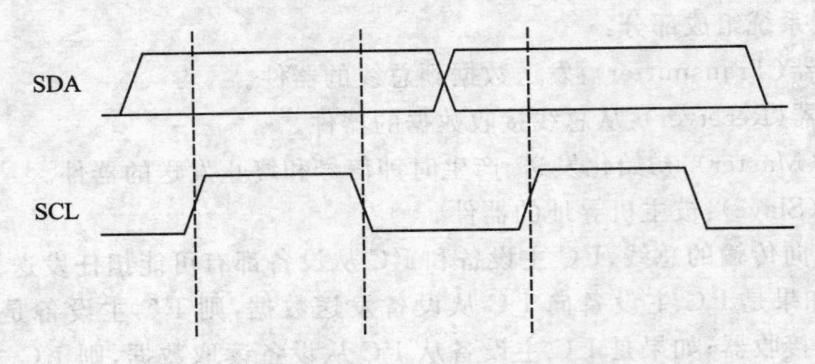

图 8.2　I^2C 总线上数据有效性的示意图

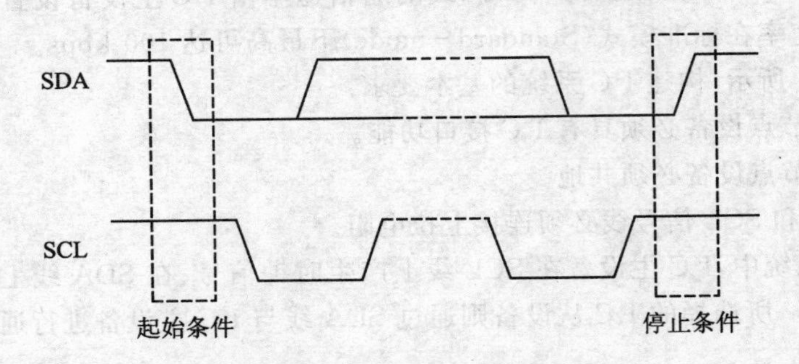

图 8.3　I^2C 起始条件和停止条件示意图

多个具有 I^2C 总线接口的器件都可以连接到同一条 I^2C 总线上，它们之间通过器件地址来区分，I^2C 总线不需要额外的地址译码器和片选信号。在 I^2C 总线系统中，I^2C 主设备是主控器件，不需要器件地址；其他器件都属于 I^2C 从设备，需要有器件地址。为了能够有效区分不同的 I^2C 从设备，必须保证同一条 I^2C 总线上所有从设备的地址都是唯一的，否则 I^2C 总线无法正常工作。

如表 8.1 所列，通常从设备地址占用 1 个字节，由 7 位地址位和一位读写标志(R/W)位组成；地址位占据字节高 7 位，读写标志位在最低位。

表 8.1　I^2C 从设备地址字节定义

D7	D6	D5	D4	D3	D2	D1	D0
DA3	DA2	DA1	DA0	A2	A1	A0	R/W

器件地址(DA3～DA0)：I^2C 器件固定地址编码，由生产厂家烧入到器件中。如 EEPROM AT24C 系列的器件地址统一为 1010。

引脚地址(A2、A1、A0)：由 I^2C 器件地址引脚 A2、A1、A0 的所接的引脚电平来确定，高电平为 1，低电平为 0。

读写控制位(R/ W)：1 表示主设备将要从从设备读取数据，0 表示主设备将要向从设备写入数据。

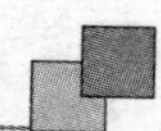

8.3　S3C2410的I²C结构分析

8.3.1　S3C2410的I²C主要结构

S3C2410的I²C主要由5部分构成:数据收发寄存器、数据移位寄存器、地址寄存器、时钟发生器、控制逻辑等部分,如图8.4所示。

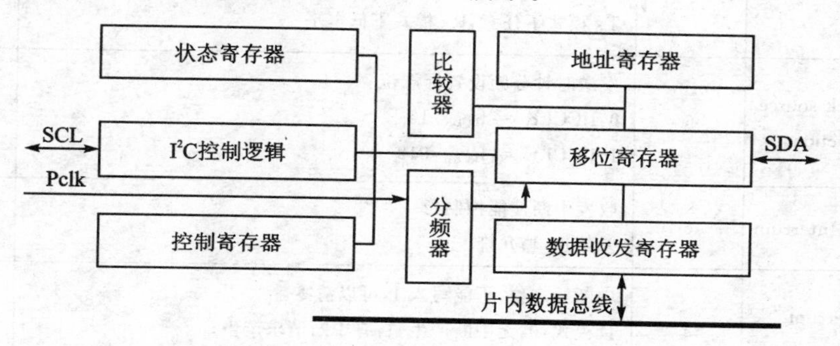

图8.4　S3C2410的I²C结构图

S3C2410的I²C总线特性:

① 具备一个I²C总线接口;

② 支持I²C总线标准传输速率100 kbps以下及高速传输速率400 kbps;

③ 支持I²C以查询方式或者中断方式运行;

④ 支持主设备和从设备两种角色,包括4种操作模式——主机发送模式、主机接收模式、从机发送模式、从机接收模式。

8.3.2　S3C2410的I²C主要寄存器

如表8.2所列,S3C2410内部有4个I²C相关专用寄存器。

表8.2　S3C2410内部I²C专用寄存器

名　称	地　址	读　写	描　述	初始值
IICCON	0x54000000	R/W	I²C总线控制寄存器	0x0X
IICSTAT	0x54000004	R/W	I²C总线控制/状态寄存器	0
IICADD	0x54000008	R/W	I²C总线地址寄存器	0xXX
IICDS	0x5400000C	R/W	I²C数据发送/接收寄存器	0xXX

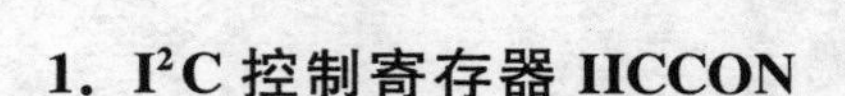

1. I^{2}C 控制寄存器 IICCON

S3C2410 的 I^{2}C 控制寄存器——IICCON，内部定义如表 8.3 所列。

表 8.3　I^{2}C 控制寄存器 IICCON 定义

字段名	位	描　述	初始值
Acknowledge generation	7	应答使能标志位 0：禁止应答，1：自动应答； 应答电平 Tx 模式下任意，Rx 模式下低电平	0
Tx clock source selection	6	发送时钟分频设置标志位 0：IICCLK = $f_{PCLK}/16$ 1：IICCLK = $f_{PCLK}/512$	0
Tx/Rx Interrupt	5	收发中断使能控制位 0：禁止，1：允许	0
Interrupt Pending flag	4	中断标志位，不能写入 1，可以清零。 读模式：0：无中断产生，1：有中断请求产生； 写模式：写 0：清除中断标志，写 1：无操作	0
Transmit clock value	3:0	发送时钟预分频值设置，根据下面公式计算： Tx clock = IICCLK/(IICCON[3:0]+1)	未定义

2. I^{2}C 控制状态寄存器(IICSTAT)

S3C2410 的 I^{2}C 控制状态寄存器—IICSTAT，内部定义如表 8.4 所列。

表 8.4　I^{2}C 控制寄存器 IICCON 定义

字段名	位	意　义	初始值
Mode selection	7:6	工作模式选择标志位 00：从机接收；01：从机发送；10：主机接收；11：主机发送	00
Busy / START STOP condition	5	I^{2}C 总线忙状态标志位 读：1 示忙；0 示闲 写：0 产生结束信号，1 产生启动信号	0
Serial output	4	I^{2}C 数据发送控制 0：禁止发送接收；1：允许发送接收	0
Arbitration Status flag	3	仲裁状态标志 0：总线仲裁成功；1：总线仲裁失败	0
Address - as - slave status flag	2	从地址匹配状态标志位 0：检测到起始/停止状态清除标志位； 1：接收到从地址与 IICADD 中地址匹配	0

续表 8.4

字段名	位	意 义	初始值
Address zero status flag	1	零地址状态标志位 0:检测到起始/停止状态清除标志位; 1:接收到地址为0x00000000b	0
Last - received bit status flag	0	最后接收位状态标志位 0:最后接收位为0,收到ACK;1:最后接收位为1,未收到ACK	0

备注:启动主设备发送0xF0;结束主设备发送0xD0;启动主设备接收0xB0;结束主设备接收0x90。

3. I²C地址寄存器(IICADD)

S3C2410的I²C地址寄存器——IICADD,内部定义如表8.5所列。

表8.5 I²C控制寄存器IICCON定义

字段名	位	意 义	初如值
Slave address	7:1	7位从设备地址	0xXX
Not mapped	0	保留	—

备注:

(1) 地址寄存器在作为从设备时有意义,为主设备时无意义。

(2) 只有在数据传输控制位IICSTAT[4]=0时可写;在任何时刻都可读。

4. I²C数据发送/接收寄存器(IICDS)

S3C2410的I²C数据发送/接收寄存器——IICDS,内部定义如表8.6所列。

表8.6 I²C控制寄存器IICDS定义

字段名	位	意 义	初始值
Data shift	7:0	8位移位接收或移位发送的数据	0xXX

备注:

(1) 在设备处于接收状态时,读取该寄存器获取的数据。

(2) 在设备发送数据时,首先要使数据传输控制位IICSTAT[4]=1,才能进行写操作,然后将所要发送的数据内容写入寄存器IICDS中。

(3) 在任何时刻,可读取该寄存器的数值。

8.4 ZLG7290B 特性

8.4.1 ZLG7290B 描述与主要特性

ZLG7290B 是广州周立功单片机发展有限公司自主设计的数码管显示驱动及键盘扫描管理芯片。ZLG7290B 能够直接驱动 8 位共阴式数码管(或 64 只独立的 LED),同时还可以扫描管理多达 64 只独立按键。其中,有 8 只按键作为功能键使用,如同 PC 键盘上的 Ctrl、Shift、Alt 等键一样。ZLG7290B 内部还设置有连击计数器,支持某按键输入按下不释放时输入连续有效。采用 I^2C 总线方式,与微控制器的接口仅需两根信号线。

ZLG7290B 芯片的主要特性如下:

- 直接驱动 8 位共阴式数码管(1 英寸以下)或 64 只独立的 LED;
- 能够管理多达 64 只按键,自动消除抖动,其中有 8 只可以作为功能键使用;
- 段电流可达 20 mA,位电流可达 100 mA 以上;
- 利用功率电路可以方便地驱动 1 英寸以上的大型数码管;
- 具有闪烁、段点亮、段熄灭、功能键、连击键计数等强大功能;
- 提供 10 种数字和 21 种字母的译码显示功能,或者直接向显示缓存写入显示数据;
- 当不接数码管而仅使用键盘管理功能时,工作电流可降至 1 mA;
- 与微控制器之间采用 I^2C 串行总线接口,只需两根信号线,节省 I/O 资源;
- 工作电压范围:+3.3~+5.5 V;
- 工作温度范围:-40~+85 ℃;
- 封装:DIP-24(窄体),SOP-24。

8.4.2 ZLG7290B 引脚功能说明

ZLG7290B 引脚功能说明如表 8.7 所列。

表 8.7 ZLG7290B 引脚功能表

引脚序号	引脚名称	功能描述
1	SC/KR2	数码管 c
2	SD/KR3	数码管 d
3	DIG3/KC3	数码管位选信号 3/键盘列信号 3
4	DIG2/KC2	数码管位选信号 2/键盘列信号 2

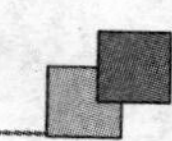

续表 8.7

引脚序号	引脚名称	功能描述
5	DIG1/KC1	数码管位选信号 1/键盘列信号 1
6	DIG0/KC0	数码管位选信号 0/键盘列信号 0
7	SE/KR4	数码管 e
8	SF/KR5	数码管 f
9	SG/KR6	数码管 g
10	DP/KR7	数码管 dp
11	GND	接地
12	DIG6/KC6	数码管位选信号 6/键盘列信号 6
13	DIG7/KC7	数码管位选信号 7/键盘列信号 7
14	INT	键盘中断请求信号,低电平/下降沿有效
15	RST	复位信号,低电平有效
16	VCC	电源,+3.3～5.5 V
17	OSC1	晶振输入信号
18	OSC2	晶振输出信号
19	SCL	I^2C
20	SDA	I^2C
21	DIG5/KC5	数码管位选信号 5/键盘列信号 5
22	DIG4/KC4	数码管位选信号 4/键盘列信号 4
23	SA/KR0	数码管 a
24	SB/KR1	数码管 b

8.4.3 ZLG7290B 寄存器说明

ZLG7290B 内部有 8 个显示缓冲寄存器 DpRam0～DpRam7,放置数码管所需要显示的内容。ZLG7290B 提供有两种显示控制方式,一种是直接向显存写入字型数据,另一种是通过向命令缓冲寄存器写入控制指令实现自动译码显示。访问这些寄存器需要通过 I^2C 总线接口来实现。ZLG7290B 的 I^2C 总线器件写操作地址是"0x70",读操作地址是"0x71"。访问内部寄存器要通过"地址"来实现。ZLG7290B 支持总计 64 个按键,其中 56 个可以定义为普通按键,8 个定义为功能键。

1. 系统寄存器 SystemReg(地址:00H)

系统寄存器的第"0"位(LSB)称作 KeyAvi,标志着按键是否有效:"0"代表没有按键被按下,"1"代表有某个按键被按下。SystemReg 寄存器的其他位暂时没有定

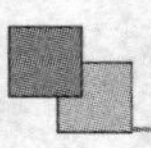

义。当按下某个键时，ZLG7290B 的 INT 引脚会产生一个低电平的中断请求信号。当读取键值后，中断信号就会自动撤销，而 KeyAvi 也同时予以反应。正常情况下，微控制器只需要判断 INT 引脚即可。

2. 键值寄存器 Key(地址:01H)

如果某个普通键被按下，从键值寄存器 Key 中读取相应的键值为 1～56。如果微控制器发现 ZLG7290B 的 INT 引脚产生了中断请求，而从 Key 中读到的键值是"0"，则表示按下的可能是功能键。键值寄存器 Key 的值在被读取后自动变成"0"。

3. 连击计数器 RepeatCnt(地址:02H)

ZLG7290B 为 56 个普通按键提供了连击计数功能。所谓连击是指按住某个普通键不释放，经过一两秒钟的延迟后(在 4 MHz 下约为 2 s)，开始连续有效，连续有效间隔时间(在 4 MHz 下约为 170 ms)在几十到几百个 ms。这一特性跟电脑上的键盘很类似。在微控制器能够及时响应按键中断并及时读取键值的前提下，当按住某个普通键一直不释放时：首先会产生一次中断信号，这时连击计数器 RepeatCnt 的值仍然是"0"；经过一两秒延迟后，会连续产生中断信号，每中断一次 RepeatCnt 就自动加 1；当 RepeatCnt 计数到"255"时就不再增加，而中断信号继续有效。

4. 功能键寄存器 FunctionKey(地址:03H)

ZLG7290B 有 8 个功能键。功能键常常是配合普通键一起使用的，就像电脑键盘上的 Shift、Ctrl 和 Alt 键。当然功能键可以单独使用，类似电脑键盘上的 F1～F12。当按下某个功能键时，在 INT 引脚也会像按普通键那样产生中断信号。功能键的键值是被保存在 FunctionKey 寄存器中的。功能键寄存器 FunctionKey 的初始值是"0xFF"，每一个位对应一个功能键，第 0 位(LSB)对应 F0，第 1 位对应 F1，以此类推，第 7 位(MSB)对应 F7。某一功能键被按下时，相应的 FunctionKey 位就清零。功能键还有一个特性就是"二次中断"，按下时产生一次中断信号，抬起时又会产生一次中断信号；而普通键只会在被按下时产生一次中断。

5. 命令缓冲区 CmdBuf0 和 CmdBuf1(地址:07H 和 08H)

通过向命令缓冲区写入相关的控制命令可以实现段寻址、下载显示数据、控制闪烁等功能。

6. 闪烁控制寄存器 FlashOnOff(地址:0CH)

FlashOnOff 寄存器决定闪烁频率和占空比，初值为"0x77"。该寄存器高 4 位表示闪烁时亮的持续时间，低 4 位表示闪烁时灭的持续时间。改变 FlashOnOff 的值，可以同时改变闪烁频率和占空比。FlashOnOff 取值"0x00"时可获得最快的闪烁速度，在 4 MHz 下，亮或灭的持续时间最短时间周期约为 280 ms。特别说明：单独设置 FlashOnOff 寄存器的值，并不会看到显示闪烁，需要配合闪烁控制命令一起使用。

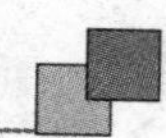

7. 扫描位数寄存器 ScanNum(地址:0DH)

ScanNum 寄存器决定扫描显示的位数,取值 0～7,对应 1～8 位。寄存器初值是"7",即数码管的 8 个位都扫描显示。实际应用中可能需要显示的位数不足 8 位,例如只显示 3 位,这时可以把 ScanNum 的值设置为 2,则数码管的第 0、1、2 位被扫描显示,而第 3～7 位不会被分配扫描时间,所以不显示。数码管的扫描位数减少后,有用的显示位由于被分配的扫描时间增多显示亮度得以提高。ScanNum 寄存器的值为 0 时,只有数码管的第 0 位在显示,亮度达到最大。

8. 显示缓冲区 DpRam0～DpRam7(地址:10H～17H)

DpRam0～DpRam7 这 8 个寄存器的取值直接决定了数码管的显示内容。每个寄存器的 8 个位分别对应数码管的 a、b、c、d、e、f、g、dp 段,MSB 对应 a,LSB 对应 dp。例如大写字母"H"的字型数据为"0x6E"(不带小数点)或"0x6F"(带小数点)。

8.4.4 ZLG7290B 控制命令详解

寄存器 CmdBuf0(地址:07H)和 CmdBuf1(地址:08H)共同组成命令缓冲区。通过向命令缓冲区写入相关的控制命令可以实现段寻址、下载显示数据、控制闪烁等功能。

1. 段寻址(SegOnOff)

ZLG7290B 段寻址指令如表 8.8 所列。

表 8.8 ZLG7290B 段寻址指令

D7	D6	D5	D4	D3	D2	D1	D0	D7	D6	D5	D4	D3	D2	D1	D0
0	0	0	0	0	0	0	1	on	0	S5	S4	S3	S2	S1	S0

在段寻址命令中,8 位数码管被看成是 64 个段,每一个段实际上就是一只独立的 LED。第 1 字节 0000,0001B 是命令字;on 表示该段是否点亮,0—灭,1—亮;S5S4S3S2S1S0 是 6 位段地址,取值 0～63。在某 1 位数码管内,各段的亮或灭的顺序按照 a、b、c、d、e、f、g、dp 进行。

2. 下载数据并译码(Download)

ZLG7290B 下载数据译码指令如表 8.9 所列。

表 8.9 ZLG7290B 下载数据译码指令格式

D7	D6	D5	D4	D3	D2	D1	D0	D7	D6	D5	D4	D3	D2	D1	D0
0	1	1	0	A_3	A_2	A_1	A_0	d_p	flash	0	d_4	d_3	d_2	d_1	d_0

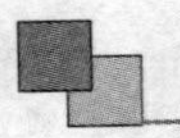
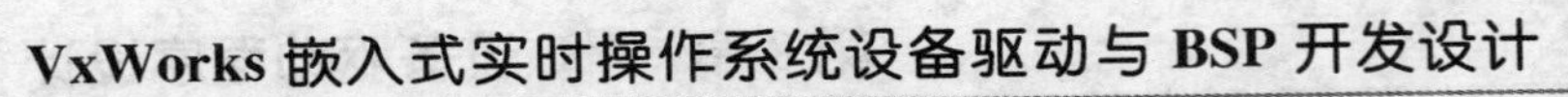

在指令格式中，高 4 位的“0110”是命令字段；$A_3A_2A_1A_0$ 是数码管显示数据的位地址（其中 A_3 留作以后扩展使用，实际使用时取 0 即可），位地址编号按从左到右的顺序依次为 0、1、2、3、4、5、6、7；d_p 控制小数点是否点亮，0—点亮，1—熄灭；flash 表示是否要闪烁，0—正常显示，1—闪烁；d4d3d2d1d0 是要显示的数据，包括 10 种数字和 21 种字母。显示数据按照表 8.10 中的规则进行译码。

表 8.10 下载数据译码命令数据表

$d_4d_3d_2d_1d_0$（二进制）					$d_4d_3d_2d_1d_0$（十六进制）	显示结果
0	0	0	0	0	00H	0
0	0	0	0	1	01H	1
0	0	0	1	0	02H	2
0	0	0	1	1	03H	3
0	0	1	0	0	04H	4
0	0	1	0	1	05H	5
0	0	1	1	0	06H	6
0	0	1	1	1	07H	7
0	1	0	0	0	08H	8
0	1	0	0	1	09H	9
0	1	0	1	0	0AH	A
0	1	0	1	1	0BH	b
0	1	1	0	0	0CH	C
0	1	1	0	1	0DH	d
0	1	1	1	0	0EH	E
0	1	1	1	1	0FH	F
1	0	0	0	0	10H	G
1	0	0	0	1	11H	H
1	0	0	1	0	12H	i
1	0	0	1	1	13H	J
1	0	1	0	0	14H	L
1	0	1	0	1	15H	o
1	0	1	1	0	16H	p
1	0	1	1	1	17H	q
1	1	0	0	0	18H	r
1	1	0	0	1	19H	t
1	1	0	1	0	1AH	U

续表 8.10

$d_4d_3d_2d_1d_0$（二进制）					$d_4d_3d_2d_1d_0$（十六进制）	显示结果
1	1	0	1	1	1BH	y
1	1	1	0	0	1CH	c
1	1	1	0	1	1DH	h
1	1	1	1	0	1EH	T
1	1	1	1	1	1FH	（无显示）

3. 闪烁控制(Flash)

表 8.11 ZLG7290B 闪烁控制

D7	D6	D5	D4	D3	D2	D1	D0	D7	D6	D5	D4	D3	D2	D1	D0
0	1	1	1	x	x	x	x	F_7	F_6	F_5	F_4	F_3	F_2	F_1	F_0

在命令格式中，高4位的“0111”是命令字段；“xxxx”表示无关位，通常取值“0000”；第2字节的 F_n($n=0\sim7$)控制数码管相应位的闪烁属性，0—正常显示，1—闪烁。复位后，所有位都不闪烁。

8.5 I²C实验设计

8.5.1 实验目的

掌握S3C2410A的I²C设备的结构配置和操作方法，以及ZLG7290芯片与LED数码管和键盘的连接原理图。

8.5.2 实验设备

(1) 硬件。

- PC 1台；
- MagicARM2410教学实验开发平台1套。

(2) 软件。

- Windows98/XP/2000系统；
- Tornado交叉开发环境。

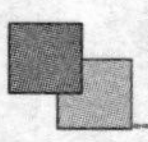

8.5.3 实验内容

LED 数码管的显示，以及键盘操作。

8.5.4 电路原理图

LED 和键盘与 I²C 总线之间的连接如图 8.5 所示。

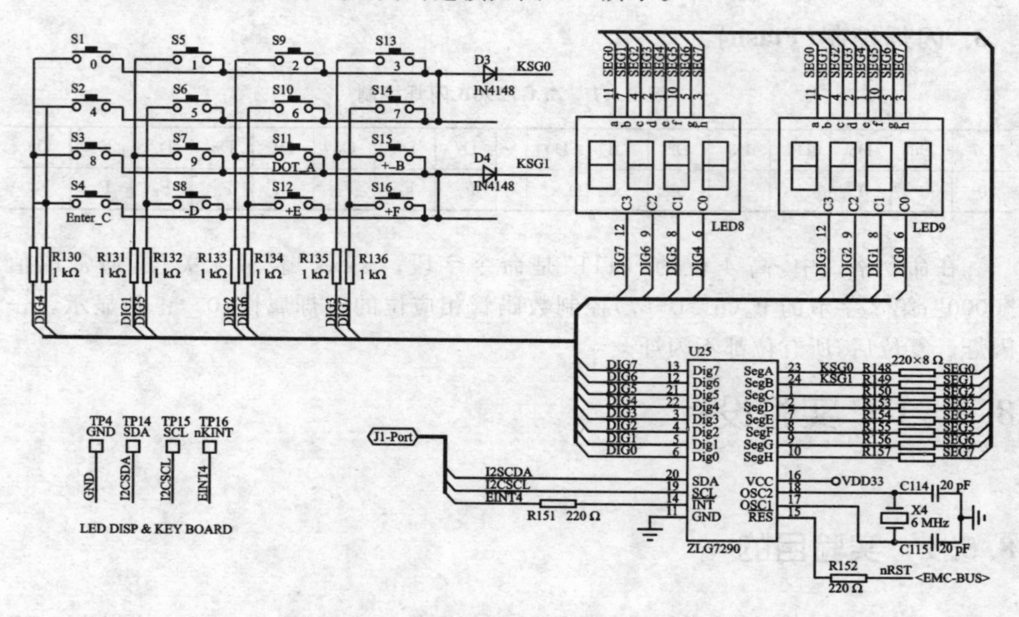

图 8.5　I²C 实验部分电路原理图

8.5.5 实验步骤

(1) 完成相关设置(主机网络地址、FTP 服务器、超级终端等)，然后系统上电，登录到 ftp 服务器上获取编译的基本映像，并且运行。

(2) 将 I²C 以及 ZLG7290 相关驱动实验的源代码(I²C 驱动目录下面)复制到 C:\Tornado2.2\target\config\MagicARM2410 目录下。

(3) 通过右键将“testiic.c”文件添加到项目文件中去，并编译，生成“.o”文件。

(4) 连接上 Target Server，并开启 Wshell。

(5) 将 Object Modules 中的“testii.o”模块下载到运行的系统中去。

(6) 在 Wshell 中输入“testiic”，观察实验结果。

(7) 全速运行程序，ZLG7290 电路控制 LED 数码管依次显示数字 0、2、4、6、8 进

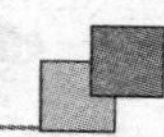

行检测，然后显示“87654321”；芯片读取按键，按下 1～8 键时相应位的字符会闪烁，按 Enter 键时蜂鸣器响一声。观察数码管和键盘按键的结果。

8.5.6 程序清单

1. s3c2410xIIC.h 文件清单

```
#ifndef __INCs3c2410xIICh
#define __INCs3c2410xIICh

#ifdef __cplusplus
extern "C" {
#endif

#ifndef _ASMLANGUAGE

#define rIICCON        0x54000000 /* IIC control */
#define rIICSTAT       0x54000004 /* IIC status */
#define rIICADD        0x54000008 /* IIC address */
#define rIICDS         0x5400000c /* IIC data shift */

#define rGPECON        0x56000040 /* Port E control */
#define rGPEDAT        0x56000044 /* Port E data */
#define rGPEUP         0x56000048 /* Pull-up control E */


/* 使能 ACK 位,IICCLK = PCLK/512 = 97656 */
#define        IICCON_DACK        ((1<<7) | (1<<6) | (1<<5) | (3<<0))
/* 中断使能(这样才能正确操作 I2C) */
#define        IICCON_DNACK        ((0<<7) | (1<<6) | (1<<5) | (3<<0))

typedef unsigned char      uint8;        /* 无符号 8 位整型变量 */
typedef signed   char      int8;         /* 有符号 8 位整型变量 */
typedef unsigned short     uint16;       /* 无符号 16 位整型变量 */
typedef signed   short     int16;        /* 有符号 16 位整型变量 */
typedef unsigned int       uint32;       /* 无符号 32 位整型变量 */
typedef signed   int       int32;        /* 有符号 32 位整型变量 */

#endif    /* _ASMLANGUAGE */

#ifdef __cplusplus
```

```
}
#endif

#endif
```

2. IICdev.h 文件清单

```
#ifndef __INCiicdevh
#define __INCiicdevh

#ifdef __cplusplus
extern "C" {
#endif

#ifndef _ASMLANGUAGE

#include "selectLib.h"
#include "iosLib.h"

#include "s3c2410xIIC.h"

/* IIC device I/O controls */

#define IIC_SLAVEADD_SET            0x1003
#define IIC_SLAVEADD_GET            0x1004

#define IIC_CLOCK_SET               0x1005
#define IIC_CLOCK_GET               0x1006

#define IIC_SUBADD_SET              0x1007
#define IIC_SUBADD_GET              0x1008


typedef  struct
{
    DEV_HDR   devHdr; /* 必须 */
    SEL_WAKEUP_LIST selWakeupList; /* 支持 SELECT 机制 */


    BOOL      readyRead;
    BOOL      readyWrite;
```

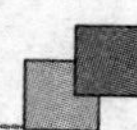

```
    BOOL        opened;

    /* 自己定义的其他成员 */
    char *pBuf;
    int  nBytes;

    UINT8 slaveAdd;
    UINT8 *pSubadd;

    int iicTxClock;   /* tx clock */

    /* iic 设备的 Rx Tx 中断 level */
    UINT8     IICintLevel;

    UINT32 *      regs;
}
IIC_DEV;

/* 其他定义 */
/**********************************************************************
** Function name: s3c2410IICInit
** Descriptions:  初始化总线(100 kHz 总线速率)
**                设 FCLK = 200 MHz,HCLK = 100 MHz,PCLK = 50 MHz
** Input:  无
** Output: 无
**********************************************************************/
extern void  s3c2410IICInit(void);

/**********************************************************************
** Function name:   ISendStr
** Descriptions:    使用硬件 I2C 发送数据
** Input:  sla      从机地址
**         suba     器件子地址(第一字节用来表示子地址字节个数)
**         s        发送数据缓冲区
**         no       发送数据个数
** Output:操作成功返回 TRUE,仲裁失败/无从机应答返回 FALSE
** Note:使用前设置好参数。程序不会更改 s、suba 缓冲区的数据
**********************************************************************/
extern int  s3c2410IICSendStr(uint8 sla, uint8 *suba, uint8 *s, uint8 no);

/**********************************************************************
** Function name: IRcvStr
```

```
** Descriptions:    使用硬件 I2C 读取数据
** Input:  sla      从机地址
**         suba     器件子地址缓冲区指针(第一字节用来表示子地址字节个数)
**         s        接收数据缓冲区
**         no       接收数据个数
** Output:操作成功返回 TRUE,仲裁失败/无从机应答返回 FALSE
** Note:使用前设置好参数。程序不会更改 s、suba 缓冲区的数据
**********************************************************************/
extern int  s3c2410IICRcvStr(uint8 sla,uint8 *suba,uint8 *s,uint8 no);
#endif

#ifdef __cplusplus
}
#endif

#endif
```

3. IICdev.c 文件清单

```
#include "vxWorks.h"
#include "iv.h"
#include "ioLib.h"
#include "iosLib.h"
#include "intLib.h"
#include "errnoLib.h"
#include "sysLib.h"

#include "s3c2410x.h"
#include "s3c2410xIIC.h"
#include "IICdev.h"
#include "s3c2410xSio.h"

/* 宏定义 */
#ifndef s3c2410x_BAUD_MIN
#define s3c2410x_BAUD_MIN           18
#endif
#ifndef s3c2410x_BAUD_MAX
#define s3c2410x_BAUD_MAX           1152000
#endif

#ifndef s3c2410x_SIO_DEFAULT_BAUD
#define s3c2410x_SIO_DEFAULT_BAUD 1152000
```

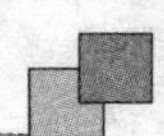

```
#endif

/* 硬件访问方法 */
#ifndef s3c2410_IIC_REG_READ
#define s3c2410_IIC_REG_READ(reg,result) \
        ((result) = *(volatile UINT32 *)(reg))
#endif

#ifndef s3c2410_IIC_REG_WRITE
#define s3c2410_IIC_REG_WRITE(reg,data) \
        (*((volatile UINT32 *)(reg)) = (data))
#endif

#ifndef s3c2410x_INT_REG_READ
#define s3c2410x_INT_REG_READ(reg,result) \
        ((result) = *(volatile UINT32 *)(reg))
#endif

#ifndef s3c2410x_INT_REG_WRITE
#define s3c2410x_INT_REG_WRITE(reg,data) \
        (*((volatile UINT32 *)(reg)) = (data))
#endif

void iicIntHandler(int iicDevId);
int  iicOpen(DEV_HDR *pDevHdr, int option, int flags);
STATUS iicDelete(char *devName);
int iicClose(int iicDevId);
int iicRead(int iicDevId, char *pBuf, int nBytes);
int iicWrite(int iicDevId, char *pBuf, int nBytes);
int iicIoctl(int iicDevId, int cmd, int arg);

LOCAL int iicDrvNum = 0;
LOCAL IIC_DEV  myiicDev;

/*****************************************************************************
 ** Function name: iicDrv
 ** Descriptions: 安装驱动例程据
 ** Input:
 ** Output:驱动安装成功返回 OK,驱动安装失败返回 ERROR
 ** Note:
 ****************************************************************************/
STATUS iicDrv(void)
```

```
{
    if(iicDrvNum > 0)
    {
        /* 驱动程序已经安装，直接返回 */
        return(OK);
    }

    if((iicDrvNum = iosDrvInstall(iicOpen, iicDelete, iicOpen, iicClose, \
        iicRead, iicWrite, iicIoctl)) == ERROR)
    {
        return (ERROR);
    }

    return (OK);
}

/*****************************************************************
** Function name: iicDevCreate
** Descriptions:  创建设备例程
** Input:   devname  所要创建的设备名称
** Output:创建设备成功返回 OK,创建设备失败返回 ERROR
** Note:
*****************************************************************/
STATUS iicDevCreate
(
    char * devName    /* 所要创建的设备名称 */
)
{
    IIC_DEV * piicDev;

    if(iicDrvNum < 1)
    {
        errnoSet(S_ioLib_NO_DRIVER);
        return (ERROR);
    }

    if ((piicDev = (IIC_DEV *)malloc(sizeof(IIC_DEV))) == NULL)
    {
        logMsg("malloc ERROR! \n",0,0,0,0,0);
        return ERROR;
    }
```

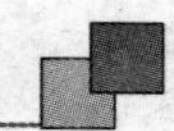

```
    /* 初始化 piicDev */
    if( iosDevAdd(&piicDev->devHdr, devName, iicDrvNum) == ERROR)
    {
        return (ERROR);
    }

    selWakeupListInit(&piicDev->selWakeupList);

    s3c2410IICInit();

    return (OK);
}

/*************************************************************
** Function name: iicOpen
** Descriptions:  打开设备
** Input:
** Output:打开设备成功返回指向设备的句柄,打开设备失败返回 ERROR
** Note:
*************************************************************/
int iicOpen
(
    DEV_HDR *pDevHdr,
    int option,
    int flags
)
{
    IIC_DEV *piicDev = (IIC_DEV *)pDevHdr;

    if(piicDev == NULL)
    {
        return (ERROR);
    }

    return ((int)pDevHdr);
}

/*************************************************************
** Function name: iicRead
** Descriptions:  读设备
** Input:
** Output:读取设备成功返回读取长度,读取设备失败返回 ERROR
```

```
** Note:
*****************************************************************/
int iicRead
(
    int iicDevId,
    char * pBuf,
    int nBytes
)
{
    IIC_DEV * piicDev = (IIC_DEV * )iicDevId;
    int ReadLength;

    ReadLength = 0;
    if(piicDev == (IIC_DEV * )NULL)
    {
        return ERROR;
    }

    s3c2410IICRcvStr(iicDevId,pBuf,nBytes);
    return (ReadLength);
}

/*****************************************************************
** Function name: iicWrite
** Descriptions:  写设备
** Input:
** Output:写设备成功返回写长度,写设备失败返回 ERROR
** Note:
*****************************************************************/
int iicWrite
(
    int iicDevId,
    char * pBuf,
    int nBytes
)
{
    IIC_DEV * piicDev = (IIC_DEV * )iicDevId;
    int WriteLength = 0;

    if(piicDev == (IIC_DEV * )NULL)
    {
        return (ERROR);
```

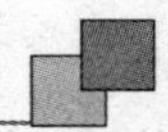

```
    }

    s3c2410IICSendStr(iicDevId,pBuf, nBytes);

    return (WriteLength);
}

/***********************************************************************
** Function name: iicIoctl
** Descriptions:  设备控制
** Input:
** Output:
** Note:
***********************************************************************/
int iicIoctl(int iicDevId, int cmd, int arg) {}

/***********************************************************************
** Function name: iicClose
** Descriptions:  关闭设备
** Input:
** Output:
** Note:
***********************************************************************/
int iicClose
(
    int iicDevId
)
{
    IIC_DEV *piicDev = (IIC_DEV *)iicDevId;

    if(piicDev == (IIC_DEV *)NULL)
    {
        return (ERROR);
    }

    piicDev->opened = FALSE;
    return (OK);
}

/***********************************************************************
** Function name: iicDelete
** Descriptions:  删除设备
```

```
** Input:
** Output:删除成功返回 OK,失败返回 ERROR
** Note:
**************************************************************/
STATUS iicDelete
(
    char * devName
)
{
    DEV_HDR * pDevHdr;
    char * pNameTail;

    pDevHdr = iosDevFind(devName, &pNameTail);
    if(pDevHdr == NULL || * pNameTail != '\0')
        return (ERROR);

    iosDevDelete(pDevHdr);
    return (OK);
}


/**************************************************************
** Function name: s3c2410IICInit
** Descriptions:  初始化总线,设 FCLK = 200 MHz,HCLK = 100 MHz,PCLK = 50 MHz
** Input:
** Output:
** Note:
**************************************************************/
void  s3c2410IICInit(void)
{
    int oldlevel;
    UINT32 tempUINT32 = 0;

    oldlevel = intLock ();

    s3c2410x_INT_REG_READ(s3c2410x_INT_CSR_SRCPND,tempUINT32);
    tempUINT32 = tempUINT32&BIT27;
    s3c2410x_INT_REG_WRITE(s3c2410x_INT_CSR_SRCPND,tempUINT32);

    s3c2410x_INT_REG_READ(s3c2410x_INT_CSR_INTPND,tempUINT32);
    tempUINT32 = tempUINT32&BIT27;
    s3c2410x_INT_REG_WRITE(s3c2410x_INT_CSR_INTPND,BIT27);
```

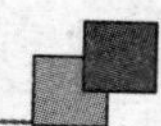

```
    s3c2410x_INT_REG_READ(s3c2410x_INT_CSR_INTMSK,tempUINT32);
    tempUINT32& = ~BIT27;
    s3c2410x_INT_REG_WRITE(s3c2410x_INT_CSR_INTMSK,tempUINT32);

    intUnlock(oldlevel);


    if(intConnect(INUM_TO_IVEC(INT_LVL_IIC), iicIntHandler, 0) == ERROR)
        return (ERROR);
    if(intEnable(INT_LVL_IIC) == ERROR)
        return (ERROR);

    /*增加按键的中断*/
    if(intConnect(INUM_TO_IVEC(INT_LVL_EINT_4_7), iicIntHandler, 0) == ERROR)
        return (ERROR);
    if(intEnable(INT_LVL_EINT_4_7) == ERROR)
        return (ERROR);

    s3c2410_IIC_REG_READ(rGPE[illegible]empUINT32);
    tempUINT32 = (tempUINT32& 0x[illegible]FFFFF) | 0xA0000000;
    s3c2410_IIC_REG_WRITE(rGPECON.tempUINT32);

    /*禁止内部上拉电阻 */
    s3c2410_IIC_REG_READ(rGPEUP,tempUINT32);
    tempUINT32 = (tempUINT32| 0xC000);
    s3c2410_IIC_REG_WRITE(rGPEUP,tempUINT32);

    /*设置 I²C 控制寄存器(使能 ACK 位时才能接收从机的应答位)*/
    s3c2410_IIC_REG_WRITE(rIICCON,IICCON_DACK);

    /*设置 I²C 为主机模式*/
    s3c2410_IIC_REG_WRITE(rIICSTAT,((3<<6)|(1<<4)));

    /*从机地址(作主机时没有用)*/
    s3c2410_IIC_REG_WRITE(rIICADD,0x10);
    s3c2410_IIC_REG_READ(rIICCON,tempUINT32);

}

/**********************************************************************
* * Function name: START_I2C
```

```
** Descriptions:启动总线,发送从机地址
** Input: sla        从机地址
** Output:操作成功返回 TRUE,仲载失败/无从机应答返回 FALSE
** Note: sla 位最低位为读写控制位
*******************************************************************/
LOCAL int  StartI2C
(
    uint8 sla
)
{
    UINT32 tempUINT32;
    UINT32  i;
    int oldlevel;

    s3c2410_IIC_REG_WRITE(rIICDS,sla);/*设置从机地址    */
    if(sla&0x01)
    {
        /*主接收模式,发送使能,启动总线*/
        s3c2410_IIC_REG_WRITE(rIICSTAT,((2<<6)|(1<<5)|(1<<4)));

        /*若是重启总线,则需要有此操作*/
        s3c2410_IIC_REG_WRITE(rIICCON,IICCON_DACK);
    }
    else
    {
        oldlevel = intLock ();

        /*主发送模式,发送使能,启动总线*/
        s3c2410_IIC_REG_WRITE(rIICSTAT,((3<<6)|(1<<5)|(1<<4)));
        s3c2410_IIC_REG_READ(rIICSTAT,tempUINT32);
        intUnlock(oldlevel);
    }
    for(;;)
    {
        for(i=0;i<100;i++);
        s3c2410_IIC_REG_READ(rIICCON,tempUINT32);
        if(tempUINT32==0xf3) break;
    };

    /*等待操作完成*/
    s3c2410_IIC_REG_READ(rIICCON,tempUINT32);
    while((tempUINT32&0x10) == 0);
```

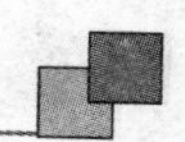

```
    /*判断操作是否成功(总线仲裁和从机应答)*/
    s3c2410_IIC_REG_READ(rIICSTAT,tempUINT32);
    if((tempUINT32&0x09) == 0)
    {
        return(TRUE);
    }
    else
    {
        /*发送结束信号*/
        if(sla&0x01)
        {
            s3c2410_IIC_REG_WRITE(rIICSTAT,((2<<6)|(0<<5)|(1<<4)));
        }
        else
        {
            s3c2410_IIC_REG_WRITE(rIICSTAT,((3<<6)|(0<<5)|(1<<4)));
        }

        s3c2410_IIC_REG_WRITE(rIICCON,IICCON_DACK);
        for(i = 0; i<5000; i++);   /*等待结束信号产生完毕 */
        return(FALSE);
    }
}

/*****************************************************************************
** Function name: I2C_SendByte
** Descriptions:发送一字节数据,并接收应答位
** Input: dat 要发送的数据
** Output:操作成功返回 TRUE,仲裁失败/无从机应答返回 FALSE
** Note:
*****************************************************************************/
LOCAL int  I2C_SendByte(uint8 dat)
{
    UINT32 tempUINT32 = 0;
    uint32  i;

    /*将数据写入 I2C 数据寄存器*/
    s3c2410_IIC_REG_WRITE(rIICDS,dat);

    /*清除中断标志,允许发送数据操作*/
    s3c2410_IIC_REG_WRITE(rIICCON,IICCON_DACK);
```

```
for(;;)
{
    for(i = 0;i<100;i ++ );

    s3c2410_IIC_REG_READ(rIICCON,tempUINT32);
    if(tempUINT32 = = 0xf3)
        break;
};

/ * 等待操作完成 * /
s3c2410_IIC_REG_READ(rIICCON,tempUINT32);
while((tempUINT32&0x10) == 0);

/ * 判断操作是否成功（总线仲裁和从机应答）* /
s3c2410_IIC_REG_READ(rIICSTAT,tempUINT32);
if((tempUINT32&0x09) == 0)
{
    return(TRUE);
}
else
{
    / * 发送结束信号      * /
    s3c2410_IIC_REG_WRITE(rIICSTAT,((3<<6)|(0<<5)|(1<<4)));

    s3c2410_IIC_REG_WRITE(rIICCON,IICCON_DACK);

    / * 等待结束信号产生完毕 * /
    for(i = 0; i<5000; i ++ );

    return(FALSE);
  }
}

/ ***************************************************************
* * Function name: I2C_RcvByteNA
* * Descriptions:接上 I²C 总线上一字节数据,并发送非应答位
* * Input: dat 用于接收数据的指针
* * Output:操作成功返回 TRUE,仲裁失败/无从机应答返回 FALSE
* * Note:
*************************************************************** /
```

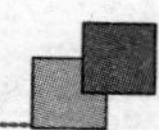

```
LOCAL int  I2C_RcvByteNA(uint8 * dat)
{
    UINT32 tempUINT32 = 0;
    uint32  i;

    /* 允许接收数据 */
    s3c2410_IIC_REG_WRITE(rIICCON,IICCON_DACK);

    for(;;)
    {
        for(i = 0;i<100;i++);

        s3c2410_IIC_REG_READ(rIICCON,tempUINT32);
        if(tempUINT32 == 0xf3)
            break;
    };

    /* 等待接收数据操作完成 */
    s3c2410_IIC_REG_READ(rIICCON,tempUINT32);
    while((tempUINT32&0x10) == 0);

    /* 判断操作是否成功（总线仲裁）*/
    s3c2410_IIC_REG_READ(rIICSTAT,tempUINT32);
    if((tempUINT32&0x08) != 0)
    {
        /* 发送结束信号      */
        s3c2410_IIC_REG_WRITE(rIICSTAT,((2<<6)|(0<<5)|(1<<4)));
        s3c2410_IIC_REG_WRITE(rIICCON,IICCON_DACK);

        /* 等待结束信号产生完毕 */
        for(i = 0; i<5000; i++);

        return(FALSE);
    }

    /* 读取数据 */
    s3c2410_IIC_REG_READ(rIICDS,tempUINT32);
    * dat = tempUINT32;

    return(TRUE);
}
```

```
/****************************************************************************
** Function name: StopI2C
** Descriptions:结束总线
** Input: send   I²C 当前模式。主发送模式时为 1,否则为 0(主接收模式)
** Output:无
** Note:
****************************************************************************/
LOCAL void  StopI2C
(
    uint8 send
)
{
    uint32  i;
    if(send)
    {
        /*发送结束信号*/
        s3c2410_IIC_REG_WRITE(rIICSTAT,((3<<6)|(0<<5)|(1<<4)));
    }
    else
    {
        /*发送结束信号*/
        s3c2410_IIC_REG_WRITE(rIICSTAT,((2<<6)|(0<<5)|(1<<4)));
    }

    s3c2410_IIC_REG_WRITE(rIICCON,IICCON_DACK);

    /*等待结束信号产生完毕*/
    for(i=0; i<5000; i++);
}


/****************************************************************************
** Function name: s3c2410IICSendStr
** Descriptions:使用硬件 I²C 发送数据
** Input: sla  从机地址
**        suba  器件子地址(第一字节用来表示子地址字节个数)
**        s   发送数据缓冲区
**        no  发送数据个数
** Output:操作成功返回 TRUE,仲裁失败/无从机应答返回 FALSE
** Note:使用前设置好参数。程序不会更改 s、suba 缓冲区的数据
****************************************************************************/
```

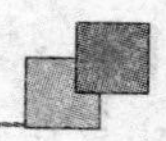

```
int  s3c2410IICSendStr
(
    uint8 sla,
    uint8 * suba,
    uint8 * s,
    uint8 no
)
{
    int  bak;
    sla = sla & 0xFE;
    if(! StartI2C(sla))
    {
        /*启动总线,发送从机地址*/
        if(! StartI2C(sla))
            return(FALSE);
    }

    /*发送器件子地址*/
    bak =  * suba ++ ;
    for(; bak>0; bak -- )
    {
        if(! I2C_SendByte( * suba ++ ))
            return(FALSE);
    }

    /*发送数据*/
    for(; no>0; no -- )
    {
        if(! I2C_SendByte( * s))
        return(FALSE);

    s ++ ;
    }

    /*结束总线*/
    StopI2C(1);

    return(TRUE);
}
```

```
/*****************************************************************
** Function name: s3c2410IICRcvStr
** Descriptions:使用硬件 I²C 读取数据
** Input: sla  从机地址
**          suba  器件子地址缓冲区指针(第一字节用来表示子地址字节个数)
**          s  接收数据缓冲区
**          no  接收数据个数
** Output:操作成功返回 TRUE,仲裁失败/无从机应答返回 FALSE
** Note:使用前设置好参数。程序不会更改 suba 缓冲区的数据
*****************************************************************/
int  s3c2410IICRcvStr
(
    uint8 sla,
    uint8 *suba,
    uint8 *s,
    uint8 no
)
{
    int  bak;

    /*子地址个数*/
    bak = *suba++;
    if(bak > 0)
    {
        sla = sla & 0xFE;
        if(! StartI2C(sla))
        {
            /*启动总线,发送从机地址(写)*/
            if(! StartI2C(sla))
                return(FALSE);
        }

        /*发送器件子地址*/
        for(; bak>0; bak--)
        {
            if(! I2C_SendByte(*suba++)) return(FALSE);
        }
    }

    /*重启总线*/
    sla = sla | 0x01;
    if(! StartI2C(sla))
```

```
        return(FALSE);          /* 启动总线,发送从机地址(读) */

    /* 读取数据     */
    for(; no>1; no--)
    {
        if(! I2C_RcvByteA(s))
            return(FALSE);

        s++;
    }

    if(! I2C_RcvByteNA(s))
        return(FALSE);

    /* 结束总线 */
    StopI2C(0);

    return(TRUE);
}
```

4. IICdev.h 清单

```
#include  "s3c2410xIIC.h"
#include "IICdev.c"
#include "zlg7290.c"
#include "stdio.h"

/* ZLG7290 控制 LED 数码管闪烁命令 */
#define     Glitter_COM      0x70
/****************************************************************************
** Function name: testiic
** Descriptions:通过 zlg7290 测试 LED 和按键
** Input:
** Output:操作成功返回 TRUE,失败返回 FALSE
** Note:
****************************************************************************/
int  testiic(void)
{
    int  i, j;
    char  disp_buf[50];  /* 定义显示缓冲区 */
    uint16  key;
```

```
/*初始化 I2C 接口*/
s3c2410IICInit();

/*所有 LED 分别显示 0、2、4、6、8 测试*/
for(i = 0; i<= 8; i = i + 2)
{
    for(j = 0; j<8; j++)
        disp_buf[j] = i;

    ZLG7290_SendBuf((uint8 *)disp_buf, 8);
    DelayNS(1000);
}

/*8 个 LED 数码管显示"87654321"*/
for(j = 0; j<8; j++)
    disp_buf[j] = j + 1;

DelayNS(1000);
ZLG7290_SendBuf((uint8 *)disp_buf, 8);

/*读取按键,设置键值对应的显示位闪烁*/
while(1)
{
    DelayNS(1);

    key = ZLG7290_GetKey();
    if((key&0xFF00) == 0)
    {
        key = key&0x00FF;
    }

    switch(key)
    {
        case  2:
            ZLG7290_SendCmd(Glitter_COM, 0x01);
            break;

        case  3:
            ZLG7290_SendCmd(Glitter_COM, 0x02);
            break;
```

```
        case 4：
            ZLG7290_SendCmd(Glitter_COM, 0x04);
            break;

        case  5：
            ZLG7290_SendCmd(Glitter_COM, 0x08);
            break;

        case 6：
            ZLG7290_SendCmd(Glitter_COM, 0x10);
            break;

        case  7：
            ZLG7290_SendCmd(Glitter_COM, 0x20);
            break;

        case  8：
            ZLG7290_SendCmd(Glitter_COM, 0x40);
            break;

        case  9：
            ZLG7290_SendCmd(Glitter_COM, 0x80);
            break;

        default：
            break;
        }
    }

  return(TRUE);
}

/*************************************************************
** Function name：DelayNS
** Descriptions：长软件延时。延时时间与系统时钟有关
** Input：dly  延时参数，值越大，延时越久
** Output：
** Note：
*************************************************************/
void  DelayNS
(
    uint32  dly
```

```
)
{
    uint32  i;

    for(; dly>0; dly--)
        for(i=0; i<50000; i++);
}
```

5. zlg7290.h 清单

```
#ifndef   __ZLG7290_H
#define   __ZLG7290_H

#ifndef   IN_ZLG7290

    #ifdef __cplusplus
    extern "C" {
    #endif

#ifndef _ASMLANGUAGE

/*********************************************************************
** Function name: ZLG7290_SendData
** Descriptions:发送数据
** Input: SubAdd   输入数据地址
**        DATA   输入值
** Output:操作成功返回 TRUE,失败返回 FALSE
** Note:
*********************************************************************/
extern unsigned char ZLG7290_SendData(unsigned char SubAdd,unsigned char Data);


/*********************************************************************
** Function name: ZLG7290_SendCmd
** Descriptions:发送命令(对子地址 7、8)
** Input: DATA1      命令 1
**        DATA2      命令 2
** Output:操作成功返回 TRUE,失败返回 FALSE
** Note:
*********************************************************************/
extern unsigned char ZLG7290_SendCmd(unsigned char Data1,unsigned char Data2);
```

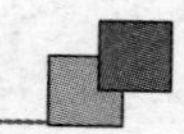

```
/********************************************************************
 ** Function name: ZLG7290_SendBuf
 ** Descriptions:向显示缓冲区发送数据
 ** Input: *disp_buf  要发送数据的起始地址
 **             num   发送个数
 ** Output:无
 ** Note:
********************************************************************/
extern void  ZLG7290_SendBuf(unsigned char *disp_buf,unsigned char num);

/********************************************************************
 ** Function name: ZLG7290_GetKey
 ** Descriptions:读取键值
 ** Input:无
 ** Output: >0 表示键值  (低 8 位为键码,高 8 位为按键次数)
 **              =0 表示无键按下
 ** Note:
********************************************************************/
extern unsigned short    ZLG7290_GetKey(void);

#endif     /* _ASMLANGUAGE */

    #ifdef __cplusplus
    }
    #endif

#endif        // IN_ZLG7290

#endif        // __ZLG7290_H
```

6. zlg7290.c 清单

```
#include     "IICdev.h"

#define  IN_ZLG7290

/*ZLG7290 的 IIC 地址 */
#define  ZLG7290      0x70


/*ZLG7290 寄存器地址(子地址)*/
#define  SubKey        0x01
```

```
#define  SubCmdBuf      0x07
#define  SubDpRam       0x10

/*****************************************************************
** Function name: DelayMS
** Descriptions:软件延时
** Input: dly 延时参数,值越大时延时越久
** Output:无
** Note:
*****************************************************************/
void DelayMS
(
    uint32 dly
)
{
    uint8  i;

    for(; dly>0; dly--)
        for(i=0; i<100; i++);
}

/*****************************************************************
** Function name: ZLG7290_SendData
** Descriptions:发送数据
** Input: SubAdd      输入数据地址
**        DATA              输入值
** Output:操作成功返回 TRUE,失败返回 FALSE
** Note:
*****************************************************************/
unsigned char ZLG7290_SendData
(
    unsigned char SubAdd,
    unsigned char Data
)
{
    unsigned char suba[2];
    if(SubAdd>0x17)
        return FALSE;

    /*子地址个数*/
    suba[0] = 1;
    /*子地址*/
```

```
    suba[1] = SubAdd;
    s3c2410IICSendStr(ZLG7290, suba, &Data, 1);

    DelayMS(10);
    return TRUE;
}


/**************************************************************
** Function name: ZLG7290_SendCmd
** Descriptions:发送命令(对子地址 7、8)
** Input: DATA1      命令 1
**        DATA2      命令 2
** Output:操作成功返回 TRUE,失败返回 FALSE
** Note:
**************************************************************/
unsigned char ZLG7290_SendCmd
(
    unsigned char Data1,
    unsigned char Data2
)
{
    unsigned char Data[2];
    unsigned char suba[2];

    Data[0] = Data1;
    Data[1] = Data2;
    suba[0] = 1;
    suba[1] = SubCmdBuf;
    s3c2410IICSendStr(ZLG7290, suba, Data, 2);
    DelayMS(10);

    returnTRUE;
}


/**************************************************************
** Function name: ZLG7290_SendBuf
** Descriptions:向显示缓冲区发送数据
** Input: *disp_buf  要发送数据的起始地址
**             num  发送个数
```

```
** Output:无
** Note:
*******************************************************************/
void  ZLG7290_SendBuf
(
    unsigned char *disp_buf,
    unsigned char num
)
{
    unsigned char  i;

    for(i = 0; i<num; i++)
    {
        ZLG7290_SendCmd(0x60 + i, *disp_buf);
        disp_buf++;
        DelayNS(100);
    }
}

/*******************************************************************
** Function name: ZLG7290_GetKey
** Descriptions:读取键值
** Input:无
** Output: >0 表示键值  (低 8 位为键码,高 8 位为按键次数)
**             =0 表示无键按下
** Note:
*******************************************************************/
unsigned short  ZLG7290_GetKey(void)
{
    unsigned char rece[2];
    unsigned char suba[2];
    rece[0] = rece[1] =  0;
    suba[0] = 1;
    suba[1] = SubKey;
    s3c2410IICRcvStr(ZLG7290, suba, rece, 2);
    DelayMS(10);

    return (rece[0] | (rece[1]<<8));
}
```

第9章 字符设备驱动程序设计实验

9.1 字符设备驱动编写概述

字符设备是指在I/O传输过程中以字符为单位进行传输的设备，如键盘、鼠标和打印机等。字符设备驱动程序一般都有7个基本I/O操作函数，但也不排除某些具体的字符设备驱动程序可能忽略其中的一个或者几个操作函数。

这7个基本I/O操作函数及其功能是：

- creat()，创建设备；
- remove()，删除设备；
- open()，打开设备；
- close()，关闭设备；
- read()，读取设备中的数据；
- write()，向设备写数据；
- ioctl()，设置设备的方式字。

驱动程序除了具有上述7种基本I/O操作函数外，还需要包括以下两种操作：

(1) 初始化函数负责在I/O系统中安装驱动程序；驱动程序将相应的设备与所需的资源关联起来；然后初始化函数再执行其他必需的硬件初始化操作。

(2) 将驱动程序加载到I/O系统中，这类函数一般称为xxDevCreate()。

对于一些简单的驱动程序，往往可以将两个函数合二为一。

在编写字符设备驱动程序时还存在一个问题：如果设备的缓冲中目前没有任何数据，但应用程序此时发起了一个read()调用，那么此时该应用程序就有可能阻塞在这个位置上，直到缓冲中有了足够的数据。如果该应用程序只是处理一个设备的数据，那么阻塞还不至于产生问题。但是，如果该应用程序要处理多个设备的数据，则阻塞在某个设备操作上就会影响正常的功能设计。要想解决类似问题，需要借助系统提供的 select 机制。实际上这个问题并不是字符设备独有的,很多设备驱动都存在同样的问题,也大都采用类似的处理办法。

支持调用 select()函数的用户驱动程序,任务能同时等待多个设备的输入,或者允许任务为设备执行指定的 I/O 操作设定最长等待时间,这需要 selectLib 函数库的支持。用户的驱动程序必须支持 select()函数的调用,设备才能实现以下的操作:

(1) 任务需要为等待一个设备执行 I/O 操作设定时间限制;

(2) 一个驱动程序同时支持多个设备,而运行的任务可能会同时等待这些设备;

(3) 任务等待某个设备的 I/O 操作,同时该设备等待其他设备的 I/O 操作;

(4) 在 select()函数执行时,驱动程序必须保存一个记载等待任务的表。当被等待的设备可用时,驱动程序激活所有等待该设备的任务。

9.2 蜂鸣器驱动设计实验

9.2.1 实验目的

熟悉 S3C2410A 处理器的 I/O 配置方法及 GPIO 输出控制。

9.2.2 实验设备

(1) 硬件。

- PC 1 台;
- MagicARM2410 教学实验开发平台 1 套。

(2) 软件。

- Windows98/XP/2000 系统;
- Tornado 交叉开发环境。

9.2.3 实验内容

控制 MagicARM2410 实验箱上的蜂鸣器报警。

9.2.4 实验原理

S3C2410A 具有 117 个通用 I/O 口,分为 A～H8 个端口。由于每个 I/O 口都有第 2 功能、甚至第 3 功能,所以需要通过设置 GPxCON 寄存器来选择 GPx 口 I/O 的功能,其中 x 可以为 A、B、C、D、E、F、G、H,表示相应的 I/O 端口,如下文所示:

- Port A(GPA):23－output port
- Port B(GPB):11－input/output port
- Port C(GPB):16－input/output port
- Port D(GPD):16－input/output port
- Port E(GPE):16－input/output port
- Port F(GPF):8－input/output port
- Port G(GPG):16－input/output port
- Port H(GPH):11－input/output port

当 I/O 设置为 GPIO 输出模式(Output 模式)时,可以通过写 GPxDAT 控制相应 I/O 口输出高电平或低电平。GPxDAT 为“1”的位对应 I/O 输出高电平,为“0”的位对应 I/O 输出低电平。

Port E 口寄存器如表 9.1 所列,Port E GPECON 寄存器说明如表 9.2 所列。

表 9.1　Port E 口寄存器表

Register	Address	R/W	Description	Reset Value
GPECON	0x56000040	R/W	Configure the pins of port E	0x0
GPEDAT	0x56000044	R/W	The data register for port E	Undefined
GPEUP	0x56000048	R/W	Pull-up disable register for port E	0x0

表 9.2　Port E GPECON 寄存器说明

GPECON	Bit	Description
GPE15	[31:30]	00 = Input　01 = Output(open drain output) 10 = IICSDA　11 = Reserved
GPE14	[29:28]	00 = Input　01 = Output(open drain output) 10 = IICSCL　11 = Reserved
GPE13	[27:26]	00 = Input　01 = Output 10 = SPICLK0　11 = Reserved
GPE12	[25:24]	00 = Input　01 = Output 10 = SPIMOSI0　11 = Reserved
GPE11	[23:22]	00 = Input　01 = Output 10 = SPIMOSO0　11 = Reserved

实验箱该部分电路原理图如图 9.1 所示。

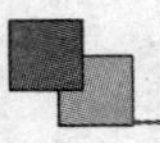

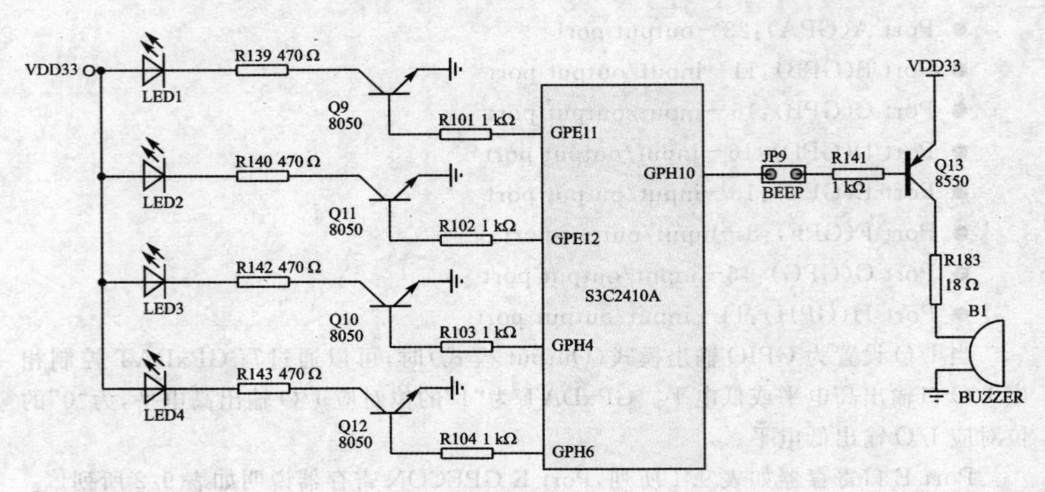

图 9.1 蜂鸣器驱动电路图

9.2.5 实验步骤

(1) 完成相关设置(主机网络地址、FTP 服务器和超级终端等),然后系统上电,登录到 ftp 服务器上获取编译的基本映像,并且运行。

(2) 复制蜂鸣器实验的源代码 BUZZ.c 到 C:\Tornado2.2\target\config\MagicARM2410 目录下。

(3) 添加 BUZZ.c 到项目文件中,编译,生成“.o”文件。

(4) 连接上 Target Server,并开启 Wshell。

(5) 将 Object Modules 中的 BUZZ.o 模块下载到运行的系统中去。

(6) 将 MagicARM2410 实验箱上的蜂鸣器跳线 JP9 短接。

(7) 在 Wshell 中输入 BUZZexamtaskStart,来执行蜂鸣器的控制任务,观察实验结果。

(8) 在蜂鸣器按照自己设定的程序工作完成后,完成了本次实验设计。

9.2.6 程序清单

```
/******蜂鸣器驱动设计实验 BUZZ.c —— only  for  MagicARM2410******/
#include "taskLib.h"

#define rGPHCON     (*(volatile unsigned *)0x56000070) //Port H control
#define rGPHDAT     (*(volatile unsigned *)0x56000074) //Port H data

// 定义蜂鸣器控制口
```

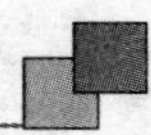

```
#define      BEEP              (1<<10)      /* GPH10 口 */
#define      BEEP_MASK         (~BEEP)

/*********************************************************************
** Function name: RunBeep
** Descriptions: 控制蜂鸣器 Be 一声音
** Input: 无
** Output: 无
*********************************************************************/
void  RunBeep(void)
{
    rGPHDAT = rGPHDAT & BEEP_MASK;              // BEEP = 0
    taskDelay(sysClkRateGet()/2);
    rGPHDAT = rGPHDAT | BEEP;                   // BEEP = 1
    taskDelay(sysClkRateGet()/2);
}

/*********************************************************************
** Function name: BUZZexamtask
** Descriptions:  控制蜂鸣器的任务
** Input: 无
** Output: 无
*********************************************************************/
void BUZZexamtask()
{
    // 初始化 I/O,rGPHCON[21:20] = 01b,设置 GPH10 为 GPIO 输出模式
    rGPHCON = (rGPHCON & (~(0x03<<20))) | (0x01<<20);

    while(1)
    {
        RunBeep();                              //蜂鸣器响一声

        taskDelay(sysClkRateGet() * 1);         //每秒执行一次 RunBeep 函数
    }

}

/*********************************************************************
** Function name: BUZZexamtaskStart
** Descriptions:  蜂鸣器任务启动函数
** Input: 无
** Output: 无
```

```
*************************************************************************/
void BUZZexamtaskStart(void)
{
    taskSpawn("BUZZexamtask",80,0,2048,
              (FUNCPTR)BUZZexamtask,0,0,0,0,0,0,0,0,0,0);
}
```

9.3 LED 流水灯驱动设计实验

9.3.1 实验目的

熟悉 S3C2410A 处理器的 I/O 配置方法及 GPIO 输出控制。

9.3.2 实验设备

(1) 硬件。

- PC 1 台；
- MagicARM2410 教学实验开发平台 1 套。

(2) 软件。

- Windows98/XP/2000 系统；
- Tornado 交叉开发环境。

9.3.3 实验内容

控制 MagicARM2410 实验箱上的 LED1～LED4 显示。将程序固化到片外 NOR FLASH 中,脱机运行程序。

9.3.4 实验原理

参见 9.2 蜂鸣器驱动实验设计。

9.3.5 实验步骤

(1) 完成相关设置(主机网络地址、FTP 服务器和超级终端等),然后系统上电,登录到 ftp 服务器上获取编译的基本映像,并且运行。

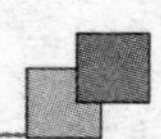

(2) 复制流水灯驱动实验源代码 led.c 到 C:\Tornado2.2\target\config\MagicARM2410 目录下。

(3) 添加 led.c 文件到项目文件中去,编译,生成".o"文件。

(4) 连接上 Target Server,并开启 Wshell。

(5) 将 Object Modules 中的 led.o 模块下载到运行的系统中去。

(6) 在 Wshell 中输入 LEDexamtaskStart,来执行 LED 的控制任务,观察实验结果。

(7) 在 4 个 LED 按照自己设定的程序工作完成后,完成了本次实验设计。

9.3.6 程序清单

```
/******LED 流水灯驱动设计实验 LED.c —— only  for  MagicARM2410******/
#include "taskLib.h"

typedef unsigned int         uint32;
#define rGPHCON     (*(volatile unsigned *)0x56000070) //Port H control
#define rGPECON     (*(volatile unsigned *)0x56000040) //Port E control
#define rGPHDAT     (*(volatile unsigned *)0x56000074) //Port H data
#define rGPEDAT     (*(volatile unsigned *)0x56000044) //Port E data

// 定义 LED 控制口(输出高电平时点亮 LED)
#define  LED1_CON       (1<<11)      /* GPE11 口 */
#define  LED2_CON       (1<<12)      /* GPE12 口 */
#define  LED3_CON       (1<<4)       /* GPH4 口  */
#define  LED4_CON       (1<<6)       /* GPH6 口  */

/*************************************************************
** Function name: LED_DispAllOn
** Descriptions: 控制 LED1~LED4 全部点亮
** Input: 无
** Output: 无
*************************************************************/
void  LED_DispAllOn(void)
{
    rGPEDAT = rGPEDAT | (0x03<<11);
    rGPHDAT = rGPHDAT | (0x05<<4);
}

/*************************************************************
```

```
** Function name: LED_DispAllOff
** Descriptions: 控制 LED1～LED4 全部熄灭
** Input: 无
** Output: 无
*******************************************************************/
void  LED_DispAllOff(void)
{
    rGPEDAT = rGPEDAT & (~(0x03<<11));
    rGPHDAT = rGPHDAT & (~(0x05<<4));
}

/*******************************************************************
** Function name: LED_DispNum
** Descriptions: 控制 LED1～LED4 显示指定 16 进制数值
                 LED4 为最高位,LED1 为最低为,点亮表示该位为 1
** Input: dat     显示数值(低 4 位有效)
** Output: 无
*******************************************************************/
void  LED_DispNum(uint32 dat)
{
    dat = dat & 0x0000000F;    // 参数过滤

    // 控制 LED4、LED3 显示(d3、d2 位)
    if(dat & 0x08)
        rGPHDAT = rGPHDAT | (0x01<<6);
    else
        rGPHDAT = rGPHDAT & (~(0x01<<6));

    if(dat & 0x04)
        rGPHDAT = rGPHDAT | (0x01<<4);
    else
        rGPHDAT = rGPHDAT & (~(0x01<<4));

    // 控制 LED2、LED1 显示(d1、d0 位)
    rGPEDAT = (rGPEDAT & (~(0x03<<11))) | ((dat&0x03) << 11);
}

/*******************************************************************
** Function name: LEDexamtask
** Descriptions:  控制 LED 的任务
** Input: 无
** Output: 无
```

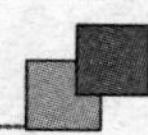

```
*****************************************************************/
void LEDexamtask()
{
    int i;
    // 初始化 I/O,rGPECON[25:22] = 0101b,设置 GPE11 和 GPE12 为 GPIO 输出模式
    rGPECON = (rGPECON & (~(0x0F<<22))) | (0x05<<22);
    // rGPHCON[13:8] = 01xx01b,设置 GPH4、GPH6 为 GPIO 输出模式
    rGPHCON = (rGPHCON & (~(0x33<<8))) | (0x11<<8);

    while(1)
    {
        for(i = 0; i<5; i++)
        {
            LED_DispAllOff();    // LED 全熄灭
            taskDelay(sysClkRateGet() * 1);
            LED_DispAllOn();     // LED 全点亮
            taskDelay(sysClkRateGet() * 1);
        }

        // 控制 LED 指示 0~F 的 16 进制数值
        for(i = 0; i<16; i++)
        {
            LED_DispNum(i);      // 显示数值 i
            taskDelay(sysClkRateGet() * 1);
        }

        taskDelay(sysClkRateGet() * 1);
    }

}

/*****************************************************************
** Function name: LEDexamtaskStart
** Descriptions: 控制 LED 的任务
** Input: 无
** Output: 无
*****************************************************************/
void LEDexamtaskStart(void)
{
    taskSpawn("LEDexamtask",80,0,2048,
            (FUNCPTR)LEDexamtask,0,0,0,0,0,0,0,0,0,0);
}
```

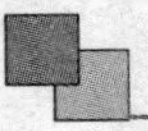

9.4 按键驱动程序设计实验

9.4.1 实验目的

掌握 3C2410A 处理器的 I/O 配置方法，能够使用 GPIO 输入模式读取开关信号。

9.4.2 实验设备

(1) 硬件。
- PC 1 台；
- MagicARM2410 教学实验开发平台 1 套。

(2) 软件。
- Windows98/XP/2000 系统；
- Tornado 交叉开发环境。

9.4.3 实验内容

读取 GPF4 口上的电平值，用来控制蜂鸣器。

9.4.4 实验原理

- 通过设置 GPxCON 寄存器来选择 GPx 口 I/O 的功能，当 I/O 设置为 GPIO 输入模式(Input 模式)时，读取 GPxDAT 寄存器即取得 I/O 口线上的电平状态。通常会使用 if(GPxDAT & (1<<n))语句来判断 GPxn 口是否为高电平；
- MagicARM2410 实验箱上使用了 S3C2410A 的 GPF4 口连接一个独立按键 KEY1。当 KEY1 键按下时，GPF4 口上的电平值为 0；当 KEY1 键放开时，由上拉电阻将 GPF4 口拉到高电平，所以其电平值为 1。

实验部分电路原理图如图 9.2 所示。

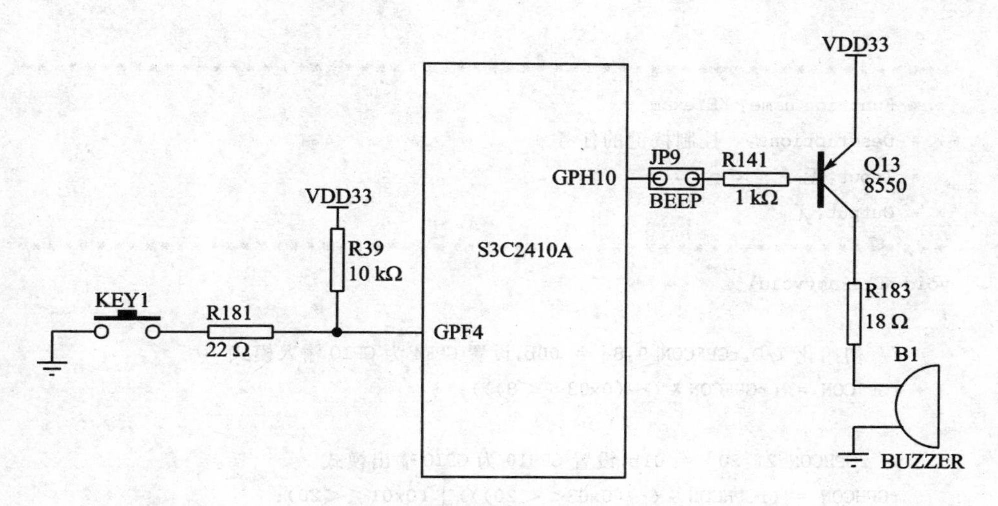

图 9.2 按键驱动电路图

9.4.5 实验步骤

(1) 完成相关设置(主机网络地址、FTP服务器和超级终端等),然后系统上电,登录到ftp服务器上获取编译的基本映像,并且运行。

(2) 复制按键驱动实验的源代码key.c到C:\Tornado2.2\target\config\MagicARM2410目录下。

(3) 添加key.c到项目文件中去,编译,生成".o"文件。

(4) 连接上Target Server,并开启Wshell。

(5) 将Object Modules中的key.o模块下载到运行的系统中去。

(6) 在Wshell中输入KEYexamtaskStart,来执行按键的控制任务,观察实验结果。

(7) 在按键状态与串口打印信息一致后,完成了本次实验设计。

9.4.6 程序清单

```
/******按键驱动程序设计实验 key.c —— only  for  MagicARM2410******/
#include "taskLib.h"

#define     KEY_CON             (1<<4)              /* GPF4 口   */

#define rGPFCON     (*(volatile unsigned *)0x56000050) //Port F control
#define rGPHCON     (*(volatile unsigned *)0x56000070) //Port H control
#define rGPFDAT     (*(volatile unsigned *)0x56000054) //Port F data
```

```
/*************************************************************
** Function name: KEYexam
** Descriptions:  控制按键的任务
** Input:无
** Output:无
*************************************************************/
void KEYexam(void)
{
   // 初始化 I/O,rGPFCON[9:8] = 00b,设置 GPF4 为 GPIO 输入模式
   rGPFCON = (rGPFCON & (~(0x03<<8)));

   // rGPHCON[21:20] = 01b,设置 GPH10 为 GPIO 输出模式
   rGPHCON = (rGPHCON & (~(0x03<<20))) | (0x01<<20);

   while(1)
   {
       if(rGPFDAT & KEY_CON)    // 读取 GPF 口线上的电平,判断 GPF4 是否为高电平
       {
            printf("\n 松开了按键!");
       }
       else
       {
            printf("\n 按下了按键!");
       }

       taskDelay(100);
   }

}

/*************************************************************
** Function name: KEYexamtaskStart
** Descriptions:  按键任务启动函数
** Input: 无
** Output: 无
*************************************************************/
void KEYexamtaskStart(void)
{
   taskSpawn("KEYexam",80,0,2048,(FUNCPTR)KEYexam,0,0,0,0,0,0,0,0,0,0);
}
```

9.5 直流电机驱动程序设计实验

9.5.1 实验目的

掌握使用PWM方式控制直流电机的转动速度。

9.5.2 实验设备

(1) 硬件。

- PC 1台;
- MagicARM2410教学实验开发平台1套。

(2) 软件。

- Windows98/XP/2000系统;
- Tornado交叉开发环境。

9.5.3 实验内容

使用S3C2410A的TOUT0口输出PWM信号控制直流电机,实现四级调速控制。通过检测按键KEY1来改变当前电机的速度级别。

9.5.4 实验原理

- S3C2410A具有4路PWM输出,输出口分别为TOUT0~TOUT3,其中两路带有死区控制功能。为了能够正确输出PWM信号,需要设置GPBCON寄存器相应的I/O为TOUTx功能。
- 通过TCFG0寄存器为PWM定时器时钟源设置预分频值,通过TCFG1寄存器选择PWM定时器时钟源。接着,通过TCNTB0寄存器设置PWM周期,通过TCMPB0设置PWM占空比。最后,通过TCON寄存器启动PWM定时器,即可输出PWM信号。

实验部分电路原理图如图9.3所示。

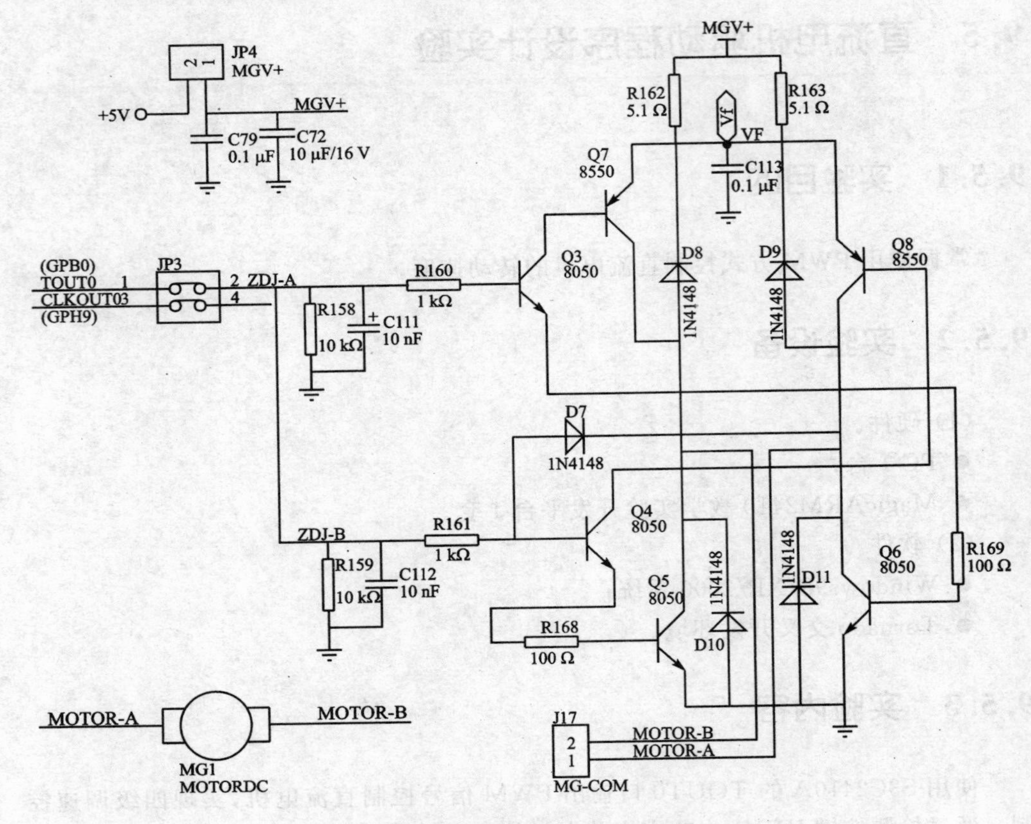

图 9.3　直流电机驱动电路图

9.5.5　实验步骤

(1) 完成相关设置(主机网络地址、FTP 服务器、超级终端等),然后系统上电,登录到 ftp 服务器上获取编译的基本映像,并且运行。

(2) 复制直流电机驱动实验源代码 DCmoto.c 到 C:\Tornado2.2\target\config\MagicARM2410 目录下。

(3) 添加 DCmoto.c 到项目文件中去,编译,生成".o"文件。

(4) 将 MagicARM2410 实验箱上的直流电机控制电路的电源跳线 JP4 短接,将直流电机控制口跳线 JP3 短接。

(5) 连接上 Target Server,并开启 Wshell。

(6) 将 Object Modules 中的 DCmoto.o 模块下载到运行的系统中去。

(7) 在 Wshell 中输入 DCmoto_taskStart,来执行按键的控制任务,观察实验结果。

(8) 全速运行程序,观察直流电机转动的速度。按下独立按键 KEY1,观察直流

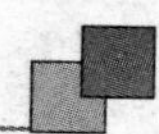

电机的转动速度是否改变。

(9) 在分析第8步原因后,完成了本次实验设计。

9.5.6 程序清单

```
/******直流电机控制实验 DCmoto.c —— only  for  MagicARM2410******/
#include "taskLib.h"

// 定义独立按键 KEY1 的输入口
#define     KEY_CON             (1<<4)          /* GPF4 口   */
typedef unsigned int        uint32;
typedef unsigned short      uint16;
#define rGPFDAT     (*(volatile unsigned *)0x56000054) //Port F data
#define rTCFG0      (*(volatile unsigned *)0x51000000) //Timer 0 configuration
#define rTCFG1      (*(volatile unsigned *)0x51000004) //Timer 1 configuration
#define rTCON       (*(volatile unsigned *)0x51000008) //Timer control
#define rTCNTB0     (*(volatile unsigned *)0x5100000c) //Timer count buffer 0
#define rTCMPB0     (*(volatile unsigned *)0x51000010) //Timer compare buffer 0
#define rGPFCON     (*(volatile unsigned *)0x56000050) //Port F control
#define rGPBCON     (*(volatile unsigned *)0x56000010) //Port B control
#define rGPBDAT     (*(volatile unsigned *)0x56000014) //Port B data
#define rGPBUP      (*(volatile unsigned *)0x56000018) //Pull - up control B
#define rGPHCON     (*(volatile unsigned *)0x56000070) //Port H control
#define rGPHDAT     (*(volatile unsigned *)0x56000074) //Port H data
#define rGPHUP      (*(volatile unsigned *)0x56000078) //Pull - up control H

/*****************************************************************
** Function name: WaitKey
** Descriptions: 等待一个有效按键,本函数有去抖功能
** Input: 无
** Output: 无
*****************************************************************/
void  WaitKey(void)
{
   uint32  i;

   while(1)
   {
      while((rGPFDAT&KEY_CON) == KEY_CON) ;      // 等待 KEY 键按下
      taskDelay(sysClkRateGet()/60);              // 延时去抖
```

```
        if( (rGPFDAT&KEY_CON) != KEY_CON) break;
    }

    while((rGPFDAT&KEY_CON) != KEY_CON);          // 等待按键放开
}

/************************************************************
** Function name: PWM_Init
** Descriptions: 初始化 PWM 定时器
** Input: cycle      PWM 周期控制值(uint16 类型)
**        duty       PWM 占空比(uint16 类型)
** Output: 无
************************************************************/
void  PWM_Init(uint16 cycle, uint16 duty)
{
    // 参数过滤
    if(duty>cycle) duty = cycle;

    // 设置定时器 0,即 PWM 周期和占空比
    // Fclk = 200MHz,时钟分频配置为 1:2:4,即 Pclk = 50 MHz
    rTCFG0 = 97;                      // 预分频器 0 设置为 98,取得 510 204 Hz
    rTCFG1 = 0;                       // TIMER0 再取 1/2 分频,取得 255 102 Hz
    rTCMPB0 = duty;                   // 设置 PWM 占空比
    rTCNTB0 = cycle;                  // 定时值(PWM 周期)
    if(rTCON&0x04)
        rTCON = (1<<1);               // 更新定时器数据(取反输出 inverter 位)
    else
        rTCON = (1<<2)|(1<<1);

    rTCON = (1<<0)|(1<<3);            // 启动定时器
}

/************************************************************
** Function name: DCmoto_task
** Descriptions:  控制直流电机的任务
** Input: 无
** Output: 无
************************************************************/
void DCmoto_task()
{
    uint16  pwm_duty;
```

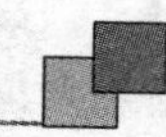

```
    // 独立按键 KEY1 控制口设置,rGPFCON[9:8] = 00b,设置 GPF4 为 GPIO 输入模式
    rGPFCON = (rGPFCON & (~(0x03<<8)));

    // TOUT0 口设置,rGPBCON[1:0] = 10b,设置 TOUT0 功能
    rGPBCON = (rGPBCON & (~(0x03<<0))) | (0x02<<0);
    // 禁止 TOUT0 口的上拉电阻
    rGPBUP = rGPBUP | 0x0001;

    // 设置 GPH9 为 GPIO 输出模式
    rGPHCON = (rGPHCON & (~(0x03<<18))) | (0x01<<18);
    rGPHDAT = rGPHDAT & (~(1<<9));                    // 输出 0 电平
    rGPHUP  = rGPHUP | (1<<9);

    // 初始化 PWM 输出,初始化占空比为 3/4
    pwm_duty = 3 * 255/4;
    //设 PWM 周期控制值为 255
    PWM_Init(255, pwm_duty);

    // 等待按键 KEY1,改变占空比
    while(1)
    {
        WaitKey();
        pwm_duty = pwm_duty + 255/4;  // 改变当前电机的速度级别
        if(pwm_duty>255)
        {
            pwm_duty = 255/4;
        }
        rTCMPB0 = pwm_duty;
    }


}

/*****************************************************************
 * * Function name: DCmoto_taskStart
 * * Descriptions:  直流电机任务启动函数
 * * Input:无
 * * Output:无
*****************************************************************/
void DCmoto_taskStart(void)
{
    taskSpawn("DCmoto - examtask",80,0,2048,
```

```
    (FUNCPTR)DCmoto_task,0,0,0,0,0,0,0,0,0,0);
}
```

9.6 步进电机驱动程序设计实验

9.6.1 实验目的

了解步进电机的控制原理，掌握 VxWorks 下电机转动控制和调速方法。

9.6.2 实验设备

(1) 硬件。

- PC 1 台；
- MagicARM2410 教学实验开发平台 1 套。

(2) 软件。

- Windows98/XP/2000 系统；
- Tornado 交叉开发环境。

9.6.3 实验内容

- 通过 4 个 GPIO 输出有序的矩形脉冲，控制 ULN2003 驱动四相步进电机实现正转，调速的功能。
- 控制的方法采用双四拍(AB — BC — CD — DA — AB)。

9.6.4 实验原理

- 步进电机是一种将脉冲转换为角位移的数据控制电机，即给它一个脉冲信号，它就按设定的方向转动一个固定的角度。用户可以通过控制脉冲的个数来控制角位移量，从而实现准确的定位操作；另外，通过控制脉冲频率来控制电机转动的速度和加速度，从而达到调速的目的。当然，对于步进电机各相绕组(即内部线圈)的控制脉冲要有一定的顺序，否则电机无法正常旋转；
- MagicARM2410 实验箱上的步进电机为四相步进电机，电机步距角为 18 度。S3C2410A 的 GPIO 驱动能力有限，必须通过 ULN2003 达林顿集成驱动芯片驱动步进电机，在步进电机和驱动电路之间连接了电阻，防止控制紊乱造成电机的损坏。

实验部分电路原理图如图 9.4 所示。

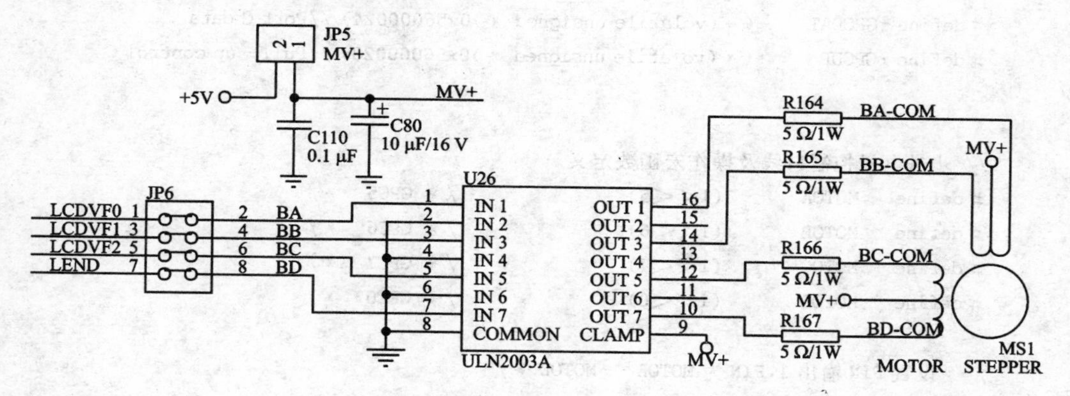

图 9.4　步进电机驱动电路图

9.6.5　实验步骤

（1）完成相关设置(主机网络地址、FTP 服务器、超级终端等),然后系统上电,登录到 ftp 服务器上获取编译的基本映像,并且运行。

（2）复制直流电机驱动实验的源代码 StepMoto. c 到 C:\Tornado2. 2\target\config\MagicARM2410 目录下。

（3）添加 StepMoto. c 到项目文件中去,编译,生成“. o”文件。

（4）将 MagicARM2410 实验箱上的步进电机控制电路的电源跳线 JP5 短接,将步进电机控制口跳线 JP6 短接。

（5）连接上 Target Server,并开启 Wshell。

（6）将 Object Modules 中的 StepMoto. o 模块下载到运行的系统中去。

（7）在 Wshell 中输入 Stepmoto_taskStart,来执行按键的控制任务,观察实验结果。

（8）全速运行程序,观察步进电机转动方向及速度。

（9）改变控制步序延时参数,观察步进电机转动的速度变化。

9.6.6　程序清单

```
/ ******直流电机控制实验 StepMoto.c——only  for  MagicARM2410 ******/
#include "taskLib.h"

typedef unsigned char        uint8;
typedef unsigned int         uint32;
```

```
#define rGPCCON     (*(volatile unsigned *)0x56000020) //Port C control
#define rGPCDAT     (*(volatile unsigned *)0x56000024) //Port C data
#define rGPCUP      (*(volatile unsigned *)0x56000028) //Pull-up control C


// 步进电机控制口线及操作宏函数定义
#define    MOTOA       (1<<5)          /* GPC5   */
#define    MOTOB       (1<<6)          /* GPC6   */
#define    MOTOC       (1<<7)          /* GPC7   */
#define    MOTOD       (1<<0)          /* GPC0   */

/* 设置 PIN 输出 1,PIN 为 MOTOA--MOTOD */
#define    GPIOSET(PIN)   rGPCDAT = rGPCDAT | PIN
/* 设置 PIN 输出 0,PIN 为 MOTOA--MOTOD */
#define    GPIOCLR(PIN)   rGPCDAT = rGPCDAT & (~PIN)

/************************************************************
** Function name: MOTO_Mode2()
** Descriptions: 步进电机双四拍程序。
**              时序控制为 AB--BC--CD--DA--AB,共控制运转 4 圈(电机步距角为
                18 度)。
** Input: dly     每一步的延时控制,值越大,延时越久
** Output: 无
************************************************************/
void MOTO_Mode2(uint8 dly)
{
    uint32 i;

    for(i=0; i<20; i++)
    {
        // AB 相有效
        GPIOSET(MOTOA);
        GPIOSET(MOTOB);
        taskDelay(sysClkRateGet()/dly);
        GPIOCLR(MOTOA);
        GPIOCLR(MOTOB);

        // BC 相有效
        GPIOSET(MOTOB);
        GPIOSET(MOTOC);
        taskDelay(sysClkRateGet()/dly);
        GPIOCLR(MOTOB);
```

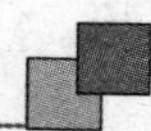

```
        GPIOCLR(MOTOC);

        // CD 相有效
        GPIOSET(MOTOC);
        GPIOSET(MOTOD);
        taskDelay(sysClkRateGet()/dly);
        GPIOCLR(MOTOC);
        GPIOCLR(MOTOD);

        // DA 相有效
        GPIOSET(MOTOD);
        GPIOSET(MOTOA);
        taskDelay(sysClkRateGet()/dly);
        GPIOCLR(MOTOD);
        GPIOCLR(MOTOA);
    }
}

/*****************************************************************
 * * Function name: Stepmoto_task
 * * Descriptions: 控制步进电机的任务
 * * Input: 无
 * * Output: 无
 *****************************************************************/
void Stepmoto_task()
{
    // 步进电机控制口设置,GPC0、GPC5 -- 7 口设置为输出
    rGPCCON = (rGPCCON & (~0x0000FC03)) | (0x00005401);

    // 禁止 GPC0、GPC5 -- 7 口的上拉电阻
    rGPCUP  = rGPCUP | 0x00E1;

    // 设置 GPC0、GPC5 -- 7 口输出低电平
    rGPCDAT = rGPCDAT & (~0x00E1);

    while(1)
    {
        MOTO_Mode2(1);                       // 控制步进电机正转
        taskDelay(sysClkRateGet() * 2);      // 停止步进电机,延时
    }

}
```

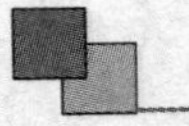


```
/************************************************************
** Function name：Stepmoto_taskStart
** Descriptions： 步进电机任务启动函数
** Input：无
** Output：无
************************************************************/
void Stepmoto_taskStart(void)
{
    taskSpawn("Stepmoto - examtask",80,0,2048,
              (FUNCPTR)Stepmoto_task,0,0,0,0,0,0,0,0,0,0);
}
```

9.7 ADC 驱动程序设计实验

9.7.1 实验目的

掌握 S3C2410A 的模/数(A/D)转换器的应用设置,进行电压信号的测量。

9.7.2 实验设备

(1) 硬件。

- PC 1 台;
- MagicARM2410 教学实验开发平台 1 套。

(2) 软件。

- Windows98/XP/2000 系统;
- Tornado 交叉开发环境。

9.7.3 实验内容

使用 AIN0 和 AIN1 测量两路直流电压,并将测量结果通过 UART0 向 PC 机发送。

9.7.4 实验原理

- S3C2410A 具有 1 个 8 通道的 10 位模数转换器(ADC),有采样保持功能,输

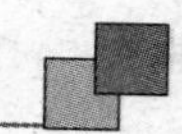

入电压范围是0～3.3 V，在2.5 MHz的转换器时钟下，最大的转换速率可达500 kSps。

- A/D转换器的AIN5和AIN7还可以与控制脚nYPON、YMON、nXPON和XMON配合，实现触摸屏输入功能。

S3C2410A的ADC功能框图如图9.5所示。

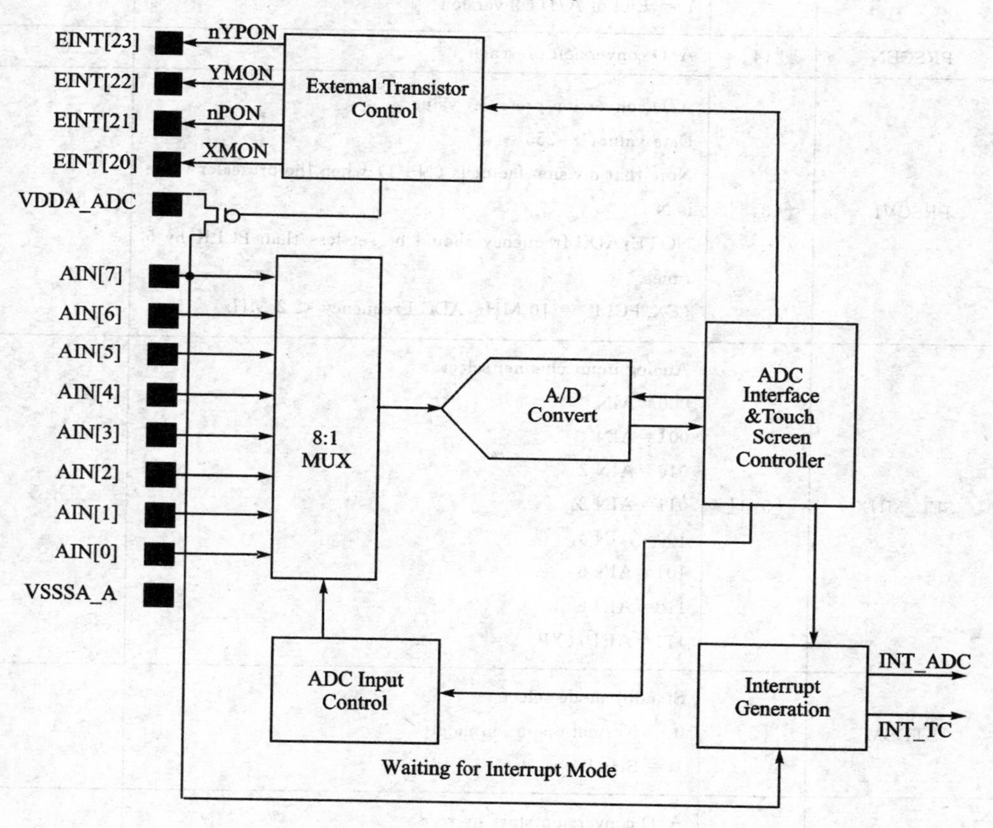

图9.5 ADC功能框图

为了正确使用A/D转换器，需要设置A/D转换器的时钟，A/D转换器的工作模式和输入通道选择，以上都是通过ADCCON寄存器来完成设置。设置完成后，置位ADCCON寄存器的ENABLE_START位来控制启动A/D转换，读ADCCON寄存器的ECFLG位来判断A/D转换是否已经结束。当一次A/D转换结束后，通过读ADCDAT0寄存器来取得A/D转换结果，寄存器的低10位数据有效。

寄存器ADCCON说明如表9.3所列。

表 9.3　寄存器 ADCCON 说明

ADCCON	Bit	Description	Initial State
ECFLG	[15]	End of conversion flag(read only) 0 = A/D conversion in process 1 = End of A/D conversion	0
PRSCEN	[14]	A/D conversion prescaler	0
PRSCVL	[13:6]	A/D conversion prescaler value Data value:1～255 Note that division factor is (N+1) when the prescaler value is N NOTE:ADC frequency should be set less than PCLK by 5 times. (EX. PCLK = 10 MHz,ADC Frequency < 2 MHz)	0xFF
SEL_MUX	[5:3]	Analog input channel select 000=AIN 0 001=AIN 1 010=AIN 2 011=AIN 3 100=AIN 4 101=AIN 5 110=AIN 6 111=AIN 7(XP)	0
STDBM	[2]	Standby mode select 0 = Normal operation mode 1 = Standby mode	1
READ_START	[1]	A/D conversion start by read 0 = Disable start by read operation 1 = Enable start by read operation	0
ENABLE_START	[0]	A/D conversion starts by setting this bit If READ_START is enabled , this value is not valid. 0 = No operation 1 = A/D conversion starts and this bit is cleared after the start-up.	0

寄存器 ADCTSC 说明如表 9.4 所列。

表9.4　寄存器ADCTSC说明

ADCTSC	Bit	Description	Initial State
Reserved	[8]	This bit should be zero	0
YM_SEN	[7]	Select output value of YMON 0 = YMON output is 0(YM = Hi－Z) 1 = YMON output is 1(YM = GND)	0
YP_SEM	[6]	Select output value of nYPON 0 = nYPON output is 0(YP = Extemal voltage) 1 = nYPON output is 1(YP is connected with AIN[5])	1
XM_SEM	[5]	Select output value of XMON 0 = XMON output is 0(XM = Hi－Z) 1 = XMON output is 1(XM = GND)	0
XP_SEM	[4]	Select output value of nXPON 0 = nXPON output is 0(XP = Extemal voltage) 1 = nXPON output is 1(XP is connected with AIN[7])	1
PULL_UP	[3]	Pull－up switch enable 0 = XP pull－up enable 1 = XP pull－up disable	1
AUTO－PST	[2]	Automatically sequencing conversion of X－position and Y－position 0 = Nomal ADC conversion 1 = Auto (Sequential) X/Y Position Conversion Mode	0
XY_PST	[1:0]	Manual measurement of X－position or Y－position 00 = No operation mode 01 = X－position measurement 10 = Y－position measurement 11 = Waiting for Interrupt Mode	0

注意：自动模式下，寄存器ADCTSC在启动读取操作之前需要进行重新配置。

寄存器ADCDAT0说明如表9.5所列。

表9.5　寄存器ADCDAT0说明

ADCDAT0	Bit	Description	Initial State
UPDOWN	[15]	Up or down state of Stylus at Waiting for Interrupt Mode. 0 = Stylus down state 1 = Stylus up state	—

续表 9.5

ADCDAT0	Bit	Description	Initial State
AUTO_PST	[14]	Automatic sequencing conversion of x－position and Y－position 0 ＝ Normal ADC conversion 1 ＝ Sequencing measurement of X－position ,Y－position	—
XY_PST	[13:12]	Manual measurement of X－position or Y－position 00 ＝ No operation mode 01 ＝ X－position measurement 10 ＝ Y－position measurement 11 ＝ Waiting for Interrupt Mode	—
Reserved	[11:10]	Reserved	
XPDATA	[9:0]	X－position conversion data value. (include Normal ADC conversion data value) Data value：0～3FF	—

实验箱实验部分电路原理图如图 9.6 所示。

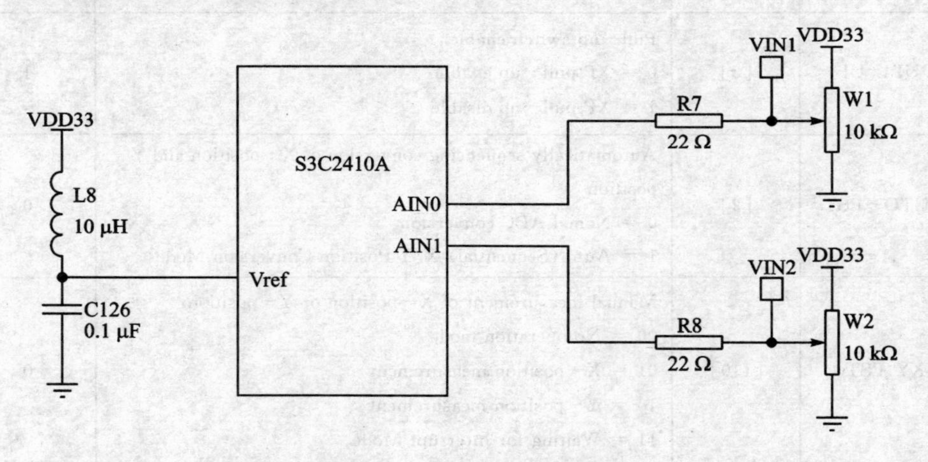

图 9.6　ADC 驱动电路图

9.7.5　实验步骤

（1）完成相关设置（主机网络地址、FTP 服务器和超级终端等），然后系统上电，登录到 ftp 服务器上获取编译的基本映像，并且运行。

（2）复制直流电机驱动实验的源代码 ADC.c 到 C:\Tornado2.2\target\config\MagicARM2410 目录下。

（3）添加 ADC.c 到项目文件中去，编译，生成.o 文件。

(4) 连接上 Target Server,并开启 Wshell。

(5) 将 Object Modules 中的 ADC.o 模块下载到运行的系统中去。

(6) 在 Wshell 中输入 ADCtaskStart,观察实验结果。

(7) 全速运行程序,调整 W1 和 W2 改变测量的电压,观察 PC 机上的“超级终端”主窗口显示电压值是否正确,如图 9.7 所示。

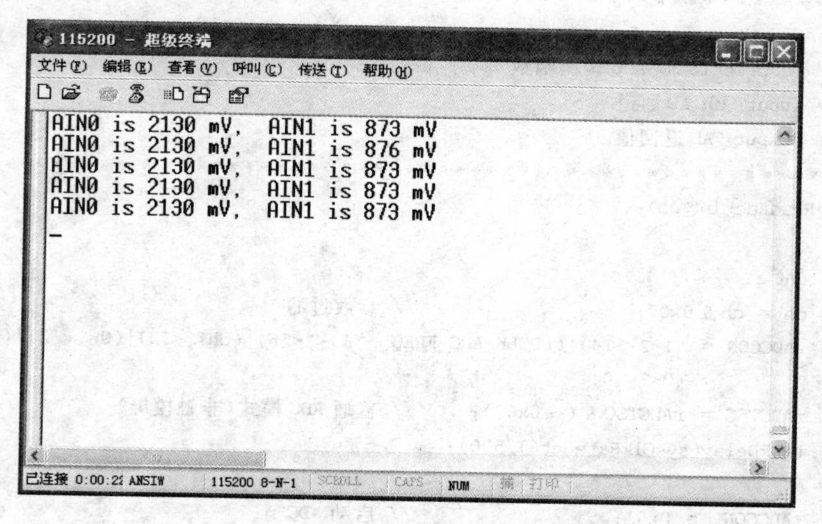

图 9.7 程序运行结果图

9.7.6 程序清单

```
/****** ADC 驱动程序设计实验 ADC.c —— only  for  MagicARM2410 ******/
#include "taskLib.h"
typedef unsigned int        uint32;

#define     FCLK        (200 * 1000000)     /* 系统时钟 */
#define     HCLK        (FCLK/2)            /* HCLK 只能为 FCLK 除以 1、2 */
#define     PCLK        (HCLK/2)            /* PCLK 只能为 HCLK 除以 1、2 */

#define rADCCON     (*(volatile unsigned *)0x58000000) //ADC control
#define rADCTSC     (*(volatile unsigned *)0x58000004) //ADC touch screen control
#define rADCDLY     (*(volatile unsigned *)0x58000008) //ADC start or Interval De-
                                                         lay
#define rADCDAT0    (*(volatile unsigned *)0x5800000c) //ADC conversion data 0
#define rADCDAT1    (*(volatile unsigned *)0x58000010) //ADC conversion data 1
```

```
// 定义用于保存 ADC 结果的变量
uint32  adc0, adc1;
// 定义 ADC 转换时钟 (2 MHz)
#define  ADC_FREQ        (2 * 1000000)

/*****************************************************************************
** Function name: ReadAdc
** Descriptions: ADC 转换函数
** Input: ch AD 通道
** Output: AD 返回值
*****************************************************************************/
int ReadAdc(int ch)
{
    int i;
    ch = ch & 0x07;                          // 参数过滤
    rADCCON = (1<<14)|((PCLK/ADC_FREQ - 1)<<6)|(ch<<3)|(0<<2)|(0<<1)|
              (0<<0);
    rADCTSC = rADCTSC & (~0x03);            // 普通 ADC 模式(非触摸屏)
    taskDelay(sysClkRateGet()/500);

    rADCCON |= 0x01;                         // 启动 ADC
    while(rADCCON & 0x01);                   // 等待 ADC 启动
    while(! (rADCCON & 0x8000));            // 等待 ADC 完成
    return (rADCDAT0 & 0x3ff);               // 返回转换结果
}

/*****************************************************************************
** Function name: ADCtask
** Descriptions:  控制 AD 的任务
** Input: 无
** Output: 无
*****************************************************************************/
void ADCtask()
{
    int   vin0, vin1;

    while(1)
    {
        // 进行 A/D 转换
        adc0 = ReadAdc(0);
        adc1 = ReadAdc(1);
```

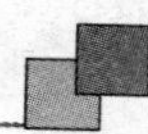

```
        // 通过串口输出显示
        vin0 = (adc0 * 3300) / 1024;      // 读算实际电压值(mV)
        vin1 = (adc1 * 3300) / 1024;
        printf("AIN0 is %d mV,  AIN1 is %d mV \n", vin0, vin1);

        // 延时
        taskDelay(sysClkRateGet() * 1);
    }

}

/*************************************************************************
** Function name: ADCtaskStart
** Descriptions: 启动 AD 的任务
** Input: 无
** Output: 无
*************************************************************************/
void ADCtaskStart(void)
{
    taskSpawn("ADCexamtask",80,0,2048,(FUNCPTR)ADCtask,0,0,0,0,0,0,0,0,0,0);
}
```

第 10 章

VxWorks 应用程序设计实验

10.1 Hello World 实验

10.1.1 实验目的

- 熟悉 Tornado 环境的使用；
- 掌握使用 Tornado 建立工程。

10.1.2 实验设备

(1) 硬件。

- PC 1 台；
- MagicARM2410 教学实验开发平台 1 套。

(2) 软件。

- Windows98/XP/2000 系统；
- Tornado 交叉开发环境。

10.1.3 实验内容

在 Tornado 中建立一个“Hello World”工程，编写应用程序，通过串口输出“Hello World”字符串。交叉编译该程序，下载到 MagicARM2410 上运行。

10.1.4 实验原理

串口在嵌入式系统当中是非常重要的数据通信接口，其本质功能是作为 CPU

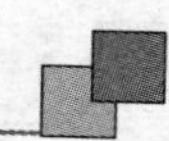

和串行设备间的编码转换器。当数据从CPU经过串行端口发送出去时，字节数据转换为串行的位；在接收数据时，串行的位被转换为字节数据。应用程序使用串口进行通信，必须在使用之前向操作系统提出资源申请要求（打开串口），通信完成后必须释放资源（关闭串口）。串口通信的优点是开发简单，在传输数据量不大、要求速度不高而传输距离较大的通信场合得到广泛应用。

在VxWorks中，对于串口的操作可以视为对一个文件的操作，而不必了解串口设备或程序驱动实现的细节。在串口通信软件的设计中，当串口初始化完成后，在使用之前利用open()函数打开相应的串口，然后进行配置。

本实验针对VxWorks内核，直接调用printf语句进行打印，将通过串口在Host上打印出相关内容。

10.1.5 实验步骤

(1) 启动Tornado开发工具，建立一个项目，建立方法如本文前面介绍。

(2) 在项目中的usrAppInit.c中编写实验代码。

(3) 在Tornado环境下用“Bulid vxWorks”编译生成vx映像。

(4) 启动Tornado开发工具所带的ftp服务器，并设定好用户和密码，以及所对应的用户登录目录（如前文所述）。

(5) 按RST键复位系统。

(6) 复位后，S3C2410A会登录到ftp服务器上获取编译的映像，并且开始执行映像。

(7) 执行映像时，观察输出。

10.1.6 程序清单

```
/*********************************************************************
** Function name: usrAppInit
** Descriptions:打印 Hello world 程序
** Input: 无
** Output: 无
*********************************************************************/
void usrAppInit (void)
    {
#ifdef     USER_APPL_INIT
    USER_APPL_INIT;          /* for backwards compatibility */
#endif
```

```
    /* add application specific code here */
    printf("Hello World form MagicArm2410 ! \n");
}
```

10.2 任务调度

10.2.1 实验目的

- 学习创建多个任务。

10.2.2 实验设备

(1) 硬件。

- PC 1 台;
- MagicARM2410 教学实验开发平台 1 套。

(2) 软件。

- Windows98/XP/2000 系统;
- Tornado 交叉开发环境。

10.2.3 实验内容

在实验 10.1 建立的 project 中,编写一段小程序:创建 10 个任务,每个任务的工作是输出其任务 ID,然后观察运行结果。

10.2.4 实验原理

VxWorks 系统的主要特点就是支持多任务运行。多任务运行环境支持实时构造多个独立任务,每个任务单独执行,并且各自拥有自己的一套资源。VxWorks 多任务内核 wind 使用中断驱动,基于优先级的任务调度机制,具有较短的上下文切换时间和中断延迟。系统中的多任务运行让外界以为是多个任务在同时执行。事实上,是系统内核通过调度算法让这些任务依次执行,而并非是同时在运行。每个任务拥有自己的上下文,包括 CPU 环境和系统资源,这些是内核调度该任务运行时所必须的。在上下文切换时,任务的上下文保存在任务控制块 TCB 中。一个任务的上下文包括:

- 代码执行点,也就是任务的程序计数器,如同中断处理;

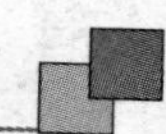

- CPU 寄存器和浮点数寄存器(如果需要);
- 动态变量和函数调用的堆栈;
- 标准输入输出 I/O 分配;
- 1 个延时定时器;
- 1 个时间片定时器;
- 内核控制结构;
- 信号处理器;
- 调试和性能监测值。

10.2.5　实验步骤

(1) 启动 Tornado 开发工具,建立一个项目,建立方法如前文介绍。

(2) 在项目中的 usrAppInit.c 中编写实验代码。

(3) 在 Tornado 环境下用"Bulid vxWorks"编译生成 vx 映像。

(4) 启动 Tornado 开发工具所带的 ftp 服务器,并设定好用户和密码,以及所对应的用户登录目录(如前文所述)。

(5) 按 RST 键复位系统。

(6) 复位后,S3C2410A 会登录到 ftp 服务器上获取编译的映像,并且开始执行映像。

(7) 执行映像时,观察输出。

(8) 如果将 taskLowB 任务优先级改为 140,观察输出。

10.2.6　程序清单

```
#include "vxWorks.h"
#include "taskLib.h"
#include "semLib.h"
#include "stdio.h"
#include "sysLib.h"
#include "loglib.h"

#define  TASK_PRI_HIGH          120        /* 高优先级 */
#define  TASK_PRI_LOW           130        /* 低优先级 */

#define  TASK_STACK_SIZE        5000       /* 任务栈 */

#define LOOP_TIMES              1000000    /* 循环的次数 */
```

```
LOCAL STATUS taskHigh ();
LOCAL STATUS taskLowA ();
LOCAL STATUS taskLowB ();

/*************************************************************
** Function name: taskPriorityDemo
** Descriptions:  创建 3 个任务，tHighPriority 优先级较高，
**                 tLowPriorityA 和 tLowPriorityB 优先级相同，
**                 3 个任务都是执行 LOOP_TIMES 次循环后退出
** Input:  无
** Output: 正确返回 OK,错误返回 ERROR
*************************************************************/
STATUS taskPriorityDemo ()
{
    /* 创建 tHighPriority */
    if (taskSpawn ("tHighPriority", TASK_PRI_HIGH, 0, TASK_STACK_SIZE,
                   (FUNCPTR) taskHigh, 0, 0, 0, 0, 0, 0, 0, 0, 0, 0)
                   == ERROR)
    {
        perror ("taskPriorityDemo: Error in spawning tHighPriority\n");
        return (ERROR);
    }
    else
    {
        printf("taskPriorityDemo:spawning tHighPriority OK\n");

    }

    /* 创建 tLowPriorityA */
    if (taskSpawn ("tLowPriorityA", TASK_PRI_LOW, 0, TASK_STACK_SIZE,
                   (FUNCPTR) taskLowA, 0, 0, 0, 0, 0, 0, 0, 0, 0, 0)
                   == ERROR)
    {
        perror ("taskPriorityDemo: Error in spawning tLowPriority\n");
        return (ERROR);
    }
    else
    {
        printf("taskPriorityDemo:spawning tLowPriorityA OK\n");

    }

    /* 创建 tLowPriorityB */
```

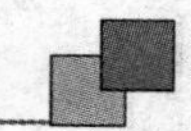

```
    if (taskSpawn ("tLowPriorityB", TASK_PRI_LOW, 0, TASK_STACK_SIZE,
                (FUNCPTR) taskLowB, 0, 0, 0, 0, 0, 0, 0, 0, 0, 0)
                == ERROR)
    {
        perror ("taskPriorityDemo: Error in spawning tLowPriority\n");
        return (ERROR);
    }
    else
    {
        printf("taskPriorityDemo:spawning tLowPriorityB OK\n");

    }

    return (OK);
}

/*************************************************************
 * * Function name: taskHigh
 * * Descriptions:  高优先级任务
 * *
 * * Input:  无
 * * Output: 正确返回 OK,错误返回 ERROR
 *************************************************************/
LOCAL STATUS taskHigh ()
{
    int i,j;

    for(i = 0; i<LOOP_TIMES; i++)
    {
      if (j++  == 500)
      {
            j = 0;
            logMsg("\n",taskIdSelf(),0,0,0,0,0);
      }
    }

    return (OK);
}

/*************************************************************
 * * Function name: taskLowA
 * * Descriptions:  低优先级任务 A
 * *
```

```
** Input: 无
** Output: 正确返回 OK,错误返回 ERROR
*********************************************************************/
LOCAL STATUS taskLowA()
{
    int i,j;

    for(i = 0; i<LOOP_TIMES; i++)
    {
        if (j++  == 500)
        {
            j = 0;
            logMsg("\n",taskIdSelf(),0,0,0,0,0);
        }
    }

    return (OK);
}

/*********************************************************************
** Function name: taskLowB
** Descriptions:  低优先级任务 B
**
** Input:  无
** Output: 正确返回 OK,错误返回 ERROR
*********************************************************************/
LOCAL STATUS taskLowB()
{
    int i,j;

    for(i = 0; i<LOOP_TIMES; i++)
    {
        if (j++  == 500)
        {
            j = 0;
            logMsg("\n",taskIdSelf(),0,0,0,0,0);
        }
    }

    return (OK);
}
```

10.3 信号量实验

10.3.1 实验目的

● 学习使用信号量实现共享资源的保护。

10.3.2 实验设备

(1) 硬件。

● PC 1 台;

● MagicARM2410 教学实验开发平台 1 套。

(2) 软件。

● Windows98/XP/2000 系统;

● Tornado 交叉开发环境。

10.3.3 实验内容

在实验一建立的 project 中,创建 2 个任务,一个任务将全局变量设为“1”;一个任务将全局变量设为“0”,观察运行结果。

10.3.4 实验原理

信号量允许多个任务相互协调其活动。任务间最直接的通信方式就是共享各种各样的数据结构。由于 VxWorks 使用单地址空间,所有任务存在于一个单一的线性地址空间中,共享数据结构也就非常容易实现。全局变量、各类缓冲、链表和指针都可以被运行在不同任务上下文的代码直接引用。然而,对于共享的数据,需要保证对其的互斥访问。VxWorks 提供了许多实现共享临界区互斥访问的机制,经过优化后,信号量已经成了任务之间最快的通信机制。VxWorks 内核 Wind 提供 3 种类型的信号量用于解决不同类型的问题:

● 二进制信号量(binary):用于互斥临界区的访问和任务之间的同步。

● 互斥信号量(mutual exclusion):属于 binary 信号量,但是具备优先级继承、安全删除和递归处理特性。局限性:不能用于 ISR,不能用 semFlush()函数。

● 计数信号量(counting):适用于保护具有多个数据的资源。

10.3.5 实验步骤

(1) 启动 Tornado 开发工具,建立一个项目,建立方法如 10.2 节所述。

(2) 在项目中的 usrAppInit.c 中编写实验代码。

(3) 在 Tornado 环境下用"Bulid vxWorks"编译生成 vx 映像。

(4) 启动 Tornado 开发工具所带的 ftp 服务器,并设定好用户和密码,以及所对应的用户登录目录(如前文所述)。

(5) 按 RST 键复位系统。

(6) 复位后,S3C2410A 会登录到 ftp 服务器上获取编译的映像,并且开始执行映像。

(7) 执行映像时,观察输出。

(8) 如果将注释为"NOTE 1"和"NOTE 2"的两条语句删除,观察输出。

10.3.6 程序清单

```
#include "VxWorks.h"
#include "taskLib.h"
#include "semLib.h"
#include "MagicARMconfig.h"

SEM_ID semBinary;
int NUM = 0;

/*****************************************************************
** Function name: NewTask
** Descriptions:  观察信号量
** Input: 无
** Output: 无
*****************************************************************/
static void NewTask(void)
{
    int i;
    for (i=0;i<10;i++)
    {
        semTake(semBinary, WAIT_FOREVER);
        printf("The new task now and the num = %d\n", ++NUM);
        semGive(semBinary);
    }
```

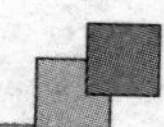

```
}

/*****************************************************************
** Function name: OldTask
** Descriptions:  观察信号量
** Input: 无
** Output: 无
*****************************************************************/
static void OldTask(void)
{
    int i;
    semGive(semBinary);
    for (i = 0;i<10;i++)
    {
        semTake(semBinary, WAIT_FOREVER);
        printf("The old task now and the num =  %d\n", --NUM);
    }
}

/*****************************************************************
** Function name: usrAppInit
** Descriptions:  信号量实验程序
** Input: 无
** Output: 无
*****************************************************************/
void usrAppInit (void)
{
#ifdef  USER_APPL_INIT
    USER_APPL_INIT;
#endif
    /* add application specific code here */

    int NewTaskID,OldTaskId;
    semBinary = semBCreate(SEM_Q_FIFO, SEM_FULL);
    semTake(semBinary, WAIT_FOREVER);

    NewTaskID = taskSpawn("NewTask",90,0x100,2000,(FUNCPTR)NewTask,0);
    OldTaskId = taskSpawn("OldTask",90,0x100,2000,(FUNCPTR)OldTask,0);

}
```

10.4 VxWorks 信号

10.4.1 实验目的

- 学习使用信号。

10.4.2 实验设备

(1) 硬件。
- PC 1 台;
- MagicARM2410 教学实验开发平台 1 套。

(2) 软件。
- Windows98/XP/2000 系统;
- Tornado 交叉开发环境。

10.4.3 实验内容

在实验一建立的 project 中,编写一段信号处理程序,将其与 SIGINT 相关连,使用 kill()发送 SIGINT 信号并调用信号处理程序,观察运行结果。

10.4.4 实验原理

信号可以用来通知任务处理特定的事件。当引起一个信号的事件发生时,信号产生(generated)。当处理事件的任务激活时,信号释放(delivered)。信号的生命期是从产生到释放之间的时间。一个已经产生但还没有释放的信号是挂起的(pending)。信号的生命期可能比较长。

任何任务都可以向一个特定任务发送信号。被信号通知的任务立即挂起它当前的执行线程,在下次任务被调度运行的时刻,指定的信号处理程序将获得处理器。甚至即便任务处于阻塞状态下,其信号处理程序仍可以被调用执行。信号处理程序是由用户提供并与特定的信号相联系,用于执行当信号发生时必要的处理工作。信号适用于错误和异常处理,很少用于任务间通信。

Wind 内核提供了两种信号接口:BSD4.3 和 POSIX 信号接口。POSIX 接口提供比 BSD4.3 接口更强大的标准接口。应用程序仅能使用其中一个。

VxWorks 提供 31 种不同的信号。程序可以调用 kill()产生一个信号,与中断和

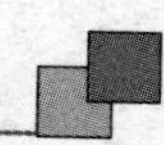

硬件异常类似;调用 sigaction()将信号与指定的信号处理程序相对应。当信号处理程序运行时,其他信号被阻塞。通过调用 sigprocmask()函数,任务可以阻止一些信号的出现,如果当信号产生时被阻塞,它的信号处理程序将在信号解除阻塞时调用。

10.4.5 实验步骤

(1) 启动 Tornado 开发工具,建立一个项目,建立方法如 10.2 节所述。

(2) 在项目中的 usrAppInit.c 中编写实验代码。

(3) 在 Tornado 环境下用“Bulid vxWorks”编译生成 vx 映像。

(4) 启动 Tornado 开发工具所带的 ftp 服务器,并设定好用户和密码,以及所对应的用户登录目录(如前文所述)。

(5) 按 RST 键复位系统。

(6) 复位后,S3C2410A 会登录到 ftp 服务器上获取编译的映像,并且开始执行映像。

(7) 执行映像时,观察输出。

10.4.6 程序清单

```
#include "vxworks.h"
#include "stdio.h"
#include "sigLib.h"
#include "taskLib.h"
#include "wdLib.h"

#define TASK_PRIORITY                   101
#define TASK_STACK_SIZE                 5000
#define NUM_OF_GIVES                    3
#define TIME_BETWEEN_INTERRUPTS  sysClkRateGet() * 2    /* 定时周期 2 秒 */

LOCAL WDOG_ID wdId = NULL;                 /* 看门狗定时器 ID */
LOCAL int syncTaskTid = 0;                 /* syncTask 的任务 ID */
LOCAL int numToGive = NUM_OF_GIVES;        /* 信号被发送的次数 */


LOCAL void syncISR(int);
LOCAL void cleanUp();
LOCAL void syncTask();
```

```
/****************************************************************
** Function name: sigDemo
** Descriptions:  主函数,创建任务 syncTaskTid,并创建和启动 watchdog 定时器
**                模拟外部异常事件
** Input:  无
** Output: OK or ERROR
****************************************************************/
STATUS sigDemo(void)
{
    /*创建看门狗定时器,用来模拟外部中断*/
    if ((wdId = wdCreate()) == NULL)
    {
        perror("wdCreate");
        cleanUp();
        return (ERROR);
    }

    /*创建一个系统任务*/
    if ((syncTaskTid = taskSpawn("tsyncTask", TASK_PRIORITY, 0,TASK_STACK_SIZE,
        (FUNCPTR) syncTask, 0,0,0,0,0,0,0,0,0,0)) == ERROR)
    {
        perror("taskSpawn");
        cleanUp();
        return (ERROR);
    }

    /*启动看门狗定时器*/
    if (wdStart(wdId, TIME_BETWEEN_INTERRUPTS,
         (FUNCPTR) syncISR, numToGive) == ERROR)
    {
        perror("wdStart");
        cleanUp();
        return (ERROR);
    }

    /* 延时设定后,结束*/
    taskDelay(sysClkRateGet() + TIME_BETWEEN_INTERRUPTS * numToGive);

    cleanUp();
    return (OK);
}
```

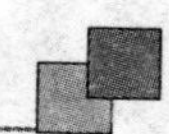

```
/*****************************************************************
 ** Function name：sigHandler
 ** Descriptions： 信号处理函数
 ** Input： 无
 ** Output：无
*****************************************************************/
LOCAL  void sigHandler
(
    int sig,       /*信号值*/
    int code,      /*附加值*/
    SIGCONTEXT *sigContext    /*任务上下文*/
)
{

    switch (sig)
    {
        case SIGUSR1:
            printErr("\nSignal SIGUSR1 received! \n");
        break;

        default:
            printErr("\nOther Signal received! \n");
        break;
    }

}

/*****************************************************************
** Function name：syncTask
** Descriptions： 运行的任务，响应异常任务发出的信号
** Input： 无
** Output：无
*****************************************************************/
LOCAL  void syncTask(void)
{

    /*绑定信号处理函数与 SIGUSR1 信号*/
    signal(SIGUSR1, (VOIDFUNCPTR)sigHandler);
    printf("setup function for SIGUSR1\n");

    /*每隔一秒在 PC 上打印"syncTask Hello"*/
    FOREVER
```

```
    {
        printf("syncTask Hello\n");
        taskDelay(sysClkRateGet());
    }

}

/*************************************************************
** Function name: syncISR
** Descriptions:  模拟的外部异常事件
** Input:  无
** Output: 无
*************************************************************/
LOCAL  void syncISR
(
    int times
)
{
    /* 向 SIGUSR1 发送信号 SIGUSR1 */
    kill(syncTaskTid,SIGUSR1);
    times--;

    if (times > 0)
        wdStart(wdId, TIME_BETWEEN_INTERRUPTS, (FUNCPTR)syncISR, times);

}

/*************************************************************
** Function name: cleanUp
** Descriptions:  删除任务和定时器
** Input:  无
** Output: 无
*************************************************************/
LOCAL  void cleanUp()
{

    if (syncTaskTid)
        taskDelete(syncTaskTid);

    if (wdId)
        wdDelete(wdId);
}
```

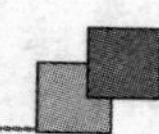

10.5 VxWorks 管道

10.5.1 实验目的

- 学习使用管道进行通信。

10.5.2 实验设备

(1) 硬件。

- PC 1 台;
- MagicARM2410 教学实验开发平台 1 套。

(2) 软件。

- Windows98/XP/2000 系统;
- Tornado 交叉开发环境。

10.5.3 实验内容

在实验一建立的 project 中,创建 2 个任务,任务 1 利用管道向任务 2 发送一条消息,任务 2 将在控制台上输出,观察运行结果。

10.5.4 实验原理

管道是消息队列的替代品,与消息队列不同的是,管道使用了 VxWorks 系统的 I/O 系统。管道是由驱动 pipeDrv 所管理的虚拟 I/O 设备,但其在底层的实现是使用了消息队列,因此可以用 VxWorks 标准的 I/O 过程 open()、read()、write()、ioctl()来访问管道。作为一个 I/O 设备,管道提供了一个消息队列所不能提供的支持,即对 select 的支持。通过使用 select 函数,任务可以同时等待来自几个管道、网口和串口的数据,这种模式能够简化编程。在创建管道设备时,需要调用 pipeDevCreate()函数,然后就可以通过标准 I/O 接口函数操作管道了。

pipeDrvCreate 函数说明如表 10.1 所列。

表 10.1　pipeDrvCreate 函数说明

pipeDrvCreate	
目标	创建一个管道设备
头文件	#include "pipeDrv.h"
函数原型	STATUS nResult = pipeDrvCreat(char * name, int nMessages, int nBytes)
参数	name 管道设备名 nMessages 队列中允许的最大消息个数 nBytes 最长消息的字节数
返回值	ERROR 错误 OK　　成功

```
#define PIPE_NAME "/pipe/pipe0" /* 管道设备的文件名 */
#define MAX_MSGNO 3 /* 最大消息数 */
#define MAX_MSGLEN sizeof (int) /* 消息大小 */
/* 创建管道 */
if (pipeDevCreate (PIPE_NAME, MAX_MSGNO, MAX_MSGLEN) == ERROR)
{
    perror ("Error in creating pipe");
}
```

10.5.5　实验步骤

(1) 启动 Tornado 开发工具，建立一个项目，建立方法如 10.2 节所述。

(2) 在项目中的 usrAppInit.c 中编写实验代码。

(3) 在 Tornado 环境下用“Bulid vxWorks”编译生成 vx 映像。

(4) 启动 Tornado 开发工具所带的 ftp 服务器，并设定好用户和密码，以及所对应的用户登录目录(如前文所述)。

(5) 按 RST 键复位系统。

(6) 复位后，S3C2410A 会登录到 ftp 服务器上获取编译的映像，并且开始执行映像。

(7) 观察输出结果。

10.5.6　程序清单

```
#include "vxWorks.h"
#include "taskLib.h"
```

```
#include "sysLib.h"
#include "stdio.h"
#include "ioLib.h"
#include "pipeDrv.h"

#define  RECEIVE_TASK_PRI          99   /*receive task 优先级*/
#define  SEND_TASK_PRI             98   /*send task 优先级*/
#define  TASK_STACK_SIZE         5000   /*任务栈大小*/

#define PIPE_NAME      "/pipe/server"   /* 管道设备文件名*/

struct msg                              /*消息结构体*/
{
    int tid;                            /*任务号*/
    int value;                          /*传递值*/
};

LOCAL int  pipeFd;                      /*管道对应的文件描述符*/
LOCAL int  MsgNum = 3;                  /*队列中最大消息个数*/
LOCAL BOOL notDone;                     /*完成标志*/

LOCAL STATUS SendTask();                /* send task */
LOCAL STATUS ReceiveTask();             /* receive task */

/***********************************************************************
** Function name: pipeDemo
** Descriptions:  利用管道通信的主程序
**                在 SendTask 和 ReceiveTask 之间创建一个管道
**                在 SendTask 通过管道发送消息
**                ReceiveTask 接收完所有消息,关闭管道
** Input:  无
** Output: 正确返回 OK,错误返回 ERROR
***********************************************************************/
STATUS pipeDemo(void)
{
    notDone = TRUE;   /*初始化全局变量*/

    /*创建管道*/
    if (pipeDevCreate(PIPE_NAME, MsgNum, sizeof (struct msg)) == ERROR)
    {
        perror("Error in creating pipe");
```

```
    }

    /* 打开管道 */
    if ((pipeFd = open(PIPE_NAME, O_RDWR, 0)) == ERROR)
    {
        perror("Error in opening pipe device");
        return (ERROR);
    }

    /* 创建 SendTask task */
    if (taskSpawn("tSendTask_pipe", SEND_TASK_PRI, 0, TASK_STACK_SIZE,
                  (FUNCPTR)SendTask, 0, 0, 0, 0, 0, 0, 0, 0, 0, 0)
                  == ERROR)
    {
        perror("SendTask: Error in spawning SendTask");
        return (ERROR);
    }

    /* 创建 ReceiveTask task */
    if (taskSpawn("tReceiveTask_pipe", RECEIVE_TASK_PRI, 0, TASK_STACK_SIZE,
                  (FUNCPTR)ReceiveTask, 0, 0, 0, 0, 0, 0, 0, 0, 0, 0)
                  == ERROR)
    {
        perror("ReceiveTask: Error in spawning ReceiveTask");
        return (ERROR);
    }

    /* 轮询等待,ReceiveTask 接收完所有消息 */
    while(notDone)
        taskDelay(sysClkRateGet());

    /* 关闭管道 */
    close(pipeFd);

    return (OK);
}

/*************************************************************
** Function name: SendTask
** Descriptions:  通过消息队列发送消息
**
```

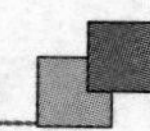

```
** Input:  无
** Output: 正确返回 OK,错误返回 ERROR
*******************************************************************/
LOCAL STATUS SendTask(void)
{
    int count;
    int value;

    struct msg sendItem;                /* 发送缓存 */

    printf("SendTask started: task ID = %#x \n", taskIdSelf());

    /* 构造 MsgNum 个消息并发送 */

    for (count = 1; count <= MsgNum; count++)
    {
        value = count * 10;

        /* 给消息结构体赋值 */
        sendItem.tid = taskIdSelf();          /* 赋值任务号 */
        sendItem.value = value;

        /* 发送消息 */
        if ((write(pipeFd, (char *)&sendItem, sizeof(sendItem))) == ERROR)
        {
            perror("Error in sending the message");
            return (ERROR);
        }
        else
        {
            printf("SendTask: task ID = %#x, send value = %d \n",
                    taskIdSelf(), value);
        }
    }


    return (OK);

}
/*******************************************************************
** Function name: ReceiveTask
** Descriptions:  通过消息队列接收消息
```

```
**
** Input:  无
** Output: 正确返回 OK,错误返回 ERROR
**********************************************************************/
LOCAL STATUS ReceiveTask (void)
{
    int count;
    struct msg ReceiveItem;              /* 接收缓存 */

    printf("\n\nReceiveTask: Started -   task ID = %#x\n", taskIdSelf());

    /* 接收 MsgNum 个消息 */
    for (count = 1; count <= MsgNum; count++)
    {
        /* 接收消息 */
        if (read(pipeFd, (char *)&ReceiveItem, sizeof (ReceiveItem)) == ERROR)
        {
            perror("Error in receiving the message");
            return (ERROR);
        }
        else
            printf ("ReceiveTask: Receiving msg of value %d from task = %#x\n",
                    ReceiveItem.value, ReceiveItem.tid);
    }

    notDone = FALSE;    /*设置 notDone,通知主任务关闭管道*/

    return (OK);

}
```

10.6 VxWorks 消息队列

10.6.1 实验目的

● 学习使用消息队列进行通信。

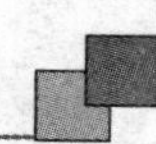

10.6.2 实验设备

(1) 硬件。

- PC 1 台；
- MagicARM2410 教学实验开发平台 1 套。

(2) 软件。

- Windows98/XP/2000 系统；
- Tornado 交叉开发环境。

10.6.3 实验内容

在实验一建立的 project 中，创建 2 个任务，任务 1 向任务 2 发送一条消息，任务 2 将在控制台上输出，观察运行结果。

10.6.4 实验原理

在单 CPU 中，VxWork 的多任务通信的主要机制是消息队列。消息队列允许以 FIFO 或基于优先级方式排队消息，消息的数目可变，消息的长度可变。任何任务都可以向消息队列发送消息，也可以从消息队列接收消息。多个任务允许从同一个消息队列收发消息。但是，两个任务间的双向通信通常需要两个消息队列，各自用于一个方向。

VxWorks 消息队列的创建、删除、发送和接收调用如下：

```
/* 生成和初始化一个消息队列 */
MSG_Q_ID msgQCreate ( int maxMsgs, int maxMsgLength, int options);

/* 删除一个消息队列 */
STATUS msgQDelete ( MSG_Q_ID msgQId );

/* 发送一个消息队列 */
STATUS msgQSend ( MSG_Q_ID msgQId, char * buffer, UINT nBytes, int timeout,
                  int priority );

/* 接收一个消息队列 */
int msgQReceive(MSG_Q_ID msgQId, char * buffer, UINT maxNBytes, int timeout );
```

消息队列库允许消息按照 FIFO 方式排队，但是也有一个例外：存在两个优先级，优先级最高的消息排在队列的头部。要创建一个消息队列可以调用 msgQCreate()。

任务调用 msgQSend()将消息发送到一个消息队列，如果没有任务在等待该队列的消息，那么这条消息增加到该队列的消息缓冲中；如果有任务在等待，那么该消息立即提供给第一个等待的任务。

任务如果需要从一个消息队列接收一条消息，它应该调用 msgQReceive()。如果该消息队列中已有消息可用，那么队列中的第一条消息立即出队，并提交给该任务；如果没有消息可用，那么该任务阻塞，并加入到等待该消息的任务队列中。等待任务队列可以按两种方式排队：基于任务优先级方式或基于 FIFO 方式。具体按哪一种方式，在消息队列创建时指定。

- timeout 参数。函数 msgQSend()和 msgQReceive()都可以指定一个超时参数，规定任务的等待时间(tick 数)：发送消息任务等待队列空间可用，接收消息任务等待消息可用。
- Urgent Messages 参数。函数 msgQSend()可以指定欲发送消息的优先级：正常 MSG_PRI_NORMAL 或紧急 MSG_PRI_URGENT。正常优先级的消息加入到消息队列的尾部，而紧急优先级的消息将添加到消息队列的头部。

10.6.5 实验步骤

(1) 启动 Tornado 开发工具，建立一个项目，建立方法如 10.2 节所述。

(2) 在项目中的 usrAppInit.c 中编写实验代码。

(3) 在 Tornado 环境下用“Bulid vxWorks”编译生成 vx 映像。

(4) 启动 Tornado 开发工具所带的 ftp 服务器，并设定好用户和密码，以及所对应的用户登录目录(如前文所述)。

(5) 按 RST 键复位系统。

(6) 复位后，S3C2410A 会登录到 ftp 服务器上获取编译的映像，并且开始执行映像。

(7) 增加另一个消息队列，使得任务 2 使用该消息队列向任务 1 发送消息，任务 1 将消息输出到控制台上。

10.6.6 程序清单

```
#include "vxWorks.h"
#include "taskLib.h"
#include "msgQLib.h"
#include "sysLib.h"
#include "stdio.h"
```

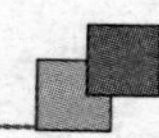

```
#define  RECEIVE_TASK_PRI        99    /* receive task 优先级 */
#define  SEND_TASK_PRI           98    /* send task 优先级 */
#define  TASK_STACK_SIZE       5000    /* 任务栈大小 */


struct msg                             /* 消息结构体 */
{
    int tid;                           /* 任务号 */
    int val;                           /* 传递值 */
};


LOCAL MSG_Q_ID msgQId;                 /*  消息队列号 */
LOCAL int  MsgNum  =  3;               /* 队列中最大消息个数 */
LOCAL BOOL notDone;                    /*  完成标志 */


LOCAL STATUS SendTask();               /* send task */
LOCAL STATUS ReceiveTask();            /* receive task */


/*************************************************************
* * Function name: msgQDemo
* * Descriptions:  利用消息队列通信的主程序
* *                 在 SendTask 和 ReceiveTask 之间创建一个消息队列
* *                 在 SendTask 通过消息队列发送消息
* *                 ReceiveTask 接收完所有消息,关闭消息队列
* * Input:  无
* * Output: 正确返回 OK,错误返回 ERROR
*************************************************************/
STATUS msgQDemo()
{
    notDone = TRUE;  /* 初始化全局变量 */

    /*  创建一个消息队列,先入先出类型 */
    if ((msgQId = msgQCreate (numMsg, sizeof (struct msg), MSG_Q_FIFO)) == NULL)
    {
        perror ("MsgQ creating error!");
        return ERROR;
    }

    /* 创建 SendTask 任务 */
```

```
    if (taskSpawn ("tSendTask", SEND_TASK_PRI, 0, TASK_STACK_SIZE,
                   (FUNCPTR) SendTask, 0, 0, 0, 0, 0, 0, 0, 0, 0, 0)
                   == ERROR)
    {
        perror ("Spawning SendTask error in msgQDemo!");
        return ERROR;
    }

    /*创建 ReceiveTask 任务*/
    if (taskSpawn ("tReceiveTask", RECEIVE_TASK_PRI, 0, TASK_STACK_SIZE,
                   (FUNCPTR) ReceiveTask, 0, 0, 0, 0, 0, 0, 0, 0, 0, 0)
                   == ERROR)
    {
        perror ("Spawning ReceiveTask error in msgQDemo!");
        return ERROR;
    }

    /*轮询等待,ReceiveTask 接收完所有消息*/
    while (notDone)
        taskDelay (sysClkRateGet ());

    /*关闭消息队列*/
    if (msgQDelete (msgQId) == ERROR)
    {
        perror ("Deleting msgQ error!");
        return ERROR;
    }

    return OK;
}

/**************************************************************
 ** Function name: SendTask
 ** Descriptions:  通过消息队列发送消息
 **
 ** Input:  无
 ** Output: 正确返回 OK,错误返回 ERROR
 **************************************************************/
LOCAL STATUS SendTask (void)
{
    int i;
```

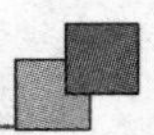

```
    int val;
    struct msg sendItem;              /*  发送缓存  */

    printf ("Into SendTask : task ID =  %#x\n", taskIdSelf ());

    /*产生 MsgNum 次发送消息*/

    for (i = 1; i <= MsgNum; i++)
    {
        val = i * 10;

        /*赋值消息结构体*/
        sendItem.tid = taskIdSelf ();           /*任务号*/
        sendItem.val = val;

        /*发送消息*/
        if ((msgQSend(msgQId, (char *) &sendItem, sizeof (sendItem),
            WAIT_FOREVER,  MSG_PRI_NORMAL)) == ERROR)
        {
            perror ("Send message error!");
            return ERROR;
        }
        else
            printf ("SendTask: tid =  %#x, send val =  %d \n",
                    taskIdSelf (), val);
    }

    return OK;
}

/**************************************************************
** Function name: ReceiveTask
** Descriptions:  通过消息队列接收消息
**
** Input:  无
** Output: 正确返回 OK,错误返回 ERROR
**************************************************************/
LOCAL STATUS ReceiveTask (void)
{
    int i;
    struct msg receiveItem;              /*  接收缓存  */
```

```
    printf ("\nInto ReceiveTask: task id = %#x\n", taskIdSelf());

    /* 接收 MsgNum 个消息 */
    for (i = 1; i <= MsgNum; i++)
    {
        /*利用消息队列接收消息*/
        if ((msgQReceive(msgQId, (char *) &receiveItem, sizeof (receiveItem),
            WAIT_FOREVER)) == ERROR)
        {
            perror ("Receive message error!");
            return ERROR;
        }
        else
            printf ("ReceiveTask: receive msg of val %d from tid = %#x\n",
                    receiveItem.val, receiveItem.tid);
    }

    notDone = FALSE;    /*设置 notDone，通知主任务关闭消息队列*/
    return OK;
}
```

10.7 VxWorks Socket 通信

10.7.1 实验目的

学习网络 Socket 通信原理。

10.7.2 实验设备

(1) 硬件。

- PC 1 台；
- MagicARM2410 教学实验开发平台 1 套。

(2) 软件。

- Windows98/XP/2000 系统；
- Tornado 交叉开发环境。

10.7.3 实验内容

在实验一建立的 project 中，编写一段 Socket 通信程序。观察运行结果。

10.7.4 实验原理

VxWorks系统和网络协议的接口靠套接字实现。Socket规范是广泛应用的、开放的和支持多种协议的网络编程接口。在VxWorks中,应用程序使用套接字接口实现访问网际协议报的功能。VxWorks套接字与UNIX BSD 4.4套接字兼容。

目前比较常使用的有两种套接字:采用流套接字的TCP协议和采用数据报套接字的UDP协议。流套接字采用TCP协议捆绑设备某一端口,提供了双向、有序、无重复且无记录边界的数据流服务。在两个TCP套接字连接后,会建立一条虚电路用以实现可靠的套接字间的通信。数据报套接字使用UDP协议捆绑某一端口,支持双向数据流,不保证可靠性,有序性和无重复性,但是保留了记录边界。与TCP相比,UDP提供了一个相对简单但适应性很强的通信方式。在UDP通信中,数据经由数据报进行传送,它们是分离、无连接的且是单独寻址的。数据报套接字使用邮信方式,每个数据报都含有目的地址和发送者地址。

10.7.5 实验步骤

(1) 启动Tornado开发工具,建立一个项目,建立方法如10.2节所述。

(2) 在项目中的usrAppInit.c中编写实验代码。

(3) 在Tornado环境下用"Bulid vxWorks"编译生成vx映像。

(4) 启动Tornado开发工具所带的ftp服务器,并设定好用户和密码,以及所对应的用户登录目录(如前文所述)。

(5) 按RST键复位系统。

(6) 复位后,S3C2410A会登录到ftp服务器上获取编译的映像,并且开始执行映像。

(7) 执行映像时,观察输出。

10.7.6 程序清单

1. udpExample.h 文件

```
/* udpExample.h                    数据套接字实验客户/服务器的头文件 */

#define SERVER_PORT         8080          /* 绑定服务器所用的端口号 */
#define REQ_MSG_SIZE        1024          /* 请求消息的最大值 */

/* 客户端请求的数据结构 */
```

```
typedef struct CLIENT_REQ
{
    int display;                            /*是否显示信息*/
    char message[REQ_MSG_SIZE];             /*定义接收缓存*/
};
```

2. udpServer. c 文件

```
/* udpServer.c - UDP服务器端程序 */

#include "vxWorks.h"
#include "sockLib.h"
#include "inetLib.h"
#include "stdioLib.h"
#include "strLib.h"
#include "hostLib.h"
#include "ioLib.h"
#include "udpExample.h"

/****************************************************************************
** Function name: udpServerProc
** Descriptions:  读取 UDP 套接字,按照设定显示客户端消息
**
** Input: 无
** Output: 正确返回 OK,错误返回 ERROR
****************************************************************************/
STATUS udpServerProc(void)
{
    struct sockaddr_in ServerAddr;              /*服务器套接字地址*/
    struct sockaddr_in ClientAddr;              /*客户端套接字地址*/
    struct CLIENT_REQ ClientRequest;            /*客户端的请求消息*/
    int socketAddSize;                          /*套接字地址数据结构大小*/
    int socketFd;                               /*套接字文件描述符*/
    char clientAdd[INET_ADDR_LEN];              /*客户端网络地址缓存*/

    /*设置本地服务器地址*/
    socketAddSize = sizeof(struct sockaddr_in);
    bzero((char *)&serverAdd , socketAddSize);
    serverAdd.sin_len = (u_char) socketAddSize;
    serverAdd.sin_family = AF_INET;
    serverAdd.sin_port = htons(SERVER_PORT);
    serverAdd.sin_addr.s_addr = htonl(INADDR_ANY);
```

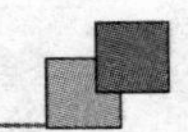

```
    /* 生成 UDP 套接字 */
    if((socketFd = socket(AF_INET,SOCK_DGRAM,0)) == ERROR)
    {
        perror("socket fail! \n");
        return ERROR;
    }

    /* 绑定 */
    if(bind(socketFd,(struct sockaddr *)&serverAdd, socketAddSize) == ERROR)
    {
        perror("bind fail! \n");
        close(socketFd);
        return ERROR;
    }

    /* 读取套接字,响应请求 */
    FOREVER
    {
        if(recvfrom(socketFd,(char *)&clientRequest,sizeof(clientRequest),0,
            (struct sockaddr *)&clientAddr, &socketAddSize) == ERROR)
        {
            perror("recvfrom fail! \n");
            close(socketFd);
            return ERROR;
        }

        /* 判断是否显示客户端消息,为真显示 */
        if(clientRequest.display)
        {
            /* 转换网络地址格式为字符型 */
            inet_ntoa_b(clientAddr.sin_addr,clientAdd);

            printf("Client message(IP - %s, PORT - %d) : %s\n", clientAdd,
            ntohs(clientAddr.sin_port), clientRequest.message);
        }
    }
}
```

3. udpServer.c 文件

```
/* udpClient.c - UDP 客户端程序 */
```

```
#include "vxWorks.h"
#include "sockLib.h"
#include "inetLib.h"
#include "stdioLib.h"
#include "strLib.h"
#include "hostLib.h"
#include "ioLib.h"
#include "udpExample.h"

/*************************************************************************
** Function name: udpClientProc
** Descriptions:  通过 UDP 套接字向服务器发送信息
**
** Input: server,服务器名或 IP 地址
** Output: 正确返回 OK,错误返回 ERROR
*************************************************************************/
STATUS udpClientProc(char * server)
{
    struct CLIENT_REQ  myRequest;                  /* 发送到服务器的请求 */
    struct sockaddr_in serverAdd;                  /* 服务器套接字地址 */
    char    display;                               /* 如果为 TRUE,服务器将打印信息 */
    int     socketAddSize;                         /* 套接字地址数据结构大小 */
    int     socketFd;                              /* 套接字文件描述符 */
    int     msgLen;                                /* 信息长度 */

    /* 生成客户端套接字 */
    if((socketFd = socket(AF_INET,SOCK_DGRAM,0)) == ERROR)
    {
        perror("socket fail! \n");
        return ERROR;
    }


    /* 生成服务器套接字地址 */
    socketAddSize = sizeof(struct sockaddr_in);
    bzero((char *)&serverAdd,socketAddSize);
    serverAdd.sin_len = (u_char)socketAddSize;
    serverAdd.sin_family = AF_INET;              /* 设置服务器 AF_INET 地址域 */
    serverAdd.sin_port = htons(SERVER_PORT);  /* 设置服务器端口号 */

    /* 设置服务器的 IP 地址 */
```

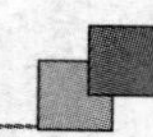

```
    if(((serverAdd.sin_addr.s_addr = inet_addr(server)) == ERROR) &&
          ((serverAdd.sin_addr.s_addr = hostGetByName(server)) == ERROR))
    {
        perror("unknown server name\n");
        close(socketFd);
        return ERROR;
    }

    /*人机交互,提示用户输入信息*/
    printf("Please input message to send: \n");
    msgLen = read(STD_IN,myRequest.message,REQ_MSG_SIZE);
    myRequest.message[msgLen - 1] = '\0';

    printf("Whether to display the message on serve(Y or N): \n");
    read(STD_IN,&display,1);
    switch(display)
    {
        case 'y':
        case 'Y':
                myRequest.display = TRUE;
                break;

        default:
                myRequest.display = FALSE;
            break;
    }

    /*发送消息到服务器*/
    if(sendto(socketFd,(caddr_t)&myRequest,sizeof(myRequest),0,
         (struct sockaddr *)&serverAdd,socketAddSize) == ERROR)
    {
        perror("sendto fail! \n");
        close(socketFd);
        return ERROR;
    }

    close(socketFd);
    return OK;

}
```

参考文献

[1] Windriver System Inc. Tornado Training Workshop, ver5. 1. May 1999
[2] Windriver System Inc. Vxworks Programmer's Guide, v5. 4. May 1999
[3] Windriver System Inc. Tornado Getting Started Guide, v2. 0. May 1999

北京航空航天大学出版社

● 博客藏经阁丛书

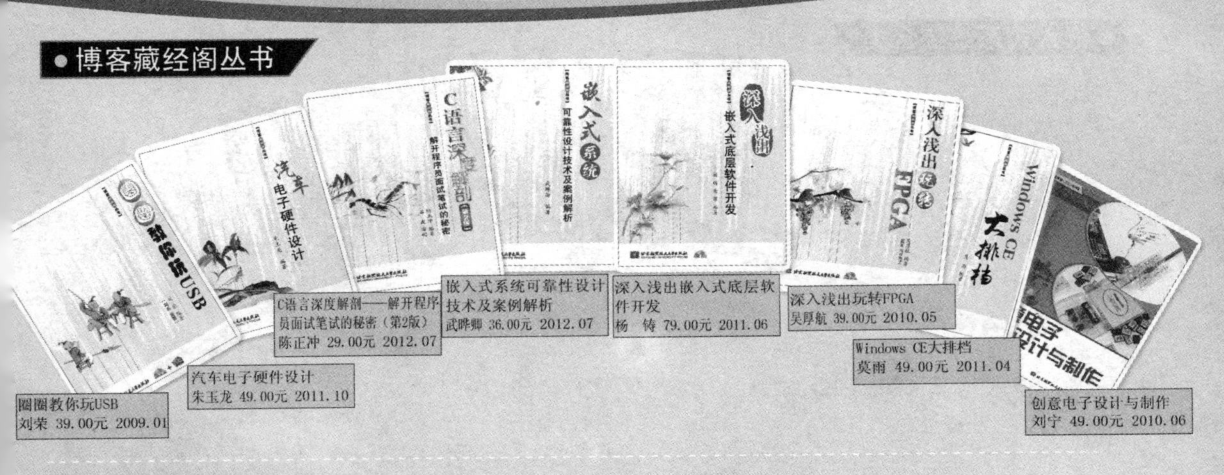

● 嵌入式系统译丛

嵌入式软件概论
沈建华 译 42.00元 2007.10

嵌入式Internet TCP/IP基础、实现及应用（含光盘）
潘琢金 译 75.00元 2008.10

嵌入式实时系统的DSP软件开发技术
郑红 译 69.00元 2011.01

ARM Cortex-M3权威指南
宋岩 译 49.00元 2009.04

链接器和加载器
李勇 译 32.00元 2009.09

● 全国大学生电子设计竞赛“十二五”规划教材

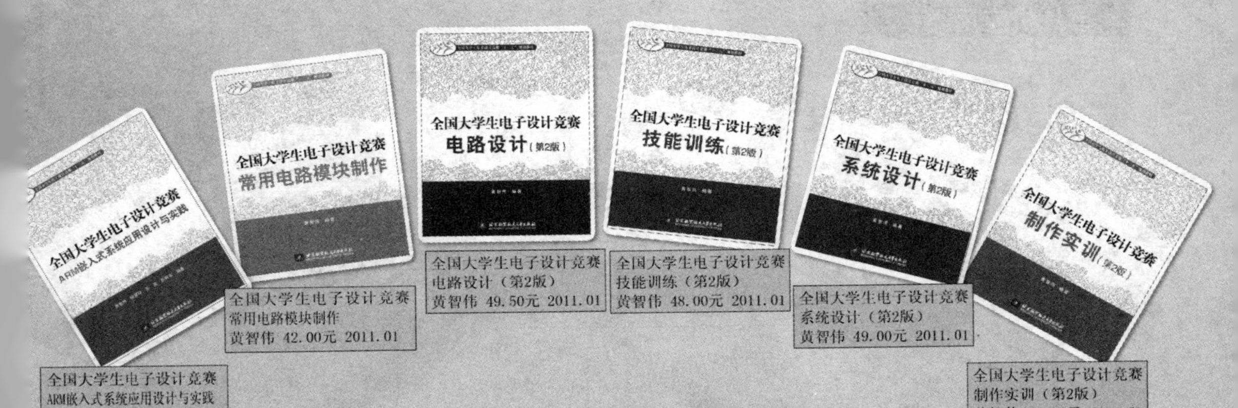